Plant Protoplasts

Editors

L. C. Fowke

Professor of Biology
Department of Biology
University of Saskatchewan
Saskatoon, Saskatchewan, Canada

F. Constabel

Head, Plant Cell Technology
Plant Biotechnology Institute
National Research Council of Canada
Saskatoon, Saskatchewan, Canada

CRC Press
Taylor & Francis Group
Boca Raton London New York

CRC Press is an imprint of the
Taylor & Francis Group, an **informa** business

First published 1985 by CRC Press
Taylor & Francis Group
6000 Broken Sound Parkway NW, Suite 300
Boca Raton, FL 33487-2742

Reissued 2018 by CRC Press

Library of Congress Cataloging in Publication Data
Main entry under title:
Plant protoplasts

Bibliography: p.
Includes index.
1. Plant protoplasts. I. Fowke, L. C.
II. Constabel, F.
QK725.P576 1985 581.87'3 84-23179
ISBN-0-8493-6473-6

A Library of Congress record exists under LC control number: 84023179

Publisher's Note
The publisher has gone to great lengths to ensure the quality of this reprint but points out that some imperfections in the original copies may be apparent.

Disclaimer
The publisher has made every effort to trace copyright holders and welcomes correspondence from those they have been unable to contact.

ISBN 13: 978-1-315-89667-0 (hbk)
ISBN 13: 978-1-351-07577-0 (ebk)

Visit the Taylor & Francis Web site at http://www.taylorandfrancis.com and the
CRC Press Web site at http://www.crcpress.com

PREFACE

Protoplasts obtained by cell wall digestion or by microsurgery have been described, analyzed, cultured, and manipulated in laboratories around the world for over 15 years. They have often become central to advances in physiology, biochemistry, and genetics, particularly somatic cell genetics. Their ever increasing importance has already resulted in numerous review articles published in scientific journals and books. In general, these reviews have focused primarily on the biotechnological potential of protoplasts. We, therefore, felt it important to highlight protoplasts as novel material to be explored for the advancement of general biology. The present book thus emphasizes the importance of plant protoplasts for fundamental research.

The first two chapters summarize the technical aspects of protoplast isolation and culture and examine the process of plant regeneration from protoplasts. Chapters 3 to 7 detail various applications of protoplasts in the area of cell biology. These chapters clearly illustrate the advantages of using protoplasts to study the plasma membrane, a vital component of plant cells which is normally inaccessible within the enclosing cell wall. They also highlight the importance of protoplasts for studies of cell growth and differentiation. Chapters 8 and 9 illustrate the major applications of protoplasts in the areas of plant physiology and virology. The final four chapters consider protoplasts as a model system for basic studies of genetic variation and mutation as well as for genetic engineering of plants via somatic hybridization or transformation.

This book is intended to be used by senior undergraduates, graduate students, and research scientists in plant biology. We hope that it will arouse their curiosity and perhaps stimulate novel experimentation with plant protoplasts.

Finally, we wish to express our sincere thanks to Ms. Pat Rennie for her important contribution to the editing of this book.

Larry Fowke
Fred Constable
Saskatoon, Canada

THE EDITORS

L. C. Fowke obtained his Ph.D. from Carleton University, Ottawa in 1968. Following a two year postdoctoral fellowship at the Australian National University in Canberra, he joined the staff of the Biology Department, University of Saskatchewan where he is currently Professor of Biology. From 1976 to 1977, Dr. Fowke returned to the Australian National University for a year as a visiting research scientist. He spent a further year abroad in 1983 to 1984 conducting research at both the Friedrich Miescher-Institute in Basel and the Institute for Physiological Botany in Uppsala.

Dr. Fowke is a cell biologist with strong interests in the structure and function of plant cells. For the past 14 years his research group has utilized plant protoplasts as a model system to study plant cells and their organelles. The work has focused on protoplast fusion and the process of cell wall regeneration by cultured protoplasts. Recent research has been directed towards understanding the role of the plasma membrane, microtubules and coated vesicles in plant cell wall formation.

Dr. Fowke's broad interests in Biology are reflected by his memberships in scientific societies. He is currently a member of the Canadian and American Societies for Cell Biology, the Canadian Society for Plant Physiologists, the Canadian Botanical Association, and the International Association for Plant Tissue Culture (IAPTC). In 1978 he served as secretary for the Fourth International Congress of the IAPTC and from 1979 to 1982 as Canadian Correspondent for the same society. Dr. Fowke has also served as advisor to *Plant Cell Reports* (1972 to 1983) and associate editor of the *Canadian Journal of Botany* (1979 to 1982).

F. Constabel obtained his Ph.D. from the University of Gottingen, Fed. Rep. Germany, in 1956. Scholarships by the French government, Maison Rothschild, and Deutsche Forschungsgemeinschaft allowed him an "apprenticeship" with Professor R. J. Gautheret at the Sorbonne in Paris (1956 to 1958) and studies of the metabolism of tissues grown in vitro with Professor R. Harder, Botanische Anlagen in Gottingen (1958 to 1959). Thereafter he joined the staff of Professor M. Steiner, Department of Pharmacognosy, University of Bonn, FRG, where he established an interest in secondary metabolism of tissues grown in vitro. He was promoted to Privatdozent for botany and pharmacognosy in 1967. After a sabbatical and work on product formation and embryogenesis at the Prairie Regional Laboratory of the National Research Council of Canada in Saskatoon with Dr. O. L. Gamborg's group in 1969 to 1970, he left Bonn as associate professor and became research officer with NRC in 1972. Since 1981 he is head of the plant cell technology project of NRC's Plant Biotechnology Institute, the former PRL, in Saskatoon.

Dr. Constabel is a plant scientist with major interest in the structure and function of cells. Most of his work is based on cell culture technology and is directed to better understanding and subsequently controlling the formation of secondary metabolites as well as plant development. Contributions in numerous papers are focused on extracellular enzymes, tannins, betalains, anthocyanins, and alkaloids, high-lighting conifer tissue culture in 1956, cell variability as early as 1967, and embryogenesis and protoplastology in the 70's. At present his primary concern is morphogenesis. Throughout his life he has remained an ardent field botanist.

Dr. Constabel is member of the Deutsche Botanische Gesellschaft, Canadian Society of Plant Physiologists, and the International Association for Plant Tissue Culture. He is co-editor for *Plant Cell Reports* and chief-editor for the *Journal of Plant Physiology,* formerly *Zeitschrift fur Pflanzenphysiologie.*

CONTRIBUTORS

Horst Binding
Professor
Botanical Institute
University of Kiel
Kiel, West Germany

Anne D. Blonstein
Doctor
Friedrich Miescher-Institut
Basel, Switzerland

Richard I. S. Brettell
Research Scientis
CSIRO
Division of Plant Industry
Canberra City, Australia

F. Constabel
Head, Plant Cell Technology
Plant Biotechnology Institute
National Research Council of Canada
Saskatoon, Saskatchewan, Canada

Adrian J. Cutler
Assistant Research Officer
Plant Biotechnology Institute
National Research Council
Saskatoon, Saskatchewan, Canada

Richard J. Daines
Associate Scientist
Pfizer Central Research
Groton, Connecticut

Tage Eriksson
Professor and Head
Institute of Physiological Botany
University of Uppsala
Uppsala, Sweden

L. C. Fowke
Professor
Department of Biology
University of Saskatchewan
Saskatoon, Saskatchewan, Canada

Alan R. Gould
Senior Research Scientist
Pfizer Central Research
Groton, Connecticut

L. R. Griffing
Reseach Associate
Department of Biology
University of Saskatchewan
Saskatoon, Saskatchewan, Canada

Christian T. Harms
Reseach Manager
CIBA-GEIGY Ltd.
Biotechnology Research Institute
Research Triangle Park, North Carolina

Patrick J. King
Doctor
Friedrich Miescher-Institut
Basel, Switzerland

H. W. Kohlenbach
Professor
Botanisches Institut der Universitat
Frankfurt am Main
Federal Republic of Germany

Philip Larkin
Research Scientist
CSIRO
Division of Plant Industry
Canberra City, Australia

Robert T. Leonard
Professor
Department of Botany and Plant Sciences
University of California, Riverside
Riverside, California

Brent G. Mersey
Associate Research Officer
National Research Council of Canada
Plant Biotechnology Institute
Saskatoon, Saskatchewan, Canada

Fusao Motoyoshi
Chief
Laboratory of Resistance Genes
Department of Molecular Biology
National Institute of Agrobiological Resources
Tsukuba Science City
Yatabe, Ibaraki, Japan

Lisa Rayder
Visiting Postdoctoral Plant Physiologist
Department of Botany and Plant Sciences
University of California, Riverside
Riverside, California
Presently at:
Biological Laboratories
Harvard University
Cambridge, Massachusetts

Presently at:
Biological Laboratories
Harvard University
Cambridge, Massachusetts

R. A. Schilperoort
Professor of Biochemistry
Department of Plant Molecular Biology
State University of Leiden
Leiden, The Netherlands

Michael A. Tanchak
Graduate Student
Department of Biology
University of Saskatchewan
Saskatoon, Saskatchewan, Canada

J. H. M. Willison
Associate Professor
Biology Department
Dalhousie University
Halifax, Nova Scotia, Canada

George Joseph Wullems
Groupleader MOLBAS Research Group
Department of Plant Molecular Biology
State University of Leiden
Leiden, The Netherlands

TABLE OF CONTENTS

Chapter 1

PROTOPLAST ISOLATION AND CULTURE

Tage R. Eriksson

TABLE OF CONTENTS

I. INTRODUCTION

Isolated protoplasts have been described as "naked" plant cells because the cell wall has been experimentally removed by either a mechanical or an enzymatic process.[1] The term protoplast was introduced by Hanstein in 1880 to designate the living matter enclosed by plant cell walls.[2] The isolated protoplast is unusual because the outer plasma membrane is fully exposed and is the only barrier between the external environment and the interior of the living cell.[1,3] The isolation of protoplasts from plant cells was first achieved by microsurgery on plasmolyzed cells.[4] Today protoplasts are isolated in large numbers by enzymatic removal of cell walls as pioneered by Cocking in 1960.[5] The plant species, the conditions of plant growth, plant age, protoplast isolation, and protoplast culture are often critical for sustained division of protoplasts. Therefore, there are no standard methods for the isolation and culture of protoplasts. While leaf mesophyll tissues of many dicotyledonous plants respond in a similar way during the protoplast isolation, the subsequent culturing might require a special protocol depending upon what species or what cultivar is used. Young, yet fully developed leaves and cell suspension cultures in the phase of logarithmic growth after subculturing generally yield the greatest number of viable protoplasts.

Plant cells are normally connected with each other via many plasmodesmata to form a multicellular tissue. When isolating protoplasts a real single cell suspension is obtained from a given tissue or cell culture. A suspension of isolated protoplasts constitutes the only proper single-cell system obtainable with higher plants. This makes them valuable and interesting as experimental material for physiological, genetic, and biochemical studies. Besides providing the possibility of applying microbiological techniques to this single-cell system with millions of individuals, protoplasts also facilitate the study of cell membranes since the plasmalemma is directly exposed to the culture medium. It is also possible to isolate good preparations of plant cell vacuoles from protoplasts. Protoplasts can also be used for the isolation of cell organelles, cell cloning, and cell fusion.

The purpose of this chapter is to present the techniques and approaches used in most studies of protoplast isolation and culture. Many workers have made modifications of techniques practiced in most laboratories. It has not been possible to include all these more or less necessary modifications.

II. PROTOPLAST ISOLATION

A. Mechanical Methods

Cell wall degrading enzymes are toxic to varying degrees and might unfavorably affect the physiology of the cells.[6,7] However, good test systems are not available for assessing the damaging components of the enzyme preparation. Nonenzymatically isolated protoplasts can be obtained, but the yields are usually insufficient.[4] In rare instances, such as with the moss *Funaria hygrometrica,* mechanically isolated protoplasts have been cultured and entire plants regenerated.[8] In a more recent approach, the starting material was new callus tissue on expanded leaves of greenhouse grown *Saintpaulia ionantha* which were grown in a specific auxin environment, thereby producing thin-walled cells. Protoplasts were isolated by gently teasing the callus tissue apart in liquid culture medium with dissecting needles.[9] In general, mechanical methods for protoplast isolation are not adequate because of low yields. However, in studies where all side effects of the wall degrading enzymes have to be avoided, the method can be used.

B. Enzymatic Methods

The enzymatic isolation of protoplasts can be performed in two different ways: the two-step (or sequential method) and the one-step method. In the two-step method, the tissue is first treated with a macerozyme or pectinase which separate the cells by degrading the middle lamella. The more or less free cells are then treated with cellulase which releases the protoplasts.[10,11] In general, the cells are exposed to the different enzymes for shorter periods than that used for the one-step method. In the one-step method the tissue is subjected to a mixture of enzymes, including macerozyme and cellulase.[12] This method generally results in higher yields from leaf tissues since both mesophyll and palisade cells release protoplasts. In some cases, however, no difference is observed between the two methods.[13] The one-step method is most frequently used today and is less labor intensive.

1. Osmotic Conditions

In isolating protoplasts, the wall pressure must be replaced by osmotic pressure in the isolation mixture and also later in the culture media. The transfer of cells and tissues to a plasmolyzing solution induces a stress situation which may have harmful effects on several important steps in cell metabolism and growth. A condensation of DNA in cell nuclei and decreased protein synthesis are two common effects of osmotic stress, although both can be reversed. Induced changes in hormone production is another effect of reduced osmotic potential in the medium. The incorporation of labeled precursors into RNA and proteins of isolated tobacco leaf protoplasts were shown to decrease with increasing osmotic pressure in the incubation medium.[14,15]

Lower (more negative) osmotic potentials are usually generated by the addition of mannitol, sorbitol, glucose, or sucrose to the enzyme mixture and to the culture medium. The properties of the various osmotica vary. Mannitol and sorbitol are the most frequently used. Mannitol is considered to be relatively inert metabolically and infuses slowly into the protoplasts. The hexitols are used separately or in combination. The protoplasts tend to sink in osmotic solutions of hexitols and glucose, but they usually float in sucrose solutions. The osmotic stabilizers need not be metabolically inert. A combination of an inert osmotic stabilizer and a metabolically active one may be of some advantage.[16] The utilization of sucrose or glucose during the early period of protoplast culture may reduce sudden osmotic shock when the regenerated cells are transferred to media with higher osmotic potential.[17] The stability, viability, and future growth of the protoplasts are closely related to the maintenance of proper osmotic conditions during isolation and subsequent culture.

The optimal osmotic potential varies with the source of the cells and tissues used. In general, a suitable osmotic potential can be generated by 0.3 to 0.7 *M* sugar solutions. When culture medium or stabilizing salts are included in the enzyme solution a reduction of the sugar concentration is recommended. Salts can readily penetrate the protoplast membranes, thereby causing difficulties in maintenance of proper osmotic potential over longer periods, especially during culture. The salts also reduce the activity of the wall degrading enzymes and should, therefore, be avoided as osmotic stabilizers. However, low concentrations (e.g., 0.1 to 0.5 m*M*) of $CaCl_2$ are sometimes recommended to stabilize membranes in solutions for preplasmolysis before enzyme treatment.[18]

2. Enzymes and Enzyme Treatment

The routine isolation of protoplasts has become possible by the availability of a number of commercial cell wall degrading enzymes. Those most extensively used are listed in Table 1. There are, however, instances where workers prefer to prepare their own enzymes.[19]

Table 1
CELL WALL DEGRADING ENZYMES MOST COMMONLY EMPLOYED IN PROTOPLAST ISOLATION

Enzyme	Manufacturer/Supplier
Cellulases and hemicellulases	
Cellulase	Calbiochem, La Jolla, Calif., U.S.
	Mayvill Chemicals Ltd., Cheshire, U.K.
	Serva Feinbiochemica, Heidelberg, W. Germany
	Sigma London Chemical Co., Dorset. U.K.
Cellulase RID	Yakult Honsha Co., Nishinomiya, Japan
Cellulase RS	Yakult Honsha Co., Nishinomiya, Japan
Cellulysin	Calbiochem, La Jolla, Calif., U.S.
Driselase	Kyowa Hakko Kogyo, Tokyo, Japan
Hemicellulase	Sigma London Chemical Co., Dorset, U.K.
Meicelase	Meiji Seika Kaisha, Tokyo, Japan
Rhozyme® HP150	Genencor Inc., Corning, U.S.
Pectinases	
Macerozyme RID	Kiki Yakult, Nishinomiya, Japan
Pectinase	Sigma Chemicals, St. Louis, U.S.
Pectinol R-10	Röhm and Haas, Philadelphia, U.S.
Pectolyase Y23	Sigma London Chemical Co., Dorset, U.K.
Rohament P	Röhm GmbH, Darmstadt, W. Germany
Other mixtures	
Glusulase	Endo Laboratories, New York, U.S.
Helicase	Industrie Biologique Francaise, Gennevilliers. France

Most of the enzymes used for protoplast isolation are extracted from fungi grown on plant cell wall constituents. However, a few have been isolated from the intestinal juice of the Roman snail *Helix pomatia* and from bacteria. The enzymes, sold under different trade names, fall into two main categories in respect to their action on plant cell wall material. The pectinases dissolve the middle lamella and in this way separate the cells from each other. The second category includes the cellulases and hemicellulases which degrade the cell walls and eventually release the protoplasts. The commercial enzymes are usually used without further purification in spite of the fact that they are most often mixtures of several enzymes. Very highly purified and crystallized enzyme preparations are often less efficient for protoplast isolation. The effectiveness of crude and less costly preparations can be improved by gel filtration.[20] Partial purification by elution through Sephadex® G-25 or BioGel® removes phenolics, salts, and inert material. It is especially important to use a purified enzyme preparation when a procedure is employed where only a few protoplasts are to be cultured.[6] Proteinaceous compounds and contaminating enzymes like ribonucleases, proteases, and peroxidases could also be responsible for the harmful effects of the commercial preparations.[17]

The viability of carrot suspension cell protoplasts was increased by desalting the enzyme.[21] Filtration through a column of BioGel® P6 was found essential for colony formation of mesophyll protoplasts of *Medicago sativa.*[22] A fivefold stimulation in plating efficiency of *Petunia parodii* mesophyll protoplasts cultured in 5 $\mu\ell$ droplets of medium was obtained when BioGel® purified Meicelase was used. Cellulysin seems to be a powerful cellulase for many types of tissues. It should, however, be used in combination with a low concentration of a pectinase like Macerozyme or Pectolyase. The latter enzyme appears to be useful in many combinations and applicable to many wall materials, difficult to degrade. With the addition of Pectolyase, higher yields of protoplasts from tobacco mesophyll[23] and tobacco suspensions were obtained.[24] Pectolyase has been found to be capable of digesting tissues which were considered impossible for protoplast isolation such as leaves and cotyledons of *Glycine max*[25,26] and leaflets of *Medicago disciformis* and *M. lupulina.*[27] Pectolyase appears to be especially effective in releasing protoplasts of soybean cotyledons when mixed with Onozuka RS and Rhozyme®. Rhozyme® has often been regarded as having properties like a hemicellulase,[28] but according to some workers, the major role of Rhozyme® should be a macerating action different from that of Macerozyme[29]

In order to stimulate release of protoplasts from mesophyll cells of *Macleaya,* Tween 80® was added.[30] To increase protoplast stability many workers include inorganic salts or complete culture medium in the enzyme solution. This is especially recommended when long treatments in the enzyme solution are used. Some workers prefer the use of potassium dextran sulfate in the enzyme solution to increase the stability of the protoplasts.[31,33] Cell organelles tend to deteriorate during the release and culture of some protoplasts. Bovine serum albumin has been used to diminish organelle deterioration in rice protoplasts and in large root nodule protoplasts.[34]

Most often, an empirical approach is required to develop an experimental protocol for protoplast isolation of specific tissues. A single enzyme might sometimes be effective. Driselase, with its cellulolytic and pectolytic activities, has successfully been used alone for protoplast isolation.[35] However, a mixture of two or more enzymes is most frequently used and many laboratories have developed standard mixtures for routine protoplast isolation. The souce of tissue used in protoplast isolation not only determines what enzymes should be used but also the enzyme concentration and the treatment time. A low enzyme concentration requires an increased treatment duration. Some cells can tolerate a long treatment period (up to 20 to 24 hr) while others cannot. It is, therefore, necessary to adjust the enzyme concentration to optimize yields and viabilities of the isolated protoplasts. Many laboratories, for practical reasons, prefer a long enzyme treatment overnight in order to have the protoplasts available in the morning. Overnight treatment can, with advantage, be performed at room temperature and with relatively low enzyme concentrations.

3. Plant and Cell Material

Protoplasts have been isolated from a variety of tissues and organs including leaves, petals, petioles, shoot apices, roots, fruits, coleoptiles, hypocotyls, stems, embryos, aleurone layers of cereal grains, root nodules, microspore mother cells, microspore tetrade, pollen grains, pollen, callus cultures, and cell suspension cultures.[17,36] The selection of plant material is important and should, of course, be dictated by the intended use of the protoplasts. Field grown plants, greenhouse grown plants, in vitro grown aseptic plants or shoots, and tissue cultures are the regular sources of plant material for protoplast isolation. Advantages and disadvantages are associated with each different source of tissue. Since the physiological condition of the source tissue markedly influences the yield and viability of isolated protoplasts, the plants or tissues must be grown under controlled conditions. Greenhouse grown plants are often readily available but light and temperature regimes, as well as nutrient supply, are often not well enough controlled to give optimal conditions. In addition, these plant sources have to be surface sterilized before use and this might reduce the viability of the protoplasts. Controlled environment chambers reduce some of the variation associated with field and greenhouse grown plants.[22,31,37-43]

Preculture of excised leaves before enzyme treatment is sometimes recommended since it increases the metabolic activity of the leaf cells and stimulates division after isolation. Callus inducing medium is most often used during this preculture period.[44,45] However, the protoplast population isolated from the cultivated leaves is not as uniform as are mesophyll protoplasts of untreated leaves. Cytokinesis induced by preculture allows high frequencies of protoplast division shortly after isolation. In lettuce, sustained protoplast division was achieved only when a preculture procedure was used.[46]

The approach of protoplast isolation by initial short-term cultivation of leaves or leaf fragments before enzyme treatment is in several ways different from techniques that are often used today for shoot cultures.[18,47,48] Surface sterilized seeds are germinated on an agar medium and shoots of these axenic plantlets are then transferred to a cultured medium without hormones or with a low concentration of cytokinin. Shoot cultures can also be started from field or pot grown plants when surface sterilization is successful. The axenic shoots are grown in jars and regularly transferred to new culture medium to ensure good and uniform growth. The advantages of shoot cultures are many, especially if the cultures are kept in controlled environment chambers. The maintenance of shoot cultures reduces large plants to manageable sizes,[49] ensures juvenile growth, eliminates annual growth cycles, and offers an axenic material for protoplast isolation.

Leaf mesophyll tissue seems to be the most commonly used as it is readily available and produces large numbers of uniform protoplasts. However, greenhouse and growth cabinet facilities are expensive and labor intensive and many workers have therefore used other sources of experimental material. Thus, protoplasts have been isolated from roots of germinating seeds,[50] from cotyledons,[51,52] from hypocotyls of young seedlings,[53,54] and from immature cotyledons of seeds taken from developing pods.[26] Low yields of protoplasts are often obtained, however, when roots and cotyledons are used for protoplast isolation.

When leaf tissues are used for protoplast isolation the lower epidermis is peeled off or the leaf is sliced into thin slices or pricked with needles followed by slicing[55] before being placed in a preplasmolyzing solution for about 1 hr prior to enzyme incubation. The preplasmolyzing solution should be of the same osmolarity as the enzyme solution and may contain the salts of culture media plus an osmotic stabilizer. Preplasmolysis reduces the amount of enzyme uptake by endocytosis during protoplast contraction. The shrinking of the protoplasts also seals the plasmodesmata, thereby probably reducing leakage of cell contents. L-arginine and L-lysine have been used in the incubation medium to reduce senescence of protoplasts and to stimulate synthesis of macromolecules.[56]

Callus and cell suspensions are frequently used for protoplast isolation and they offer aseptic conditions from the beginning. Age of the cultures influences the complexity of the cell wall structure and reduces the protoplast yield from old cultures. Therefore, it is important to subculture every 3 to 8 days in order to reduce the number of old and dead cells in the culture. The protoplast viability and the protoplast yield are greater when protoplasts are isolated from a culture in the early log phase of the growth cycle.[57] The isolation efficiency can be further improved by increasing levels of auxin, reducing sugar concentrations, and by the addition of sulfur-containing amino acids or mercaptioethanol to the cell culture medium.[58-60]

4. Protoplast Purification

Protoplasts are usually purified by a combination of filtration, centrifugation, and washing. The enzyme solution containing the protoplasts is filtered through a stainless steel or nylon mesh (50 to 100 μm) to remove larger portions of undigested tissues and cell clumps. Highly vacuolated protoplasts from mesophyll cells or cell suspensions are then most often floated on a sucrose solution containing 15 to 20% sucrose as plasmolyticum. The filtered protoplast-enzyme solution is mixed with a suitable volume of a sucrose solution to give a final sucrose concentration of about 15 to 20%. The exact procedure is determined by the osmoticum used in the enzyme solution. Centrifugation at about 100 $\times$ g for 7 to 10 min gives a band of protoplasts floating on the top. The band is easily sucked off with a Pasteur pipette. Some workers prefer specially constructed centrifuge tubes with graduated neck to facilitate the collection and quantification of floated protoplasts.[44,61] The remaining protoplasts often float free of most of the debris but they are still contaminated by the enzyme. Washing is performed by resuspending the protoplasts in solutions which allow pelleting after centrifugation. Here osmotic solutions made up with mannitol, sorbitol, or salts are often used. The washing of the protoplasts is generally repeated two or three times. If Ca^{++} ions have not been included earlier, addition of 0.05 to 0.1*M* $CaCl_2$ to the wash solution to stabilize the protoplast membranes is recommended.[62-64] Some workers use seawater for washing and pelleting protoplasts.[65-67] To avoid large losses during centrifugation, some workers have used denisty buffers of polyethylene glycol (Molecular weight 6000), dextran (Molecular weight 40,000), sodium phosphate, and sorbitol. In these systems the debris and broken protoplasts are left suspended in the lower phase while the intact protoplasts are collected at the interface.[68] Good purification has been obtained with lymphoprep, a density buffer containing 0.6% sodium metriosoate and 5.6% Ficoll.[69] Iso-osmotic gradients of sucrose or Percoll with increasing concentrations of a salt like KCl to compensate for the decreased osmolarity in less dense bands at the top have also been used.[70,71] Density gradients have also been used for protoplasts from cotyledons of *Brassica napus*,[72] and for embryoid protoplasts of carrot.[35] The resuspension of the protoplasts must be carried out with great care in order to avoid injury. After the protoplasts have been examined for density and viability, they are ready for culture.

5. Protoplast Viability and Density

Before testing a protoplast preparation in an experiment, it is often necessary to determine the viability of the protoplasts. Fluorescein diacetate, a dye which only accumulates inside the plasmalemma of viable protoplasts, can be detected by fluorescence microscopy.[69] Evans blue[68,73] and phenolsafranine[74-77] have also been used as stains to determine protoplast viability. Intact viable protoplasts are capable of excluding these stains while broken and dead protoplasts are permeable to them. Cyclosis or protoplasmic streaming has been reported to be a measure of viability.[78] With methylene blue, protoplasts appear bright yellow if they have sufficient reducing power, while dead protoplasts retain the original blue color.[79] Neutral red is concentrated in cells if they are metabolically active and can thus be used in viability tests.[80]

Protoplasts have both maximum as well as minimum plating densities for growth. The optimum plating efficiency for tobacco protoplasts is about 5×10^4 protoplast per milliliter and the protoplasts fail to divide when plated at ten times this concentration.[81] Other protoplasts have similar ranges. It is possible, however, to reduce the protoplast density when very complex and undefined culture media are used. The concentration of protoplasts in a given preparation can be determined by the use of a hemocytometer. A type with a field depth of at least 0.2 mm is necessary since mesophyll protoplasts may be too large for the hemocytometers designed for counting red blood cells.

III. PROTOPLAST CULTURE

Isolated protoplasts are usually cultured in either liquid or semisolid media. Liquid culture allows easy dilution and transfer but has the disadvantage of not permitting isolation of colonies derived from one parent cell. Protoplasts immobilized in semisolid media give rise to cell clones and allow accurate determination of plating efficiency. Once immobilized, however, hand manipulations are required for transfer to other culture media.

The methods used for protoplast culture are basically the same as those employed for tissue and cell culture. However, protoplasts are fragile and, therefore, cannot be shaken too much and the need for adjustment of the osmolarity in the culture medium makes protoplast cultivation less reproducible than cell cultivation. In principle, protoplasts can be cultured in liquid media or on solid agar nutrient media. Different culture systems have been developed to meet special requirements. A range of plastic vessels with excellent optical properties are available today. The pattern and the aim of the study as well as the size of the protoplast population determine the type of culture vessels and culture system to be used.

A. Culturing in Liquid Media

1. Liquid Cultures

Liquid cultures in petri dishes of various sizes seem to be the most commonly used method today. The protoplasts are suspended in a small volume of liquid culture medium at an appropriate density and placed in petri dishes which are then sealed with Parafilm® to reduce the loss of water from the culture medium. It is important to maintain the protoplasts in a shallow layer of medium (approximately 1 mm) to provide adequate oxygen supply.[18,47,82-85] Protoplasts have also been cultured in liquid medium on a fabric support[86] or on filter paper.[87,88] This method offers a convenient way of changing the culture media and allows a better growth of some protoplasts.[67,88,89] One important advantage of liquid culture is that it allows a gradual change of the osmolarity of the culture medium and in this way promotes rapid cell regeneration. The methods described above require relatively large volumes of a protoplast suspension. However, small volumes of protoplasts can be cultured in liquid medium in one of the following ways.

2. Drop Cultures

Drop cultures involve placing small drops (40 to 100 $\mu\ell$) of a protoplast suspension on the inner side of the lid of a petri dish. When the lid is applied to the dish the culture drops are suspended from the lid.[90] Erect drops have also been employed on the bottom of a petri dish.[17,20,37,91] Fresh medium can be added in small drops when required.

3. Microchamber Cultures

Microchamber cultures are similar to hanging drop cultures. Some workers have used cavity slides and microchambers in order to follow the development of individual protoplasts. Drop cultures in cavity slides are mounted by the placement of a drop of the protoplast suspension on a sterile cover glass which is then inverted on the cavity slide and sealed with mineral oil.[92] Microchambers serve the same purpose as cavity slides but they generally offer an optically better view since the depth of the chambers can be kept at a minimum.[93] Special arrangements have to be made to avoid drying out and to ensure adequate oxygen supply in microchambers.

4. Multiple Drop Array Technique

Generally, when the drop culture technique is used, five to ten relatively large drops (up to 200 $\mu\ell$) are placed on the bottom of an ordinary petri dish. In the multiple drop technique the drop size is reduced to 40 $\mu\ell$ and this makes it possible to place up to 50 drops per petri dish.[94] The technique allows testing of a large number of slightly different media compositions in a few petri dishes.

5. Microdroplet Cultures

For this technique the size of drops is reduced to 0.25 to 0.50 $\mu\ell$ so that each droplet contains only one protoplast.[95] The microdroplet cultures have been made possible by the use of special Cuprak petri dishes. These dishes enable drops to be placed in separated numbered wells. Improved culture media and preculturing of the protoplasts is recommended before placing them into microcultures. The technique has an advantage when one or a few protoplasts are to be cultured.

B. Culturing on Semisolid Media

1. Agar as Gelling Agent

Agar of different qualities is most frequently used to solidify the culture media. The protoplast suspension at double the required final density is gently mixed with an equal volume of an agar medium kept molten at about 43 to 45°C. The concentration of agar should be chosen to give a soft agar gel when mixed with the cell suspension. The warm agar medium should be added to the protoplast suspension which is kept at room temperature. The agar plating technique was originally used for the plating of cell suspension cultures[96] and the method was later modified and applied to tobacco mesophyll protoplasts.[11,97]

2. Agarose or Alginate as Gelling Agent

More recently, there has been interest in other gelling agents besides agar, such as alginate which has been shown to give similar plating efficiencies as agar.[98] Gelling agents have been investigated for culture of cells in beads for secondary product production. Agar, agarose, K-carrageenan, alginate, gelatin, and polyacrylamide have been used. Agarose has given the best results in terms of retention of viability and secondary product production.[99] Several workers report improved protoplast culture efficiency by solidification of media with agarose instead of agar.[100-102] These studies show that the agar component plays an important role in the solidified protoplast culture media. The superiority of agarose is probably not only due to the absence of contaminating toxic substances. One major difference between agar and agarose is that agar is more negatively charged.[103] Seaplaque® agarose, is, in fact, a chemically modified agarose with additional hydroxyethyl groups introduced into the agarose molecule. The superior ability of agarose to support protoplast culture may relate to the essentially neutral characteristic of the polymer.[100] Protoplasts are plated in thin layers of agarose on top of already poured and solidified media in petri dishes. The use of superior gelling products, such as agarose gels, may have special advantages with new or recalcitrant protoplast systems.

C. Combination of Liquid and Solid Media

1. Gel Embedded Protoplast Cultures

The combination of agarose plating and bead culture has been shown to dramatically improve plating efficiencies of protoplasts in many species tested. In bead-type culture, the protoplasts are incorporated into the whole depth of the medium when plated. The gelled agar or agarose is then cut into blocks which are transferred to large volumes of liquid culture medium and placed on a shaker. In these studies agarose was shown to be superior to agar as a gelling agent. The cultures were also able to grow from lower densities in agarose.[101]

2. Semisolid Media for Liquification

Procedures for the culture of plant protoplasts in semisolid media with the potential for liquification and large-scale recovery of microcalli and transfer to new media have been reported.[99,104] Alginate was used as a gelling agent in the first study of liquification of a semisolid medium.[99] The method was improved when Sea Prep^{TM} agarose was used as a gelling agent.[104] This method combines the advantages of cell immobilization with those of liquid culture. However, it remains to be shown that this technique, which involves remelting of the culture medium at 40°C for 1 to 2 hr, can be used generally without inducing alterations in the protoplasts or selection of heat tolerant cells.

D. Different Feeder Techniques

Several modifications of the feeder techniques have been reported. Only the principal modifications are discussed below.

1. Feeder Layers

In many types of studies it is desirable to reduce the plating density under the minimum plating density for a given protoplast preparation. This is especially important when particular mutant or hybrid cells are to be selected on agar plates. A feeder layer consists of X-irradiated nondividing but living protoplasts plated in agar medium in petri dishes. Protoplasts are plated on this feeder layer at low density in a thin layer of agar medium.[105-107] The feeder layer effect is not species specific and can be obtained even when a cellophane is separating the two layers. The use of a cellophane sheet above the feeder layer allows rapid and convenient transfers of growing protoplasts to new media which enhances growth and development of the colonies.

2. Nurse Cultures

In fusion studies where only a few heterokaryons are produced, it might be necessary to use nurse cultures where one of the cell lines is supporting the growth of the fusion products.[108] The nitrate reductase deficient mutant of tobacco has been used to nurse heterokaryons between *Solanum tuberosum* and *S. stenotomum*.[109]

3. Reservoir Media

The use of quadrant petri dishes in which two of the four quadrants containing fresh medium are connected to two quadrants containing cultured protoplasts is another useful modification in the plating technique.[110,111] The advantage of such X-plates when compared to traditional culturing methods is probably due to the continuous leakage of nutrients to the developing protoplasts. When activated charcoal was added to the reservoir medium protoplast survival was significantly increased. When activated charcoal was used less browning of the developing protoplasts occurred and this technique might be of help in protoplast cultures where browning is a problem.[102] The culturing in X-plates as well as culturing in a very thin layer over solid media allows high protoplast densities without reducing plating efficiencies.

4. Filter Paper Discs on Agar Media

Another recent culture modification consists of laying a filter paper disc on the agar medium before pouring a thin layer of protoplast suspension over the agar medium.[88] The filter paper provides a physical support which promotes the growth of the protoplasts and facilitates the transfer to new media.[37,67,89] Filter paper also stimulates gas exchange and adsorbs the brownish substances.[37]

5. Agar Drop Technique

It is also possible to embed protoplasts in solid drops of agar. Small drops of liquid agar medium containing protoplasts at desired density are placed in petri dishes. The gelled agar drops are then overlayered with more agar medium. This technique allows a comparison of several media and reduces loss of moisture.[37]

E. Nutritional Requirements

1. The Needs of Protoplasts

Protoplasts from different species and even from a different source of the same species differ in their nutritional requirements and these must be empirically determined. The nutritional requirements of protoplasts differ also from the requirements of isolated cells of the same source. This could be explained by the fact that the membranes of the protoplasts are directly exposed to the culture medium and the protoplasts are also exposed to more rapid volume changes than the cultured cells. Therefore, special care must be taken at the beginning of protoplast culture to control the flow of metabolites across the plasmalemma. The loss of compounds can probably be partly compensated for by the use of complex extracts such as coconut milk and casein hydrolysate. An excessive uptake of nutrients can also become deleterious to the protoplasts. This is especially a problem when protoplasts are transferred to a medium with low concentration of osmoticum.

2. Culture Media

Generally, the basic constituents in the most frequently used protoplast media are similar to those used for cell cultures. Many minor modifications of the most widely used tissue culture media have been demonstrated to improve protoplast survival and colony formation. It is possible to separate most protoplast culture media into two categories, chemically defined and complex culture media, sometimes referred to as regular and rich culture media, respectively. Examples in the first category are B5 and WB while KM8p can be placed in the second category (Table 2). A highly enriched culture medium is often supplemented with coconut milk and/or casein hydrolysate or yeast extract which make these media chemically undefined. They were developed for culturing protoplasts at low density and will support growth of protoplasts from a large number of species.[113] The actual requirements of isolated protoplasts in terms of the composition and nature of the culture medium is complex and poorly understood.

3. Inorganic Elements

The nitrogen source and/or the total amount of nitrogen are often varied in culture media. Some protoplasts grow better in media with a high proportion of reduced nitrogen while others prefer high nitrate concentrations.[113] For some protoplast cultures ammonium ions seem to be toxic.[31,55,112,114] Therefore, a common modification of the MS-medium is to reduce the ammonium content and replace it with nitrate and/or amino acids. There are, however, studies showing that ammonium nitrate is beneficial for protoplast regeneration.[115-117] Ammonium nitrate has also been shown to be somewhat deleterious for protoplast regeneration of some species.[114,118-121] It seems well established that calcium is beneficial for protoplast regeneration.[115-124] The microelements seem to be employed in satisfactory levels.[125] For tobacco protoplasts cultured at low density a threefold increase in the concentration of microelements was found beneficial.[126] Potassium iodide was found to stimulate growth of cultured pea mesophyll protoplasts.[115]

4. Organic Elements

A suitable carbon source is of great importance in a protoplast culture medium. Sucrose and glucose are the two sugars most often recommended for protoplast culture media. In studies of effects of different sugars, there are reports favoring sucrose[57,63,115] and those in favor of glucose.[16,28,91,120,127,128] It has been shown that glucose can serve both as an osmoticum and a carbon source. Sucrose, on the other hand, often reaches a toxic level when used as osmoticum. Several workers have included other sugars in the protoplast culture media (Table 2) because they may have an important role in the synthesis of other cell wall components, like hemi-celluloses and pectins.[59,91,129,130] Organic acids such as citric acid, malic acid, and fumaric acid have been shown to enhance protoplast growth and have, therefore, been included.[126,131,132]

Table 2
SOME COMMONLY USED MEDIA FOR PROTOPLAST CULTURE

	V 47[149]	KM 8p[131]	V-KM[150]	MS[151a]	B5[152]	WB[137]
Macro elements (mg/*l*)						
NH_4NO_3	144	600	1444	1650	134	384
KNO_3	1480	1900	1480	1900	2500	3000
$CaCl_2 \cdot 2\,H_2O$	735	600	735	440	150	780
$MgSO_4 \cdot 7\,H_2O$	984	300	934	370	250	500
$(NH_4)_2SO_4$	—	—	—	—	134	—
$NaH_2PO_4 \cdot H_2O$	—	—	—	—	150	150
KH_2PO_4	68	170	68	170	—	—
KCl	—	300	—	—	—	—
$FeSO_4 \cdot 7\,H_2O$	28	—	28	28	28	28
Na_2EDTA	37	—	37	37	37	37
Sequestrene 330 Fe	—	28	—	—	—	—
Micro elements (mg/*l*)						
H_3BO_3	2	3	3	6.2	3	3
$MnSO_4 \cdot H_2O$	5	10	10	22.3	10	10
$ZnSO_4 \cdot 7\,H_2O$	1.5	2	2	8.6	2	2
$NaMoO_4 \cdot 2\,H_2O$	0.1	0.25	0.25	0.25	0.25	0.25
$CuSO_4 \cdot 5\,H_2O$	0.015	0.025	0.025	0.025	0.025	0.025
$CoCl \cdot 6\,H_2O$	0.015	0.025	0.025	0.025	0.025	0.025
KI	0.25	0.75	0.75	0.83	0.75	0.75
Sugars (g*l*)						
Sucrose	9.0	0.25	0.25	30	34	20
Mannose	—	0.25	0.25	—	—	—
Glucose	9.0	68.4	108.9	—	—	—
Fructose	—	0.25	02.5	—	—	—
Ribose	—	0.25	0.25	—	—	—
Xylose	—	0.25	0.25	—	—	—
Rhamnose	—	0.25	0.25	—	—	—
Cellobiose	—	0.25	0.25	—	—	—
Sorbitol	—	0.25	0.25	—	—	45
Mannitol	81.9	—	0.25	—	—	45
Vitamins and organic acids (mg/*l*)						
Inositol	100	100	100	100	100	100
Nicotinic acid	4	—	1	0.5	1	1
Nicotinic amid	—	1	—	—	—	—
Pyrridoxin-HCl	0.7	1	1	0.5	1	1
Thiamine-HCl	4	10	1	0.1	10	10
D-Ca-Pantothenate	—	1	1	—	—	—
Folic acid	0.4	0.4	0.4	—	—	—
p-Aminobenzoic acid	—	0.02	0.02	—	—	—
Biotin	0.04	0.01	0.01	—	—	—
Cholinchloride	—	1	1	—	—	—
Ascorbic acid	—	2	2	—	—	—
Vitamin A	—	0.01	0.01	—	—	—
Vitamin D_3	—	0.01	0.01	—	—	—
Vitamin B_{12}	0.02	0.02	—	—	—	—
Glycine	—	2	—	—	—	—
Na-pyruvate	—	20	20	—	—	—
Citric acid	—	40	40	—	—	—
Malate	—	40	40	—	—	—
Fumarate	—	40	40	—	—	—
Organic supplements (mg/*l*)						
Edamin	—	—	—	1000	—	—
Casein hydrolysate	—	250	250	—	—	250

Table 2 (continued)
SOME COMMONLY USED MEDIA FOR PROTOPLAST CULTURE

	V 47[149]	KM 8p[131]	V-KM[150]	MS[151a]	B5[152]	WB[137]
Coconut milk	—	—	20 ml/l	—	—	—
Hormones (mg/l)						
2,4-Dichlorophenoxyacetic acid	—	0.2	—	—	2	2
2-Isopentenyladenine	—	—	—	—	—	0.2
α-Naphthaleneacetic acid	1.5	1	1.5	—	—	—
Indole-3-acetic acid	—	—	1	—	—	—
Zeatin	—	0.5	—	—	—	—
Kinetin	—	—	—	3	0.75	1
6-Benzylaminopurine	0.4	—	0.4	—	—	—

[a] MS and B5 are designed as tissue culture media and are therefore lacking osmotic stabilizers.

Detailed studies on the role of vitamins in a protoplast culture medium are lacking but the presence of some vitamins is beneficial for good protoplast growth.[50,97,115,126,133] Amino acids have been reported to enhance protoplast division for some special protoplast cultures.[44,86,119,134]

Casein hydrolysate,[131,135-137] yeast extract,[131,138-140] and potato extract[141] are also beneficial for protoplast growth. Coconut milk is included today in the protoplast culture media by several workers. These supplements are, however, not always positive or necessary for successful growth of protoplasts. Media containing them are chemically undefined and vary with the quality of the supplements.

5. Growth Factors

The majority of protoplast culture media contain one or more auxins plus one or two cytokinins to stimulate protoplast division and growth. Only a few protoplast systems grow without added growth regulators in the culture media.[142] By including a synthetic auxin transport inhibitor, a phytotropin, in the culture medium, mesophyll protoplasts of *Nicotiana debneyi* were able to form colonies and plants without exogenous auxin.[143] The authors concluded that the protoplasts were self-sufficient in their auxin requirements. In general, protoplast culturing starts with a relatively high, 1 to 3 mg/l, concentration of NAA or 2,4-D along with a lower, 0.1 to 1.0 mg/l, of BAP or zeatin. When protoplast division has started, it is often recommended to change the exogenous hormone supply by feeding the cultures with culture medium of changed hormone content. The hormone concentration can be especially critical in low density cultures.[126] A change in the auxin to cytokinin balance is the most often used method for stimulating morphogenesis in the callus formed. In some cases a reduction of the auxin content in the culture medium allows plant regeneration. However, most protoplast systems require more complicated and stepwise changes over several weeks. The effects of growth regulators on cell wall formation, cell division, and plant regeneration will be discussed further in later chapters.

F. Environmental Factors

1. Light

Generally, high light intensity inhibits protoplast growth when applied from the beginning of the culture. The inhibition by high light intensity is not clearly understood but might be related to the rapid bleaching of the chloroplasts. Many workers initiate protoplast culturing in darkness or dim light for a few days and later transfer the cultures to light of about 2000 to 5000 lux.[94,102,126] There are also reports of better protoplast growth when the cultures are kept in continuous darkness.[22,37] In contrast, it has been shown that for legume species light was necessary for initiating protoplast division.[67] The sensitivity to light seems to be genetically controlled and some species are light sensitive while others show light tolerance.[84]

2. Temperature and pH

Protoplast cultures are generally cultured at temperatures ranging between 20 to 28°C. However, some tomato[146] and cotton[116] protoplasts even grow well at higher temperatures. Chilling of freshly isolated tomato mesophyll protoplasts at 7°C for 12 hr increased the division frequency.[147] A similar effect has been obtained by chilling cotyledons of the legume *Cyamopsis tetragonoloba*, before isolating protoplasts from them.[148]

A pH in the range of 5.5 to 5.9 is recommended for most protoplast culture media and it seems to be satisfactory for most protoplasts. Somewhat higher pH values also enhance protoplast division. The pH of the culture medium has been found to be important in studies of the toxicity of exogenously supplied auxin.[142]

IV. CONCLUSIONS

Protoplast isolation and culture is a well-established technique for many plant species. The choice of source material for protoplast isolation and the importance of its physiological status are emphasized by many workers. Shoot cultures seem to be an excellent material which gives high and reproducible yields of protoplasts. The selection of culture techniques is also of considerable importance. The use of agarose as a gelling agent or for embedded cultures seems to be an important improvement for many protoplasts which are difficlut to culture. The nutritional requirements differ with different species and must be empirically determined. A majority of the plants which can successfully be used for protoplast isolation belong to the family Solanaceae. Unfortunately, most plants of legumes and cereals are still difficult or impossible to culture from protoplasts. The reason for this is unknown, but most probably many factors together are responsible for the failure in culturing these protoplasts. The procedure for protoplast isolation works satisfactorily today even for many legumes and a few cereals when mesophyll tissues are used. The culturing of these protoplasts is, however, still difficult or, in the case of the cereals, impossible. In many cases cell wall formation and/or cell division are inhibited. Basic research is required to clarify the mechanisms involved in the regulation of cell wall formation and cell division in protoplasts. The nutritional requirements of cultured protoplasts are poorly understood and an improvement in culturing can certainly be obtained by using more complete culture media including appropriate growth regulator supply. The protoplast source, especially for cereals, seems to be important in determining the success or failure in the culturing of protoplasts. Meristematically active tissues such as immature embryos, immature leaves, and immature inflorescences are most likely to respond to treatment. There has, however, been significant progress in culturing protoplasts isolated from various sources within the forage legumes during the last few years, but with the grain legumes the progress has been slower.[144] Cells competent for forming cell walls and dividing seem to stay competent only a short period of time. It is desirable to find ways of prolonging this period of competence or ways of inducing competence for cell wall formation and cell division. We are, however, far from understanding how this can be accomplished. The possibility of selecting and culturing morphogenetic cell lines from recalcitrant plant species indicates that totipotent cells are present in the cereals[145] and that it is possible to keep competent cells in cultures. Pretreatment of the source material as well as selection of special genotypes might improve the culture success with recalcitrant species.

REFERENCES

1. Cocking, E. D., Plant cell protoplasts, isolation and development, *Ann. Rev. Plant Physiol.*, 23, 29, 1972.
2. Constabel, F.,Isolation and culture of plant protoplasts, in *Plant Tissue Culture Methods*, Wetter, L. R. and Constabel, F., Eds., N. R. C. C., Prairie Regional Laboratory, Saskatoon, Sask., Canada, 1982, 38.
3. Evans, P. K. and Cocking, E. C., Isolated plant protoplasts, in *Plant Tissue and Cell Culture*, 2nd ed., Street, H. E., Ed., Blackwell Scientific, Oxford, 1977, 103.
4. Klercker, J., Eine Methode zur Isolterung lebender Protoplasten, *Svensk. Vet'Akad. Forh., (Stockholm)*, 9, 463, 1892.
5. Cocking, E. C., A method for the isolation of plant protoplasts and vacuoles, *Nature (London)*, 187, 927, 1960.
6. Patnaik, G., Wilson, D., and Cocking, E. C., Importance of enzyme purification for increased plating efficiency and plant regeneration from single protoplasts of *Petunia parodii, Z. Pflanzenphysiol.*, 102, 199, 1982.
7. Pilet, P. E., Transaminase activities of root protoplasts, *Experientia*, 28, 638, 1972.
8. Binding, H., Regeneration un Verschmelzung nackter Laubmoosprotoplasten, *Z. Pflanzenphysiol.*, 55, 305, 1966.
9. Bilkey, P. C. and Cocking, E. C., A non-enzymatic method for the isolation of protoplasts from callus of *Saintpaulia ionantha* (African violet), *Z. Pflanzenphysiol.*, 105, 285, 1982.
10. Takebe, I., Otsuki, Y., and Aoki, S., Isolation of tobacco mesophyll cells in intact and active state, *Plant Cell Physiol.*, 9, 115, 1968.
11. Nagata, T. and Takebe, I., Cell wall regeneration and cell division in isolated tobacco mesophyll protoplasts, *Planta*, 92, 301, 1970.
12. Power, J. B. and Cocking, E. C., Isolation of leaf protoplasts: macromolecular uptake and growth substance response, *J. Exp. Bot.*, 21, 64, 1970.
13. Zieg, R. G. and Outka, D. E., The isolation, culture, and callus formation of soybean pod protoplasts, *Plant Sci. Lett.*, 18, 105, 1980.
14. Premecz, G., Ruzicska, P., Olah, T., and Farkas, G. L., Effect of "osmotic stress" on protein and nucleic acid synthesis in isolated tobacco protoplasts, *Planta*, 141, 33, 1978.
15. Kaur-Sawhney, R., Altman, A., and Galtson, A. W., Dual mechanisms in polyamine-mediated control of ribonuclease activity in oat leaf protoplasts, *Plant Physiol.*, 62, 158, 1978.
16. Lu, C. Y., Vasil, V., and Vasil, I. K., Isolation and culture of protoplasts of *Panicum maximum* Jacq. (Guinea Grass) — somatic embryogenesis and plantlet formation, *Z. Pflanzenphysiol.*, 104, 311, 1981.
17. Vasil, I. K. and Vasil, V., Isolation and culture of protoplasts, in *Perspectives in Plant Cell and Tissue Cultures*, Vasil, I. K., Ed., Academic Press, New York, 1980 *(Int. Rev. Cytol.*, 1, Suppl. 11B).
18. Binding, H., Nehls, R., Kock, R., Finger, J., and Mordhorst, G., Comparative studies on protoplast regeneration in herbaceous species of the *Dicotyledoneae* class, *Z. Pflanzenphysiol.*, 101, 119, 1981.
19. Engler, D. E. and Grogan, R. G., Isolation, culture and regeneration of lettuce leaf mesophyll protoplasts, *Plant Sci. Lett.*, 28, 223, 1982.
20. Kao, K. N., Gamborg, O. L., Miller, R. A., and Keller, W. A., Cell division in cells regenerated from protoplasts of soybean and *Haplopappus gracilis, Nature (New Biol.)*, 232, 124, 1971.
21. Slabas, A. R., Powell, A. J., and Lloyd, C. W., An improved procedure for the isolation and purification of protoplasts from carrot suspension culture, *Planta*, 147, 283, 1980.
22. Santos, A. V. P. dos, Outka, D. E., Cocking, E. C., and Davey, M. R., Organogenesis and somatic embryogenesis in tissues derived from leaf protoplasts and leaf explants of *Medicago sativa, Z. Pflanzenphysiol.*, 99, 261, 1980.

23. Nagata, T. and Ishill, S., A rapid method for the isolation of mesophyll protoplasts, *Can. J. Bot.*, 57, 1820, 1979.
24. Nagata, T., Okada, K., Takebe, I., and Matsui, C., Delivery of tobacco mosaic virus RNA into plant protoplasts mediated by reverse-phase evaporation vesicles (liposomes), *Mol. Gen. Genet.*, 184, 161, 1981.
25. Schwenk, W., Pearson, A., and Roth, M. K., Soybean mesophyll protoplasts, *Plant Sci. Lett.*, 23, 153, 1981.
26. Lu, D. Y., Cooper-Bland, S., Pental, D., Cocking, E. C., and Davey, M. R., Isolation and sustained division of protoplasts from cotyledons of seedlings and immature seeds of *Glycine max* L., *Z. Pflanzenphysiol.*, 111, 389, 1983.
27. Johnson, L. B., Stuteville, D. L., Higgins, R. K., and Douglas, H. L., Pectolyase Y-23 for isolating mesophyll protoplasts from several *Medicago* species, *Plant Sci. Lett.*, 26, 133, 1982.
28. Kao, K. B. and Michayluk, M. R., Plant regeneration from mesophyll protoplasts of alfalfa, *Z. Pflanzenphysiol.*, 96, 135, 1980.
29. Patnaik, G. and Cocking, E. C., A new enzyme mixture for the isolation of leaf protoplasts, *Z. Pflanzenphysiol.*, 107, 41, 1983.
30. Lang, H. and Kohlenbach, H. W., Differentiation of alkaloid cells in cultures of *Macleaya* mesophyll protoplasts, *Planta Med.*, 46, 78, 1982.
31. Sink, K. C. and Niedz, R. P., Factors controlling plating efficiency in tomato, in *Proc. 5th Int. Cong. Plant Tissue and Cell Cultures,* Fujiwara, A., Ed., IAPTC, Maruzen Co., Tokyo, Japan, 1982, 583.
32. Rao, I. V. R., Mehta, U., and Ram, H. Y. M., Whole plant regeneration from cotyledonary protoplasts of *Crotolaria juncea*, in *Proc. 5th Int. Cong. Plant Tissue and Cell Cultures,* Fujiwara, A., Ed., IAPTC, Maruzen Co., Tokyo, Japan, 1982, 595.
33. Mii, M. and Cheng, S.-M., Callus and root formation of mesophyll protoplasts of carnation, in *Proc. 5th Int. Cong. Plant Tissue and Cell Cultures,* Fujiwara, A., Ed., IAPTC, Maruzen Co., Tokyo, Japan, 1982, 585.
34. Lia, K.-L and Liu, L.-F., Correlation between fine structure and viability of rice protoplasts, *Proc 5th Int. Cong. Plant Tissue and Cell Cultures,* Fujiwara, A., Ed., IAPTC, Maruzen, Tokyo, Japan, 1982, 603.
35. Nomura, K., Nitta, T., Fukei, K., Fujimura, T., and Komamine, A., Isolation and characterization of protoplasts from carrot somatic embryos, in *Proc. 5th Int. Cong. Plant Tissue and Cell Cultures,* Fujiwara, A., Ed., IAPTC, Maruzen Co., Tokyo, Japan, 1982, 587.
36. Vasil, I. K., The progress, problems, and prospects of plant protoplast research, in *Advances in Agronomy,* Academic Press, New York, 1976, 28, 119.
37. Arcioni, S., Davey, M. R., Santos, A. V. P. dos, and Cocking, E. C., Somatic embryogenesis in tissues form mesophyll and cell suspension protoplasts of *Medicago coerulea* and *M. glutinosa, Z. Pflanzenphysiol.*, 106, 105, 1982.
38. Saxena, P. K., Gill, R., Rashid, A., and Maheshwari, S. C., Plantlet formation from isolated protoplasts of *Solanum melongena* L., *Protoplasma,* 106, 355, 1981.
39. Bhatt, D. P. and Fassuliotis, D., Plant regeneration from mesophyll protoplasts of eggplant *Z. Pflanzenphysiol.*, 104, 81, 1981.
40. Gunn. R. E. and Shepard, J. F., Regeneration of plants from mesophyll derived protoplasts of British potato (*Solanum tuberosum* L.), *Plant Sci. Lett.*, 22, 97, 1981.
41. Cassels, A. C. and Barlass, M., Method for the isolation of stable mesophyll protoplasts from tomato leaves throughout the year under standard conditions, *Physiol. Plant.*, 42, 236, 1978.
42. Cassels, A. C. and Cocker, F. M., Seasonal and physiological aspects of the isolation of tobacco protoplasts, *Physiol. Plant.*, 56, 69, 1982.
43. Shepard, J. F. and Totten, R. D., Isolation and regeneration of tobacco mesophyll cell protoplasts under low osmotic conditions, *Plant Physiol.*, 55, 689, 1975.
44. Donn, G., Cell division and callus regeneration from leaf protoplasts *of Vicia narbonensis, Z. Pflanzenphysiol.*, 86, 65, 1978.
45. Gatenby, A. A. and Cocking, E. C., Callus formation from protoplasts of marrow stem kale, *Plant Sci. Lett.*, 8, 275, 1977.
46. Berry, S. F., Lu, D. Y., Pental, D., and Cocking, E. C., Regeneration of plants from protoplasts of *Lactuca sativa* L., *Z. Pflanzenphysiol.*, 108, 31, 1982.
47. Binding, H., Cell cluster formation by leaf protoplasts from axenic cultures of haploid *Petunia hybrid* L., *Plant Sci. Lett.*, 2, 185, 1974.
48. Binding, H., Jörgensen, J., Krumbiegel-Schroeren, G., Finger, J., Mordhorst, G., and Suchowiat, G., Culture of apical protoplasts from shoot culture in the orders *Fabales, Rosales,* and *Caryophyllales,* in *Poster Proc., 6th Int. Protoplast Symp.*, Potrykus, I., Harms, C. T., Hinnen, A., Hütter, R., King, P. J., and Shillito, R. D., Eds., Birkhäuser Verlag, Boston, 1983, 34.
49. Smith, M. A. L. and McCowan, B. H., A comparison of source tissues for protoplast isolation from three woody plant species, *Plant Sci. Lett.*, 28, 149, 1983.

50. Xu, Z.-H., Davey, M. R., and Cocking, E. C., Isolation and sustained division of *Phaseolus aureus* (mung bean) root protoplasts, *Z. Pflanzenphysiol.*, 104, 289, 1981.
51. Lu, D. Y., Pental, D., and Cocking, E. C., Plant regeneration from seedling cotyledon protoplasts, *Z. Pflanzenphysiol.*, 107, 59, 1982.
52. Burger, E. W. and Hackett, W., The isolation, culture and division of protoplasts from *Citrus* cotyledons, *Physiol. Plant.*, 56, 324, 1982.
53. Ahuja, P. S., Hadiuzzaman, S., Davey, M. R., and Cocking, E. C., Prolific plant regeneration from protoplast derived tissues of *Lotus corniculatus* L., *Plant Cell Rep.*, 2, 101, 1983.
54. Glimelius, K., and Ottosson, A., Improved culture ability of the genus *Brassica* by using hypocotyls as the source for protoplasts, in *Poster Proc. 6th Int. Protoplast Symp.*, Potrykus, I., Harms, C. T., Hinnen, A., Hütter, R., King, P. J., and Shillito, R. D., Eds., Birkhäuser Verlag, Boston, 1983, 64.
55. Bokelmann, G. S. and Roest, S., Plant regeneration from protoplasts of potato (*Solanum tuberosum* cv. Bintje), *Z. Pflanzenphysiol.*, 109, 259, 1983.
56. Altman, A., Ravindar, K.-S., and Galston, A. W., Stabilization of oat leaf protoplasts through polyamine-mediated inhibition of senescence, *Plant Physiol.*, 60, 570, 1977.
57. Uchimiya, H. and Murashige, T., Evaluation of parameters in the isolation of viable protoplasts from cultured tobacco cells, *Plant Physiol.*, 54, 936, 1974.
58. Douglas, G. C., Keller, W. A., and Setterfield, G., Somatic hybridization between *Nicotiana rustica* and *N. tabacum*. I. Isolation and culture of protoplasts and regeneration of plants from cell cultures of wild type and chlorophyll deficient strains, *Can. J. Bot.*, 59, 208, 1981.
59. Simmonds, J. A., Simmonds, D. A., and Cumming, B. G., Isolation and cultivation of protoplasts from morphogenetic callus cultures of *Lilium*, *Can. J. Bot.*, 57, 512, 1979.
60. Wallin, A., Glimelius, K., and Eriksson, T., Pretreatment of cell suspensions as a method to increase the protoplast yield of *Haplopappus gracillis*, *Physiol. Plant.*, 40, 307, 1977.
61. Bornman, J. F., Bornman, C. H., and Björn, L.-O., Effects of ultraviolet radiation on viability of isolated *Beta vulgaris* and *Hordeum vulgare* protoplasts, *Z. Pflanzenphysiol.*, 105, 297, 1982.
62. Nagata, T. and Takebe, I., Cell wall regeneration and cell division in isolated tobacco mesophyll protoplasts, *Planta*, 92, 301, 1970.
63. Arnold, S. von, and Eriksson, T., Factors, influencing the growth and division of pea mesophyll protoplasts, *Physiol. Plant.*, 36, 193, 1976.
64. Potrykus, I., Harms, C. T., Lörz, H., and Thomas, E., Callus formation from stem protoplasts of corn (*Zea mays* L.) *Mol. Gen. Genet.*, 156, 347, 1977.
65. Binding, H. and Nehls, R., Regeneration of isolated protoplasts of *Vica faba* L., *Z. Pflanzenphysiol.*, 88, 327, 1978.
66. Krumbiegel, G. and Schieder, O., Selection of somatic hybrids after fusion of protoplasts *Datura innoxia* Mill and *Atropa belladonna* L., *Planta*, 145, 371, 1979.
67. Oelck, M. M., Bapat, V. A., and Schieder, O., Protoplast culture of three legumes: *Arachis hypogaea, Melilotus officinalis, Trifolium resupinatum*, *Z. Pflanzenphysiol.*, 106, 173, 1982.
68. Kanai, R., and Edwards, G. E., Purification of enzymatically isolated mesophyll protoplasts from C_3, C_4 and crassulacean acid metabolism plants using an aqueous dextran polyethylene glycol two-phase system, *Plant Physiol.*, 52, 484, 1973.
69. Larkin, P. J., Purification and viability determinations of plant protoplasts, *Planta*, 128, 213, 1976.
70. Harms, C. T. and Potrykus, I., Enrichment for heterokaryocytes by the use of iso-osmotic density gradients after plant protoplast fusion, *Theor. Appl. Genet.*, 53, 49, 1978.
71. Scowcroft, W. R. and Larkin, P. J., Isolation, culture and plant regeneration from protoplasts of *Nicotiana debnevi*, *Aust, J. Plant Physiol.*, 7, 635, 1980.
72. Alexander, R. and Dubert, F., The isolation, purification, and culture of *Brassica napus* cv. Lingot cotyledon protoplasts, in *Poster Proc., 6th Int. Protoplast Symp.*, Potrykus, I., Harms, C. T., Hinnen, A., Hütter, R., King, P. J., and Shillito, R. D., Eds., Birkhäuser Verlag, Boston, 1983, 62.
73. Glimelius, K., Wallin, A., and Eriksson, T. C., Agglutinating effects of Concanavalin A on isolated protoplasts of *Daucus carota*, *Physiol. Plant.*, 31, 225, 1974.
74. Ferrari, T. E., Palmer, J. E., and Widholm, J., Monitoring protoplast production from plant cells, *Plant Sci. Lett.*, 4, 145, 1975.
75. Hughes, B. G., White, F. G., and Smith, M. A., Effect of plant growth, isolation, and purification conditions on barley protoplast yield, *Biochem. Physiol. Pflanzen.*, 172, 67, 1978.
76. Ulrich, T. H., Chowdhury, J. B., and Widholm, J. M., Callus and root formation from mesophyll protoplasts of *Brassica rapa*, *Plant Sci. Lett.*, 19, 347, 1980.
77. Karanaratne, S. M. and Scott, K. J., Mitotic activity in protoplasts isolated from *Sorghum bicolor* leaves, *Plant Sci. Lett.*, 23, 11, 1981.
78. Raj, B. and Herr, J. M., Isolation of protoplasts from the placental cells of *Lycopersicon pimpinellifolium*, *Exp. Cell Res.*, 64, 469, 1970.
79. Hooley, R., Protoplasts isolated from aleurone layers of wild oat (*Avena fatua* L.) exhibit the classic response to gibberellic acid, *Planta*, 154, 29, 1982.

80. De La Roche, A. I., Keller, W. A., Sing, J., and Siminovitch, D., Isolation of protoplasts from unhardened tissues of winter rye and wheat, *Can. J. Bot.*, 55, 1181, 1977.
81. Evans, P. K. and Cocking, E. C., Isolated plant protoplasts, in *Plant Tissue and Cell Culture*, Street, H. E., Ed., Blackwell Scientific, Oxford, 1977, 103.
82. Potrykus, I. and Durand, J., Callus formation from single protoplasts of *Petunia Nature (New Biol.)*, 237, 286, 1972.
83. Schieder, O., Regeneration of haploid and diploid *Datura innoxia* Mill. mesophyll protoplasts to plants, *Z. Pflanzenphysiol.*, 76, 462, 1975.
84. Banks, M. S. and Evans, P. K., A comparison of the isolation and culture of mesophyll protoplasts from several *Nicotiana* species and their hybrids, *Plant Sci. Lett.*, 7, 409, 1976.
85. Lörz, H., Wernicke, W., and Potrykus, I., Culture and plant regeneration of *Hyoscyamus* protoplasts, *Planta Med.*, 36, 21, 1979.
86. Kirby, E. G. and Cheng, T. Y., Colony formation from protoplasts derived from douglas fir cotyledons, *Plant Sci. Lett.*, 14, 145, 1979.
87. Partanen, C. R., Power, J. B., and Cocking, E. C., Isolation and division of protoplasts of *Pteridium aquilinum*, *Plant Sci. Lett.*, 17, 333, 1980.
88. Partanen, C. R., Filter paper as a support and carrier for plant protoplast cultures, *In Vitro*, 17, 77, 1981.
89. Santos, A. V. P. dos, Dutka, D. E., Cocking, E. C., and Davey, M. R., Organogenesis and somatic embryogenesis in tissues derived from leaf protoplasts and leaf explants of *Medicago sativa*, *Z. Pflanzenphysiol.*, 99, 261, 1980.
90. Kao, K. N., Keller, W. A., and Miller, R. A., Cell division in newly formed cells from protoplasts of soybean, *Exp. Cell. Res.*, 62, 338, 1970.
91. Vasil, V., and Vasil, I. K., Isolation and culture of cereal protoplasts. I. Callus formation from pearl millet (*Pennisetum americanum*) protoplasts, *Z. Pflanzenphysiol.*, 92, 379, 1979.
92. Bawa, S. B. and Torrey, J. G., "Budding" and nuclear division in cultured protoplast of corn, *Convolvulus* and onion, *Bot. Gaz.*, 132, 420, 1971.
93. Vasil, V. and Vasil, I. K., Regeneration of tobacco and Petunia plants from protoplasts and culture of corn protoplasts, *In Vitro*, 1D, 83, 1974.
94. Potrykus, I., Harms, C. T., and Lörz, H., Multiple-drop array (MDA) technique for the large-scale testing of culture media variations in hanging microdrop cultures of single cell systems. I. The technique, *Plant Sci. Lett.*, 14, 231, 1979.
95. Gleba, Y. Y., Microdroplet cultures: tobacco plants from single mesophyll protoplasts, *Naturwissenschaften*, 65, 158, 1978.
96. Bergmann, L., Growth and division of single cells of higher plants *in vitro*, *J. Gen. Physiol.*, 43, 841, 1960.
97. Nagata, T. and Takebe, I., Plating of isolated tobacco mesophyll protoplasts on agar medium, *Planta*, 99, 12, 1971.
98. Adaoha-Mbanaso, E. N. and Roscoe, D. H., Alginate: an alternative to agar in plant protoplast culture, *Plant Sci. Lett.*, 25, 61, 1981.
99. Brodelius, P. and Nilsson, K., Entrapment of plant cells in different matrixes. A comparative study, *FEBS Letts.*, 122, 312, 1980.
100. Lörz, H., Larkin, P. I., Thomson, I., and Scowcroft, W. R., Improved protoplast culture and agarose media, *Plant Cell, Tissue Organ Cult.*, 2, 217, 1983.
101. Shillito, R. D., Paszkowski, J., and Potrykus, I., Agarose plating and a bead type culture technique enable and stimulate development of protoplast-derived colonies in a number of plant species, *Plant Cell Rep.*, 2, 244, 1983.
102. Carlberg, I., Glimelius, K., and Eriksson, T., Improved culture ability of potato protoplasts by use of activated charcoal, *Plant Cell Rep.*, 2, 223, 1983.
103. Duckworth, M. and Yaphe, W., The structure of agar, *Carbohydr. Res.*, 16, 435, 1971.
104. Adams, T. L. and Townsend, J. A., A new procedure for increasing efficiency of protoplast plating and clone selection, *Plant Cell Rep.*, 2, 165, 1983.
105. Raveh, D., Huberman, E., and Galun, E., *In vitro* culture of tobacco protoplasts: use of feeder techniques to support division of cells plated at low densities, *In Vitro*, 9, 216, 1973.
106. Raveh, D. and Galun, E., Rapid regeneration of plants from tobacco protoplasts plated at low densities, *Z. Pflanzenphysiol.*, 76, 76, 1975.
107. Cella, R. and Galun, E., Utilization of irradiated carrot cell suspensions as feeder layer for cultured *Nicotiana* cells and protoplasts, *Plant Sci. Lett.*, 19, 243, 1980.
108. Menczel, L., Lazar, G., and Maliga, P., Isolation of somatic hybrids by cloning *Nicotiana* heterokaryons in nurse cultures, *Planta*, 143, 29, 1978.
109. Hein, T., Prezewozny, T., and Schieder, O., Culture and selection of somatic hybrids using an auxotrophic cell line, *Theor. Appl. Genet.*, 64, 119, 1983.

110. Shepard, J. F., Bidney, D., and Shahin, E. A., Potato protoplasts in crop improvement, *Science*, 208, 17, 1980.
111. Shepard, J. F., Mutant selection and plant regeneration from potato mesophyll protoplasts, in *Genetic Improvements of Crops*, Rubenstein, I., Gegenbach, B., and Green, C. E., Eds., University of Minnesota Press, Minneapolis, 1980, 185.
112. Pental, D., Cooper-Bland, S., Harding, K., Cocking, E. C., and Muller, E. J., Cultural studies on nitrate reductase deficient *Nicotiana tabacum* mutant protoplasts, *Z. Pflanzenphysiol.*, 105, 219, 1982.
113. Binding, H., Nehls, R., and Jörgensen, I., Protoplast regeneration in higher plants, in *Proc. 5th Int. Cong. Plant Tissue and Cell Culture*, Tokyo, Fujiwara, A., Ed., IAPTC, Maruzen, Tokyo, Japan, 1982, 575.
114. Shepard, J. F. and Totten, R. E., Mesophyll cell protoplasts of potato — isolation, proliferation, and plant regeneration, *Plant Physiol.*, 60, 313, 1979.
115. Arnold, S. von, and Eriksson, T., A revised medium for growth of pea mesophyll protoplasts, *Physiol. Plant.*, 39, 257, 1977.
116. Bhojwani, S. S., Power, J. B., and Cocking, E. C., Isolation, culture, and division of cotton callus protoplasts, *Plant, Sci. Lett.*, 8, 85, 1977.
117. Nehls, R., Isolation and regeneration of protoplasts from *Solanum nigrum* L., *Plant Sci. Lett.*, 12, 183, 1978.
118. Upadhya, M. D., Isolation and culture of mesophyll protoplasts of potato (*Solanum tuberosum* L.), *Potato Res.*, 18, 438, 1975.
119. Boyes, C. J. and Sink, K. C., Regeneration of plants from callus derived protoplasts of *Salpiglossis sinuata*, *J. Am. Hortic. Sci.*, 106, 42, 1981.
120. Zapata, F. J., Sink, K. C., and Cocking, E. D., Callus formation from leaf mesophyll protoplasts of theree *Lycopersicon* species: *L. esculentum* cv. Walter, *L. pimpinellifolium* and *L. hirsutum* f. *glabratum*, *Plant Sci. Lett.*, 23, 41, 1981.
121. Kao, K. N., Gamborg, O. L., Michayluk, M. R., Keller, W. A., and Miller, R. A., The effects of sugars and inorganic salts on cell regeneration and sustained division in plant protoplats, in Protoplasts et Fusion de Cellules Somatiques Vegetales, Colloq. Int., CNRS 212, Centre National de la Recherche Scientifique, Paris, 1973, 207.
122. Pelcher, L. E., Gamborg, O. L., and Kao, K. N., Bean mesophyll protoplasts: production, culture, and callus formation, *Plant Sci. Lett.*, 3, 107, 1974.
123. Bourgin, J. P. and Missionier, C., Culture of haploid mesophyll protoplasts from *Nicotiana alata*, *Z. Pflanzenphysiol.*, 87, 55, 1978.
124. Xuan, L. T. and Menczel, L., Improved protoplast culture and plant regeneration from protoplast-derived callus in *Arabidopsis thaliana*, *Z. Pflanzenphysiol.*, 96, 77, 1980.
125. Meyer, Y. and Abel, W. O., Budding and cleavage division of tobacco mesophyll protoplasts in relation to pseudo-wall and wall formation, *Planta*, 125, 1, 1975.
126. Caboche, M., Nutritional requirements of protoplast derived, haploid tobacco cells grown at low cell densities in liquid medium, *Planta*, 149, 7, 1980.
127. Brar, D. S., Rambold, S., Constabel, F., and Gamborg, O. L., Isolation, fusion, and culture of *Sorghum* and corn protoplasts, *Z. Pflanzenphysiol.*, 96, 269, 1980.
128. Mülbach, H. P., Different generation potentials of mesophyll protoplasts from cultivated and a wild species of tomato, *Planta*, 148, 89, 1980.
129. Constabel, F., Kirkpatrick, J. W., and Gamborg, O. L., Callus formation from mesophyll protoplasts of *Pisum sativum*, *Can. J. Bot.*, 51, 2105, 1973.
130. Shahin, E. A. and Shepard, J. F., Cassava mesophyll protoplasts: isolation, proliferation, and shoot formation, *Plant Sci. Lett.* 17, 459, 1980.
131. Kao, K. N. and Michayluk, M. R., Nutrient requirements for growth of *Vicia hajastana* cells and protoplasts at very low population density in liquid media, *Planta*, 126, 105, 1975.
132. Negrutiu, I. and Mousseau, J., Protoplast culture from *in vitro* grown plants in *Nicotiana sylvestris* Spegg. and Comes., *Z. Pflanzenphysiol.*, 100, 373, 1980.
133. Brenneman, F. M. and Galston, A. W., Experiments on the cultivation and calli formation of agriculturally important plants. I. Oat (*Avena sativa* L.) *Biochem. Physiol. Pflanzen.*, 168, 453, 1975.
134. Gamborg, O. L. Shyluk, J. P., and Kartha, K. K., Factors affecting the isolation and callus formation in protoplasts from the shoot apices of *Pisum sativum* L., *Plant Sci. Lett.*, 4, 285, 1975.
135. Poirier-Hamon, S., Rao, P. S., and Harada, H., Culture of mesophyll protoplasts and stem segments of *Antirrhinum majus* (snapdragon): growth and organization of embryoids, *J. Exp. Bot.*, 25, 752, 1974.
136. Galun. E. and Raveh, D., *In vitro* culture of tobacco protoplasts: survival of haploid and diploid protoplasts exposed to X-ray radiation at different times after isolation, *Radiat. Bot.*, 15, 79, 1975.
137. White. D. W. R. and Bhojwani, S. S. Callus formation from *Trifolium arvense* protoplasts derived cells plated at low density, *Z. Pflanzenphysiol.*, 102, 257, 1981.

138. Kameya, T. and Uchimya, H., Embryoids derived from isolated protoplasts of carrot, *Planta*, 103, 356, 1972.
139. Maretzki, A. and Nickell, L. G., Formation of protoplasts from sugarcane cell suspensions and the regeneration of cell cultures from protoplasts, in Protoplastes et Fusion de Cellules Somatiques Vegetales, Colloq. Intern., CNRS 212, Centre National de la Recherche Scientifique, Paris, 1973, 51.
140. Dudits, D., Kao, K. N., Constabel, F., and Gamborg, O. L., Embryogenesis and formation of tetraploid and hexaploid plants from carrot protoplasts, *Can. J. Bot.*, 54, 1063, 1976.
141. Li, X., Plantlet regeneration from mesophyll protoplasts of *Digitalis lanata* Ehrh., *Theor. Appl. Genet.*, 60, 345, 1981.
142. Davey, M. R., Recent development in the culture and regeneration of plant protoplasts, in *6th Int. Protoplast Symp.*, Potrykus, I., Harms, C. T., Hinnen, A., Hütter, R., King, P. J., and Shillito, R. D., Eds., Birkhaüser Verlag, Boston, 1983, 19.
143. Larkin, P. J., Scowcroft, W. R., Geissler, A. E., and Katekar, G. F., Phytotropins. IV. Effects of phytotropins on cultured plant cells and protoplasts. *Aus. J. Plant. Physiol.*, 9, 297, 1982.
144. Dale, P. J., Protoplast culture and plant regeneration of cereals and other recalcitrant crops, in *6th Int. Protoplast Symp.*, Potrykus, I., Harm, C. T., Hinnen, A., Hütter, R., King, P. J., and Shillito, R. D., Eds., Birkhäuser Verlag, Boston, 1983, 31.
145. Vasil, V. and Vasil, I. K., Isolation and culture of cereal protoplasts. II. Embryogenesis and plantlet formation from protoplasts of *Pennisetum americanum*, *Theor. Appl. Genet.*, 56, 97, 1980.
146. Zapata, F. J., Evans, P. K., Power, J. B., and Cocking, E. C., The effect of temperature on the division of leaf protoplasts of *Lycopersicon esculentum* and *Lycospersicon peruvianum*, *Plant Sci. Lett.*, 8, 119, 1977.
147. Mühlback, H. P. and Thiele, H., Response to chilling of tomato mesophyll protoplasts, *Planta*, 151, 399, 1980.
148. Saxena, P. K., Gill, R., Rashid, A., and Maheshawari, S. C., Colony formation by cotyledonary protoplasts of *Cyamopsis tetragonoloba* L., *Z. Pflanzenphysiol.*, 106, 277, 1982.
149. Binding, H., Regeneration von haploiden und diploiden Pflanzen aus Protoplasten von *Petunia hybrida* L., *A. Pflanzenphysiol.*, 74, 327, 1974.
150. Binding, H. and Nehls, R., Regeneration of isolated protoplasts in *Solanum dulcamara* L., *Z. Pflanzenphysiol.*, 85, 279, 1977.
151. Murashige, T. and Skoog, F. A., A revised medium for rapid growth and bioassays with tobacco tissue cultures, *Physiol. Plant.*, 15, 473, 1962.
152. Gamborg, O. L., Miller, R. A., and Ojima, K., Nutrient requirements of suspension cultures of soybean root cells, *Exp. Cell Res.*, 50, 151, 1968.

Chapter 2

REGENERATION OF PLANTS

Horst Binding

TABLE OF CONTENTS

I. INTRODUCTION

Isolated protoplasts have become most useful for studies in plant physiology, genetics and plant breeding. The preservation of organogenic potential as well as the differentiation of adventitious embryos and shoots are fascinating features in developmental biology; organ formation and plant regeneration, indeed, are essential steps in experimentation aimed at cloning and genetic modification. This chapter is devoted to developmental phenomena in embryophytes. Observation and results of plant regeneration from callus and cell suspensions which did not originate from protoplasts will not be considered here, even though some of the processes and conditions are widely congruent.

The term "regeneration" already applies to the formation of a cell by synthesis of a new cell wall. It is used here in the sense of reorganization of shoots and plants. According to this definition regeneration has been obtained with 98 species (Table 1). First reports of bryophytes date from 1966[14] and those of spermatophytes from 1971.[113] Plant regeneration has not yet been achieved in 40% of those species in which callus or cell suspensions have been obtained with protoplasts (65 out of 163 species; Table 2). In a number of species, plant regeneration occurred occasionally; sometimes results could not be repeated by other investigators or even in the same laboratory. In other cases, success was limited to only one genotype so that techniques employed in plant regeneration were not applicable to a given species in general.

The dilemma in this field of research is the lack of knowledge of the basic conditions and processes leading to certain organogenic activities. A few apparently general features, however, may be concluded from available experimental results. Common observations will be presented without or with only a few exemplary references.

Three different pathways led to regeneration: moss cells formed protonemata from which gametophores were differentiated; in seed plants, either the formation of adventitious shoots was followed by rooting of cuttings or by grafting, or embryo-like structures were organized which then developed directly into plants.

II. REGENERATION OF GAMETOPHYTES IN BRYOPHYTES

With liverworts, regeneration was obtained only in *Sphaerocarpos.*[123] Polarity was established in the regenerating cells after transfer from isotonic to hypotonic seawater. Gametophyte thalli developed either directly from a short tubal outgrowth of the cells or after formation of a rhizoid.

Regeneration in mosses followed a common pattern. Polarity was visible in protoplast-derived cells by the first day of culture. High intensities of red or blue light were needed for this process in *Physcomitrella.*[65] The process was delayed by 2 days after the transfer to low osmotic media in *Funaria hygrometrica* and *Physcomitrium eurystomum* when the cells were grown in seawater.[14] Protonemata, buds, and gametophytes developed on low osmotic media which were usually employed for in vitro propagation.

III. REGENERATION VIA SHOOT FORMATION

A. Processes of Shoot Formation

Shoots arise from meristems after organization of a polar center. The establishment of polarity was a distinct developmental step, especially in cases in which cell colonies were transferred from medium 8p[134] or modifications of this medium rich in organic nutrients. In other cases it was a more gradual, also easily reversible, process. The organization of shoot primordia was frequently detectable as areas of darker green

Table 1
SPECIES OF EMBRYOPHYTES IN WHICH PLANTS HAVE BEEN REGENERATED FROM PROTOPLASTS (OR AT LEAST PLANTLETS OR SHOOTS)

Bryophyta

Anoectangium thomsonii (95)
Funaria hygrometrica (14,30,11)
Physcomitrella patens (112,6,37,65)
Physcomitrium eurystomum (14)
Polytrichum juniperinum (51)
Sphaerocarpos donnellii (123)

Spermatophyta

Angiospermae
- Magnoliatae (Dicotyledoneae)
  - Ranunculaceae
    - *Nigella arvensis* (23)
    - *Ranunculus sceleratus* (41)
  - Rosaceae
    - *Fragaria ananassa* (26)
  - Fabaceae
    - *Clianthus formosus* (27,28)
    - *Lotus corniculatus* (2)
    - *Medicago coerulea* (5)
    - *Medicago glutinosa* (5)
    - *Medicago sativa* (69,93,66,82,7)
    - *Onobrychis viciifolia* (3)
    - *Trifolium repens* (59,127,3)
    - *Trigonella corniculata* (81,94)
    - *Trigonella foenum-graecum* (130,104)
    - *Vigna aconitifolia* (105)
  - Rutaceae
    - *Citrus aurantium* (118)
    - *Citrus limon* (118)
    - *Citrus reticulata* (118)
    - *Citrus sinensis* (117)
  - Linaceae
    - *Linum usitatissimum* (26,8)
  - Geraniaceae
    - *Geranium sp.* (67)
  - Euphorbiaceae
    - *Manihot esculenta* (103)
  - Apiaceae
    - *Daucus carota* (58,68,42,24)
  - Brassicaceae
    - *Arabidopsis thaliana* (132,24)
    - *Brassica napus* (70,115,71,74,76,131,4,56)
    - *Brassica oleracea* (82,131,56)
    - *Sinapis alba* (26)
    - *Sinapis arvensis* (26)
  - Resedaceae
    - *Reseda luteola* (24)
  - Solanaceae
    - *Atropa belladonna* (57,77)
    - *Browallia viscosa* (90)
    - *Capsicum annuum* (97)
    - *Datura innoxia* (99,82)
    - *Datura metel* (100)
    - *Datura meteloides* (100)
    - *Hyoscyamus muticus* (79,126,125)
    - *Lycopersicon esculentum* (83)
    - *Lycopersicon peruvianum* (133,1)
    - *Nicotiana acuminata* (33)
    - *Nicotiana alata* (31,1,48)
    - *Nicotiana debneyi* (102)
    - *Nicotiana forgetiana* (91)
    - *Nicotiana glauca* (33)
    - *Nicotiana langsdorffii* (33)
    - *Nicotiana longiflora* (33)
    - *Nicotiana nesophila* (45)
    - *Nicotiana otophora* (9)
    - *Nicotiana paniculata* (33)
    - *Nicotiana plumbaginifolia* (52)
    - *Nicotiana repanda* (45)
    - *Nicotiana rustica* (53)
    - *Nicotiana sanderae* (89)
    - *Nicotiana stocktonii* (45)
    - *Nicotiana suaveolens* (33)
    - *Nicotiana sylvestris* (9,32,84,46)
    - *Nicotiana tabacum* (113,88,72,16,92, 78,54,50,62,125,82)
    - *Petunia axillaris* (91)
    - *Petunia hybrida* (40,43,49,15,119,75,17,18)
    - *Petunia inflata* (91)
    - *Petunia parodii* (63)
    - *Petunia parviflora* (111)
    - *Petunia violacea* (91)
    - *Salpiglossis sinuata* (34,35)
    - *Solanum brevidens* (87)
    - *Solanum chacoense* (38)
    - *Solanum dulcamara* (20,19)
    - *Solanum etuberosum* (10)
    - *Solanum fernandezianum* (10)
    - *Solanum khasianum* (75)
    - *Solanum luteum* (24,86)
    - *Solanum melongena* (13,96,55)
    - *Solanum nigrum* (85,24,86,25,26)
    - *Solanum phureja* (101)
    - *Solanum tubersoum* (109,22,124,106, 110,60,114,107,108,116,1,29,39,73)
    - *Solanum xanthocarpum* (98)
  - Scrophulariaceae
    - *Digitalis lanata* (128)
    - *Nemesia strumosa* (64)
    - *Rehmannia glutinosa* (129)
  - Lamiaceae
    - *Majorana hortensis* (26)
  - Asteraceae
    - *Cichorium endivia* (23)
    - *Cichorium intybus* (24)
    - *Gaillardia grandiflora* (23)
    - *Helianthus annuus* (23)
    - *Lactuca sativa* (12,44)
    - *Senecio jacobaea* (24)

Table 1 (continued)
SPECIES OF EMBRYOPHYTES IN WHICH PLANTS HAVE BEEN REGENERATED FROM PROTOPLASTS (OR AT LEAST PLANTLETS OR SHOOTS)

Senecio silvaticus (24)
Senecio vernalis (26)
Senecio viscosus (24)
Senecio vulgaris (21)

Liliatae (Monocotyledoneae)
Liliaceae
Asparagus officinalis (36)

Amaryllidaceae
Hemerocallis cv. (47)
Poaceae (plantlets only)
Bromus inermis (70)
Panicum maximum (80)
Pennisetum americanum (120)
Pennisetum purpureum (121,122)

Note: Numbers indicate references, ordered by year of publication, alphabetically with a year; only selected references are given for frequently investigated species, e.g., *Nicotiana tabacum.*

pigmentation. Here the cell surfaces were integrated into flat or hemispherical structures and leaves soon developed. In some cases leaves were formed on the callus apparently without the preceding step of apex organization. This observation, however, would require further confirmation by detailed investigations, because it may be explained by disorganization of apices which had already produced leaf primordia, as repeatedly documented for sunflower[28] and somatic hybrid clones of *Hyoscyamus* (x) *Nicotiana.*[135]

Shoots developed either directly from the isolated protoplasts or, more commonly, from cell colonies. The adventitious origin of the shoots was clearly recognized when plant material without preexisting apices was employed. When apex-derived protoplasts were present, their direct development to shoot primordia may be assumed. Such a pathway would lead to the formation of primary shoots without the need for de- and redifferentiation. This could have been the case when regeneration occurred within a short period of time. For example, shoots emerged in *Petunia hybrida* and *Solanum nigrum* 14 days after isolation of protoplasts from shoot apices.[28] The protoplast-derived cells were similar to apical meristem cells and unlike the more common parenchyma-type cells of callus cultures. The meristem-type growth which continued during prolonged culture in medium V-KM[20] was observed also with several other species in which no shoot primordia have been obtained.[24] On the assumption that meristem-type cell colonies represent true meristems, initiation of a shoot apex would require one developmental step only, i.e., the initiation of the characteristic polarity of a shoot primordium.

B. Plant Material

In general, in the capability of shoot formation from protoplasts of dicotyledons has been found to occur widely with respect to the taxonomic order, as well as to the type of organ used as protoplast donor material. In monocotyledons, a parallel finding has not been reported except for *Asparagus.*[36] Plants have been regenerated from all parts of embryos and seedlings,[2,4,131] from leaves as the most common source, from petals,[28,48] shoot apices,[22-27] roots,[130,131] as well as from callus and cell suspension cultures. The results led to the conclusion that totipotency of plant cells is not limited to primary meristems and that cells do not lose this potential during differentiation. Doubts regarding totipotency first emerged in the course of investigations of cereal cells.[136] This particular issue will be discussed in Section V. Loss of totipotency has frequently been observed in the course of in vitro culture. This became especially apparent with cultures of parenchyma-type cells from which protoplasts often could not be regenerated to plants.

Table 2
PLANTS FROM PROTOPLASTS OF EMBRYOPHYTES (AT LEAST GAMETOPHYTES IN BRYOPHYTA)

	Callus	Plant
Bryophyta		
Marchantiatae	1	6
Sphaerocarpaceae	—	1
Marchantiaceae	1	—
Bryatae		
Pottiales — Pottiaceae	—	1
Funariales — Funariaceae	—	3
Polytrichales — Polytrichaceae	—	1
Spermatophyta		
Gymnospermae	3	—
Pinaceae		
Angiospermae		
Magnoliatae	53	85
Ranunculales — Ranunuculaceae	3	2
Caryophyllales — Cactaceae	1	—
Rosales — Rosaceae	3	1
Fabales — Fabaceae	16	9
Myrtales — Onagraceae	1	—
Rutales — Rutaceae	1	4
Sapindales — Aceraceae	1	—
Geraniales		
Linaceae	—	1
Geraniaceae	—	1
Rhamnales — Vitaceae	1	—
Euphorbiales — Euphorbiaceae	—	1
Araliales — Apiaceae	1	1
Capparales — Brassicaceae	4	5
Resedaceae	—	1
Cucurbitales — cucurbitaceae	2	—
Malvales — Malvaceae	1	—
Dipsacales — Valerianaceae	1	—
Gentianales — Apocynaceae	1	—
Solanales — Solanaceae	7	45
Convolvulaceae	2	—
Scrophulariales — Scrophulariaceae	2	3
Lamiales — Lamiaceae	—	1
Asterales — Asteraceae	5	10
Liliatae		
Liliales	8	6
Liliaceae	1	1
Amaryllidaceae	—	1
Poales — Poaceae	7	4

Note: Numbers indicate species or hybrid cultivars (in *Rosa*). CallUs means that only callus or continuously proliferating cell suspensions were obtained some of which differentiated roots or proembryo-like structures.

C. Growth Conditions for Shoot Formation

It may be concluded from a number of experiments that as a rule shoot formation is most likely to occur when the conditions for shoot development are applied comparatively early. Delay in addition of regeneration media, for instance, drastically reduced the degree of shoot formation of callus of *Solanum nigrum*.[25] The organogenic capacity was also lost in *Petunia* regenerants after a short growth phase on callus-promoting NT medium[137] with NAA.

1. Culture at High Osmolarities

It is clear that the composition of the culture medium plays a decisive role in regulating shoot formation. Protoplast media which are of high osmolarity were primarily chosen for their suitability to promote mitotic acitivity and were usually not designed to initiate differentiation (for composition of protoplast culture media see Chapter 1). Still, some criteria for the suitability of certain factors of protoplast culture media for the preservation or induction of organized growth can be deduced from investigations made. Meristem-type cells formed particularly in the presence of rich media. In a series of experiments with V-KM medium, for instance, differentiation of meristem cells occurred in six species of Solanaceae as well as in the genera *Arabidopsis, Reseda, Majorana, Cichorium, Hellianthus,* and *Senecio.*[24] The protoplasts were obtained from shoot tips; in the Solanaceae, also from leaves. Predominantly parenchyma-type cells resulted form using the poor medium V-47.[15] Colonies of meristem-type cells differentiated shoots a few days after transfer from 8p medium and modifications of it to low osmotic media or had occurred already in the protoplast culture medium. Shoot formation from colonies of parenchyma-type cells transplanted from V-47 was delayed and less effective in *Petunia.*[28] Early reduction of the auxin concentration in media of high osmolarity was proposed to improve the yield of organogenic colonies. The duration of culture of colonies on high osmotic media appeared not to be critical within 3 and 7 weeks in an experiment with *Clianthus,* as long as fresh medium was added once a week.[28]

2. Culture in Regeneration Media of Low Osmolarity

Reduction of the osmolarity of culture media was performed either continuously or stepwise, resulting in sugar concentrations adequate for nutrition only. Retaining an osmoticum in addition to nutritional carbohydrates may be advantageous as shown with 0.2 *M* mannitol in *Solanum tuberosum* cultures.[1,114] When the osmolarities were lowered the composition of basal media NT, MS[138] and B5[139] remained unchanged. Some protoplast culture media, e.g., V-47, proved not to be useful when employed at low osmotic concentration. Basal MS and B5 media enriched by certain organic additives were most commonly used for the development of shoots. Single inorganic components have rarely been evaluated with respect to organogenesis.

Special attention has been paid to the action of phytohormones on protoplast-derived calli. A cytokinin was required for shoot formation in most of the taxa. Concentrations of 0.03 to 10 μM were usually applied. 6-Benzyladenine, zeatin, zeatin ribose, or kinetin were adequate. Development of shoots was suppressed at higher concentrations, e.g., in *Lotus.*[2] Also, combinations of benzyladenine and zeatin were reported to promote shoot formation. Low concentrations of auxins were added to regeneration media for protoplast-derived cells of a number of species. Most frequently NAA, but also 1AA and 2,4-D, have been applied. Organization of shoot apical meristems was commonly impeded by higher auxin concentrations (above 10 μM). In *Petunia* the capability of shoot formation was irreversibly lost by a 7-d growth period on agar medium with 2 mg/ℓ NAA.[17,18] A longer period of competence for shoot formation in the presence of 2,4-D[28] was found in *Clianthus*. This corresponds with the observation that regeneration was obtained from protoplasts which were prepared from cell suspensions grown in the presence of 2,4-D.[132] Gibberellic acid has also been included in regeneration media, for instance with potato.[29,76]

Regeneration media were further fortified with vitamins. These were employed as for plant tissue culture. Various combinations have been proposed. However, reports did not specify whether a certain formulation reflected a requirement for differentiation only, a requirement by all cell cultures of the particular species under investigation, or some other requirement. Nutritional sugars were applied either at the same concentrations as in protoplast culture media (2 or 3%) or at slightly lower concentrations (1%), or they were drastically reduced, as when osmotic stabilization had been established by a nutritional sugar, for instance, in 8p media.

Good results have been obtained by the addition of organic nitrogen sources. Glycine has been most widely used, as well as arginine, glutamine, and asparagine.[74,102,127] Some media were further enriched by additional amino acids, casein hydrolyzate, or liquid coconut endosperm. Adenine sulfate was also included in regeneration media.

The most important factor of the physical environment appeared to be illumination. Changes from dark or low light intensities to higher intensities, sometimes in combination with the transfer from continuous illumination to a day/night regime have been advantageous. Light is probably essential for the induction of polarity at least in most of the investigated species, comparably to the moss cell system. Interestingly, in *Solanum khasianum* initiation of regeneration was established in light, whereas well-organized shoots developed only in subsequent continuous darkness.

Some observations indicated that the stimulus for the commencement of a certain developmental pattern was provided by the changing of the growth conditions rather than by the prolonged application of these conditions. This led to the hypothesis that the polar organization of the meristem is induced by a short-term polarity of applied nutritional and/or physical factors.

D. Rooting, Transfer to Soil, and Grafting

Usually, adventitious roots have been induced easily at cuttings of regenerated shoots on media which contained either no hormones or only auxin, preferably IAA. The process has been promoted frequently by reduction of the sucrose concentration to 1%, in some cases by addition of cytokinin at low concentrations, or by combinations of two or three auxins. The young plantlets survived transfer to soil if handled cautiously. Good results were obtained with 20 species[28] if the leaflets were small and the roots measured not more than 5 mm in length and if the root meristems were prevented from drying by rapid transfer into sopping wet soil. Thereafter, the humidity of the greenhouse (about 80%) was sufficient for continued growth. Other investigators prefer intermediate steps including passages through other substrates such as vermiculite, or gradual adaptation to the lower humidity of the greenhouse.

In a few cases, rooting was nearly or completely impossible, or the plantlets did not survive transfer to soil. The only way to grow these protoplast-derived plants to maturity was by grafting. This approach has been successful in interspecific hybrids of *Datura*,[140] for instance.

IV. REGENERATION VIA EMBRYO-LIKE STRUCTURES

A. Definition and Process

Embryos are young organisms enclosed in maternal tissue. In spermatophytes, they are further characterized by a certain developmental pattern leading to the bipolar organization of a root and a shoot apex. In the typical case, embryos are derived from a zygote, but adventitious (somatic) embryos are also formed in nature in the nucellus of several species. Adventitious embryo-like structures also were observed in callus and cell suspension cultures, and on embryo explants. The terms "embryoid" or "embryo" are used for these structures. Although they occur outside maternal tissue, the similarity of the developmental patterns of somatic with sexual embryos may justify the application of the term "embryo". Structures showing early developmental stages up to the organization of bipolarity are commonly referred to as "proembryos". In cases in which the development terminates with proembryo-like structures, misinterpretation cannot be excluded. This may have been the case with *Antirrhinum*[141] in which dark-green globular bodies differentiated a root only.[24]

In *Brassica napus* the development of embryo formation started immediately from cultured cells.[76] Usually, globular stages emerged in cell colonies. In cases in which the original cells were differentiated, the stimulus for the initiation of embryogenesis was clearly set after protoplast isolation (cf. Section IV,B). With *Daucus* protoplasts from embryogenic cell cultures, even if these appeared unorganized, embryogenesis was already established but arrested in early stages.[142]

B. Plant Material

Regeneration of plantlets via embryogenesis was obtained in the genera *Ranunculus*,[41] *Medicago*,[5,69] *Trigonella*,[81] *Vigna*,[105] *Citrus*,[117,118] *Daucus*,[28,42,68] *Brassica*,[76,115] *Atropa*,[57] *Hyoscyamus*,[79] *Lycopersicon*,[133] *Nicotiana*,[46,78] *Solanum*,[55] *Hemerocallis*,[47] *Bromus*,[70] *Panicum*,[80] and *Pennisetum*.[120-122] Embryo-like structures appeared also in *Glycine max*[143] and *Antirrhinum majus*[141] but did not grow to plantlets. Protoplasts were isolated from somatic and sexual embryos,[4,69,74,82] leaves,[41,46,76,78,79,81] shoot apices,[28] and from embryogenic cell suspensions.[5,47,57,70,80,117,118,120-122]

C. Growth Conditions for Embryo Formation

The most important factor for the induction and preservation of embryogenesis in monocoytledons as well as in dicotyledons was the initial culture of protoplasts and cell colonies in the presence of 2,4-D. Subsequent reduction of the 2,4-D concentrations then led to the development of embryos and plantlets. Other auxins at lower concentrations and cytokinins were added to the protoplast media for embryo formation in cell cultures of various members of the class Magnoliatae. Independence of 2,4-D was recently observed in cell colonies of *Trigonella*. These produced embryos on MS medium with NAA and 6-BA; plants developed after transfer to hormone-free medium. NAA alone was sufficient for embryogenesis in *Atropa*.[57]

A number of regenerated embryos continued to grow and formed plantlets which could be transferred to soil by applying the same techniques as described for regenerated shoots (see Secion III.C). In *Solanum melongena*, development of embryos was followed by a phase in which multiple embryo formation took place. The secondary embryos were then grown to maturity.[55]

V. GENETIC CONTROL OF PLANT REGENERATION

No indication of genetic pecularities with respect to polarity and filamentous outgrowth can be deduced from the few experiments with protoplast-derived cells of bryophytes. In atypical protenemata (rhizoidonema) of *Funaria* the production of buds was impeded, apparently by polyploidy.[14,30]

Investigations of plant formation from protoplasts of angiosperms revealed some indication of the genetic control. Embryogenesis was obtained with both classes of angiosperms by application of rather similar procedures; shoots developed reliably only in the dicotyledons. No basic taxonomic limitations for one or the other pathway of plant regeneration existed; so far, plant formation occurred in orders across all subclasses. Recalcitrant taxa, i.e., those with which callus or cell suspensions could be established while plant formation failed, also appeared to occur in diverse orders.

Several authors discussed the occurrence of basic taxonomic differences in the response of protoplasts and cell colonies with respect to plant regeneration. The family Solanaceae, for instance, was claimed to be extraordinarily conditioned, whereas legumes were classified as recalcitrant. Recent successes in the Fabaceae, however, disprove the validity of this opinion: abundant shoot formation and embryo differentiation both were obtained in three legume species. However, until now more recalcitrant species have been found with Fabaceae than with Solanaceae. Monocotyledons responded negatively to most of the experimental approaches; shoot regeneration failed except in *Asparagus.*[36] As regards trees, *Citrus* is the only regenerated species so far. With monocotyledons and trees reliable plant regeneration required protoplasts which were prepared from embryogenic cell cultures. Success obtained in one laboratory, however, has often not been repeated by other research groups indicating that more efforts are needed to make technologies generally applicable.

Protoplast regeneration has been tried in some families which so far have not been mentioned because not even callus could be obtained from protoplasts. These are, for example, the economically important Orchidaceae and Chenopodiaceae. Callus from various types of explants showed a low degree or lack of organized growth so that one may expect regeneration to be difficult from protoplast-derived cell colonies, also.

An interesting conclusion may be drawn from the co-occurrence of good response and failure of regeneration in closely related taxa. One may say that not basal metabolism but certain products of secondary differentiation are responsible for the variable behavior of the protoplasts and their derivatives. Such opinion may be supported by the observation, for example, with several Asteraceae, that regeneration was easily obtained when isolated cells retained their meristematic character up to the onset of shoot growth.[24]

Control of response to conditions of plant regeneration by a single or few genes can be deduced from genotypic differences within a given species. This has been stressed repeatedly and is suggested by restoration of organogenic potential in somatic hybrids of two *Nicotiana* strains.[144] Obviously, regeneration is highly controlled by phytohormones; metabolism of these compounds probably plays a decisive role in the expression of intraspecific genotypic peculiarities with respect to regeneration. Other genetically controlled factors are also suggested by experimental results of regeneration. In *Solanum tuberosum* and some Asteraceae, for instance,[24] regeneration did not occur unless meristem protoplasts were prepared in the absence of differentiated tissue. The protoplasts were probably affected by particular secondary metabolites. Species-specific responses to environmental factors of various nature also indicate the existence of gene-controlled pathways which, however, are far from being understood. In fact, these considerations seem to be in accordance with a common basic congruity of the processes involved in regeneration. However, some findings in somatic hybridization experiments do not fit this picture. Intraspecific and intrageneric hybrids have been regenerated frequently (see Chapter 12). Plants also have been obtained from protoplast fusion of the closely related genera *Lycopersicon* and *Solanum*;[145] however, in hybrids from more distantly related genera,[135,146-150] formation of well-organized shoots was retarded, mitoses exhibited reduced chromosome numbers, and the phenotypes of the shoots tended to resemble one of the parents. This can be explained only by different physiological (and also genetic) bases for the accomplishment of differentiation in the parental species. These differences appeared to be more relevant in interfamilial hybrid clones in which plants have never been obtained.

The influence of some types of genotypic changes on totipotency has frequently been investigated or suggested but not documented. Haploidy[1,15,22,31,71,76,79,100,115] as well as polyploidy[42] were found not to affect totipotency in protoplast-derived cell colonies. Aneuploidy has been suggested as one of the factors restricting organized growth in long-term cultures. Single gene mutations which were significant for organogenic response have not been found.

VI. PROPERTIES OF REGENERATED PLANTS

The genetic constitution of regenerated plants reflects the nature of the protoplast donor cells as well as the consequences of alterations of the isolated protoplasts and mutations in the course of regeneration. The phenotypes may also have changes due to physiological irregularities during plant development.

A. Phenotypic Normalization

Polyploid protonemata of *Funaria* showed gradual phenotypic normalization with respect to increasing formation of buds. This process is not easily explained because the chromosome numbers were stable.

A common peculiarity in regenerants of dicotyledons was a watery appearance of the leaves due to lack of gas-containing intercellular spaces. Normalization occurred spontaneously or as a result of transfer to larger culture vessels or media with reduced hormone contents. As another peculiarity, young plants on soil repeatedly formed irregular leaves and some other developmental abnormalities which subsequently returned to normal. These features may have been caused by physical or physiological factors (for example, humidity) or by the mosaic nature of the plantlets resulting from overgrowth of one type of tissue by more vigorous cell types.

B. Aspects of Cloning via Protoplasts

Regeneration from protoplasts has been proposed as a useful method for clonal propagation as well as for detection and segregation of somaclonal variation of the donor material. Predominantly homogeneous populations have been regenerated in a number of species. Separation of the mosaic components of donor material have been obtained from somatic hybrids[25] and transformed cell lines.[151]Segregation from mosaics may be counteracted occasionally by spontaneous fusion of different types of cells. Regeneration from such mixed cells, however, appeared to be a rare event as concluded from experiments with fusion hybrids and mutant-derived periclinal chimeras.[26]

Regeneration from heterokaryotic and cybrid protoplasts may lead to hybrid, cybrid, and uniparental cells aggregated in one cell colony. The regeneration of homogeneous or mosaic plants depends on several events. The later that organized growth is initiated and the earlier that the cell organelles segregate, the higher is the probability of the formation of genetically pure plants. However, when an apex is organized from more than one cell, mosaics may arise despite prior termination of organelle segregation. With protoplasts from suspensions of mixed cells it therefore seems more appropriate to avoid mosaics by employing embryogenesis, which supposedly starts from single cells. Subcloning of the regenerants via single cells is the most reliable way to obtain pure segregants from experiments which started with mixed protoplasts.

C. Somaclonal Variation by Mutation During Regeneration

It has been stated earlier that, usually, rather homogeneous populations of regenerants have been obtained, sometimes involving phenotypic normalization of development. However, stable variants of the parental phenotypes have also been found[110] (see also Chapter 10). In cases in which somaclonal variation in the original plant material is ruled out, the most likely explanation would be mutation in the course of regeneration; but in most cases clear-cut discrimination between mutant and epigenetic modification is not possible before sexual progenies are analyzed.

A number of morphological aberrants can be ascribed to the occurrence of polyploid or aneuploid tissue. Developmental disorders are increased if tissues of different chromosome numbers are associated in mosaic or periclinal chimeral material.

Mutations of single genes have apparently resulted in chlorophyll deficiencies in regenerated plants.[25] The situation is less clear for peculiar characters found in potato regenerants.[110]

VII. CONCLUDING REMARKS

Several processes are involved in the various pathways of regeneration from protoplasts. A prominent feature concerns conservation of developmental processes already initiated before the protoplast stage. Dedifferentiation of cultured cells to real meristematic cells and induction of polarity may also be important. Discrimination between initiation of plant regeneration on the one hand and resumption of regeneration which had already been initiated on the other hand turned out to be difficult in a number of cases. This uncertainty impedes the solution of the basic question of whether development and differentiation of plant cells are directed only by factors of their environment including intercellular correlations, or whether a certain developmental pattern may, once initiated, be continued by an intracellular, more or less irreversible, mechanism.

It has already been mentioned that reorganization of polar growth via embryos or adventitious shoots could not be established in a number of genotypes, species, genera, families, or even orders as a whole. Especially in the latter cases, the question has been raised as to whether the totipotency of meristematic cells is lost during their development to other differentiated forms, for example, in plant organs or nonembryogenic cell or callus cultures. Establishment of embryogenic cell clones from different types of explants in the family Poaceae, for instance, cannot be taken as an unequivocal disproval of this thesis. All explants were taken from immature organs probably containing cells which had not yet lost their totipotency. No clear-cut neoformation of either embryogenic or shoot apical differentiation from protoplasts has been found in this family, so far.

Organized growth may also be suppressed or abolished by secondary metabolites or by environmental factors. Alternatively, a stimulus for one or another step in the pathway of regeneration must be provided. It may be considered, for instance, in the *Daucus* system, that development is arrested in the meristematic stage until a stimulus is provided for formation of polarity.[142]

Variable as the pathways of regeneration and the factors and processes involved in the achievement of regeneration are diverse as the experimental approaches have been, especially in so-called recalcitrant taxa, success has been obtained by assaying different genotypes, various types of organs, and cell cultures, as well as modifications of the culture conditions for the protoplasts and their derivatives. The method most widely applied is apparently based on the uninterrupted preservation of meristematic differentiation from the donor cell to the establishment of polarity. This pathway may either be integrated in regeneration via embryos as has been established, for instance, in some Poaceae and Fabaceae, or it may lead to unipolar growth of shoots as occurred in several species of the Asteraceae.

ACKNOWLEDGMENTS

The author thanks Dr. Gabriela Krumbiegel-Schroeren, Dr. Reinhard Nehls, and Karin Binding for the help with the search for reference data and for critical reading of the manuscript.

REFERENCES

1. Adams, T. L. and Townsend, I. A., A new procedure for increasing efficiency of protoplast plating and clone selection, *Plant Cell Rep.*, 2, 165, 1983.
2. Ahuja, P. S., Hadiuzzaman, S., Davey, M. R., and Cocking, E. C., Prolific plant regeneration from protoplast-derived tissues of *Lotus corniculatus* L. (Birdsfoot Trifoil), *Plant Cell Rep.*, 2, 101, 1983.
3. Ahuja, P. S., Lu, D. Y., Cocking, E. C., and Davey, M. R., An assessment of the cultural capabilities of *Trifolium repens L. (white clover) and Onobrychis viciifolia* Scop. (sainfoin) mesophyll protoplasts, *Plant Cell Rep.*, 2, 269, 1983.
4. Alexander, R. and Dubert, F., The isolation, purification and culture of *Brassica napus* cv. Lingot cotyledon protoplasts, in *Protoplasts 1983 — Poster Proceedings,* Potrykus, I., Harms, C. T., Hinnen, A., Hütter, R., King, P. J., and Shillito, R. D., Eds., Birkhäuser Verlag, Basel, 1983, 62.
5. Arcioni, S., Davey, M. R., Santos, A. V. P. Dos, and Cocking, E. C., Somatic embryogenesis in tissues from mesophyll and cell suspension protoplasts of *Medicago coerulea* and *M. glutinosa, Z. Pflanzenphysiol.*, 106, 105, 1982.
6. Ashton, N. W., and Cove, D. J., The isolation and preliminary characterization of auxotrophic and analogue resistant mutants of the moss *Physcomitrella patens, Mol. Gen. Genet.*, 154, 87, 1977.
7. Atanassov, A. J. and Brown, D. C. W., Plant regeneration from suspension culture protoplasts and mesophyll protoplasts of alfalfa, in *Protoplasts 1983 — Poster Proceedings,* Potrykus, I., Harms, C. T., Hinnen, A., Hütter, R., King, P. J., and Shillito, R. D., Eds., Birkhäuser Verlag, Basel, 1983, 40.
8. Bakarat, M. N. and Cocking, E. C., Towards somatic hybridization in the genus *Linum* (flax), in *Protoplasts 1983 — Poster Proceedings,* Potrykus, I., Harms, C. T., Hinnen, A., Hütter, R., King, P. J., and Shillito, R. D., Eds., Birkhäuser Verlag, Basel, 1983, 78.
9. Banks, M. S. and Evans, P. K., A comparison of the isolation and culture of mesophyll protoplasts from several *Nicotiana* species and their hybrids, *Plant Sci. Lett.* 7, 409, 1976.
10. Barsby, T. and Shepard, J. F., Regeneration of plants from mesophyll protoplasts of *Solanum* species of the *Etuberosa* group, *Plant. Sci. Lett.*, 31, 101, 1983.
11. Batra, A. and Abel, W. O., Development of moss plants from isolated and regenerated protoplasts, *Plant. Sci. Lett.*, 20, 183, 1981.
12. Berry, S. F., Lu, D. Y., Pental, D., and Cocking, E. C., Regeneration of plants from protoplasts of *Lactuca sativa* L., *Z. Pflanzenphysiol.*, 108, 31, 1982.
13. Bhatt, D. P. and Fassuliotis, G., Plant regeneration from mesophyll protoplasts of eggplant, *Z. Pflanzenphysiol.*, 104, 81, 1981.
14. Binding, H., Regeneration und Verschmelzung nackter Laubmoosprotoplasten, *Z. Pflanzenphysiol.*, 55, 305, 1966.
15. Binding, H., Regeneration of haploid and diploid plants from protoplasts of *Petunia hybrida* L., *Z. Pflanzenphysiol,* 74, 327, 1974; *de Bruijn, F., transl., Plant Molec. Biol. Newsl.*, 1, 77, 1980.
16. Binding, H., Reproducibly high plating efficiencies of isolated protoplasts from shoot cultures of tobacco, *Physiol. Plant.*, 35, 225, 1975.
17. Binding, H. and Krumbiegel-Schroeren, G., Protoplast regeneration, in *Petunia,* Sink, K. C., Ed., *Theor. Appl. Genet.*, 9, 123, 1984.
18. Binding, H. and Krumbiegel-Schroeren, G., Isolation and culture of protoplasts. *Petunia,* in *Cell Culture and Somatic Cell Genetics of Plants,* Vol. 1, Vasil, I. K., Eds., Academic Press, New York, 1984, 345.
19. Binding, H. and Mordhorst, G., Haploid *Solanum dulcamara:* androgenesis, shoot culture and plant regeneration from isolated protoplasts, *Plant Sci. Lett.*, 35, 77, 1984.
20. Binding, H. and Nehls, R., Regeneration of isolated protoplasts to plants in *Solanum dulcamara* L., *Z. Pflanzenphysiol.*, 85, 279, 1977.
21. Binding, H. and Nehls, R., Protoplast regeneration to plants in *Senecio vulgaris* L., *Z. Pflanzenphysiol.*, 99, 183, 1980.
22. Binding, H., Nehls, R., Schieder, O., Sopory, S. K., and Wenzel, G., Regeneration of mesophyll protoplasts isolated from dihaploid clones of *Solanum tuberosum, Physiol. Plant.*, 43, 52, 1978.
23. Binding, H., Nehls, R., and Kock, R., Versuche zur Protoplastenregeneration dikotyler Pflanzen unterschiedlicher systematischer Zugehörigkeit, *Ber. Dtsch. Bot. Ges.*, 93, 667, 1980.
24. Binding, H., Nehls, R., Kock, R., Finger, J., and Mordhorst, G., Comparative studies on protoplast regeneration in herbaceous species of the Dicotyledoneae class, *Z. Pflanzenphysiol.*, 101, 119, 1981.
25. Binding, H., Jain, S. M., Finger, J., Mordhorst, G., Nehls, R., and Gressel, J., Somatic hybridization of an atrazine resistent biotype of *Solanum nigrum* with *Solanum tuberosum.* I. Clonal variation in morphology and in atrazine sensitivity, *Theor.* Appl. Genet., 63, 273, 1982.
26. Binding, H., Nehls, R., and Jörgensen, J., Protoplast regeneration in higher plants, in *Plant Tissue Culture,* Fujiwara, A., Ed., Maruzen Co., Tokyo, 1982, 575.

27. Binding H., Jörgensen, J., Krumbiegel-Schroeren, G., Finger, J., Mordhorst, G., and Suchowiat, G., Culture of apical protoplasts from shoot cultures in the orders Fabales, Rosales, and Caryophyllales, in *Protoplasts 1983 — Poster Proceedings*, Potrykus, I., Harms, C. T., Hinnen, A., Hütter, R., King, P. J., and Shillito, R. D., Eds., Birkhäuser Verlag, Basel, 1983, 90.
28. Binding, H., Finger, J., Jörgensen, J., Krumbiegel-Schroeren, G., Mordhorst, G., Nehls, R., Suchowiat, G., unpublished results; a paper on *Clianthus* in preparation.
29. Bokelman, G. S. and Roest, S., Plant regeneration from protoplasts of potato *Solanum tuberosum* cv. Bintje), *Z. Pflanzenphysiol.*, 109, 259, 1983.
30. Bopp, M., Zimmerman, S., and Knoop, B., Regeneration of protonema with multiple DNA content from isolated protoplasts of the moss *Funaria hygrometrica, Protoplasma*, 104, 119, 1980.
31. Bourgin, J. P. and Missionier, C., Culture of haploid mesophyll protoplasts from *Nicotiana alata, Z. Pflanzenphysiol.*, 87, 55, 1978.
32. Bourgin, J. P., Missonier, C., and Chupeau, Y., Culture de protoplastes de mesophylle de *Nicotiana sylvestris* Spegazzini et Comes haploide et diploide, *C. R. Acad. Sci. Paris Sér. D*, 282, 1853, 1976.
33. Bourgin, J. P., Chupeau, Y., and Missonier, C., Plant regeneration from mesophyll protoplasts of several *Nicotiana* species, *Physiol. Plant.*, 45, 288, 1979.
34. Boyes, C. J. and Sink, K. C., Regeneration of plants from callus derived protoplasts of *Salpiglossis sinuata, J. Am. Hortic. Sci.*, 106, 42, 1981.
35. Boyes, C. J., Zapata, F. J., and Sink, K. C., Isolation, culture and regeneration to plants of callus protoplasts of *Salpiglossis sinuata* L., *Z. Pflanzenphysiol.*, 99, 471, 1980.
36. Bui-Dang-Ha, D. and Mackenzie, I. A., The division of protoplasts from *Asparagus officinalis* L. and their growth and differentiation, *Protoplasma*, 78, 215, 1973.
37. Burgess, J. and Linstead, P. J., Studies on the growth and development of protoplasts of the moss *Physcomitrella patens* and its control by light, *Planta*, 151, 331, 1981.
38. Butenko, R. G., Kuchko, A. A., Vitenko, A. A., and Avetisov, V. A., Obtaining and cultivation of leaf mesophyll protoplasts of *Solanum tuberosum L. and Solanum chacoense* Bitt., *Soviet Plant Physiol.*, 24, 660, 1977.
39. Carlberg, I., Glimelius, K., and Eriksson, T., Improved culture ability of potato protoplasts by use of activated charcoal, *Plant Cell Rep.*, 2, 223, 1983.
40. Donn, G., Hess, D., and Potrykus, I., Wachstum und Differenzierung in aus isolierten Protoplasten von *Petunia hybrida* entstandenem Kallus, *Z. Pflanzenphysiol.*, 69, 423, 1973.
41. Dorion, N., Chupeau, Y., and Bourgin, J. P., Isolation, culture and regeneration into plants of *Ranunculus sceleratus* L. leaf protoplasts, *Plant Sci. Lett.*, 5, 325, 1975.
42. Dudits, D., Kao, K. N., Constabel, F., and Gamborg, O. L., Embryogenesis and formation of tetraploid and hexaploid plants from carrot protoplasts, *Can. J. Bot.*, 54, 1063, 1976.
43. Durand, J., Potrykus, I., and Donn, G., Plants from protoplasts of *Petunia, Z. Pflanzenphysiol.*, 69, 26, 1973.
44. Engler, D. E. and Grogan, R. G., Isolation, culture and regeneration of lettuce leaf mesophyll protoplasts, *Plant Sci. Lett.*, 28, 223, 1983.
45. Evans, D. A., Chromosome stability of plants regenerated from mesophyll protoplasts of *Nicotiana* species, *Z. Pflanzenphysiol.*, 95, 459, 1979.
46. Facciotti, D. and Pilet, P. E., Plants and embryoids from haploid *Nicotiana sylvestris* protoplasts, *Plant Sci. Lett.*, 15, 1, 1979.
47. Fitter, M. S. and Krikorian, A. D., Recovery of totipotent cells and plantlet production from daylily protoplasts, *Ann. Bot.*, 48, 591, 1981.
48. Flick, C. E. and Evans, D. A., Isolation, culture, and plant regeneration from protoplasts isolated from flower petals of ornamental *Nicotiana* species, *Z. Pflanzenphysiol.*, 109, 379, 1983.
49. Frearson, E. M., Power, J. B., and Cocking, E. C., The isolation, culture and regeneration of *Petunia* leaf protoplasts, *Dev. Biol.*, 33, 130, 1973.
50. Gamborg, O. L., Shyluk, J. P., Fowke, L. C., Wetter, L. R., and Evans, D., Plant regeneration from protoplasts and cell cultures of *Nicotiana tabacum* sulfur mutant (Su/Su), *Z. Pflanzenphysiol.*, 95, 255, 1979.
51. Gay. L., The development of leafy gametophytes from isolated protoplasts of *Polytrichum juniperinum* Willd., *Z. Pflanzenphysiol.*, 79, 33, 1976.
52. Gill, R., Rashid, A., and Maheshwari, S. C., Regeneration of plants from mesophyll protoplasts of *Nicotiana plumbaginifolia* Viv., *Protoplasma*, 96, 375, 1978.
53. Gill, R., Rashid, A., and Maheshwari, S. C., Isolation of mesophyll protoplasts of *Nicotiana rustica* and their regeneration into plants flowering in vitro, *Physiol. Plant.*, 47, 7, 1979.
54. Gleba, Y. Y., Microdroplet culture: tobacco plants from single mesophyll protoplasts, *Naturwissenschaften*, 65, 158, 1978.
55. Gleddie, S. C., Keller, W. A., and Setterfield, G., Somatic embryogenesis and plant regeneration from protoplasts of eggplant (*Solanum melongena* L.), in *Protoplasts 1983 — Poster Proceedings*, Potrykus, I., Harms, C. T., Hinnen, A., Hütter, R., King, P. J., and Shillito, R. D., Eds., Birkhäuser Verlag, Basel, 1983, 66.

56. Glimelius, K. and Ottosson, A., Improved culture ability of the genus *Brassica* by using hypocotyls as the source for protoplasts, in *Protoplasts 1983 — Poster Proceedings*, Potrykus, I., Harms, C. T., Hinnen, A., Hütter, R., King, P. J., and Shillito, R. D., Eds., Birkhäuser Verlag, Basel, 1983, 64.
57. Gosch, G., Bajaj, Y. P. S., and Reinert, J., Isolation, culture, and induction of embryogenesis in protoplasts from cell-suspension of *Atropa belladonna*, *Protoplasma*, 86, 405, 1975.
58. Grambow, H. J., Kao, K. N., Miller, R. A., and Gamborg, O. L., Cell division and plant development from protoplasts of carrot cell suspension cultures, *Planta*, 103, 348, 1972.
59. Gresshoff, P. M., In vitro culture of white clover: callus, suspension, protoplast culture and plant regeneration, *Bot. Gaz.*, 14, 157, 1980.
60. Gunn, R. E. and Shepard, J. F., Regeneration of plants from mesophyll derived protoplasts of British potato (*Solanum tubersoum* L.) cultivars, *Plant Sci. Lett.*, 22, 97, 1981.
61. Gwoze, E. A. and Waliszewska, B., Regeneration of enzymatically isolated protoplasts of the moss *Funaria hygrometrica* (L.) Sibth., *Plant Sci. Lett.*, 15, 41, 1979.
62. Harms, C. T., Lörz, H., and Potrykus, I., Multiple-drop array (MDA) technique for the large-scale testing of culture media variations in hanging microdrop cultures of single cell systems. II. Determination of phytohormone combinations for optimal division response in *Nicotiana tabacum* protoplast cultures, *Plant Sci. Lett.*, 14, 237, 1979.
63. Hayward, G. and Power, J. P., Plant production from leaf protoplasts of *Petunia parodii*, *Plant Sci. Lett.*, 4, 407, 1975.
64. Hess, D. and Leipoldt, G., Regeneration of roots and shoots from isolated mesophyll protoplasts of *Nemesia strumosa*, *Biochem. Physiol. Pflanzen*, 174, 411, 1979.
65. Jenkens, G. I. and Cove, D. J., Light requirements for regeneration of protoplasts of the moss *Physcomitrella patens*, *Planta*, 157, 39, 1983.
66. Johnson, L. B., Stuteville, D. L., Higgins, R. K., and Skinner, D. Z., Regeneration of alfalfa from protoplasts of selected Regen S clones, *Plant Sci. Lett.*, 20, 297, 1981.
67. Kameya, T., Culture of protoplasts from chimeral plant tissue of nature, *Jpn. J. Genet.*, 50, 417, 1975.
68. Kameya, T. and Uchimiya, H., Embryoids derived from isolated protoplasts of carrot, *Planta*, 103, 356, 1972.
69. Kao, K. N. and Michayluk, M.R., Plant regeneration from mesophyll protoplasts of alfalfa, *Z. Pflanzenphysiol.*, 96, 135, 1980.
70. Kao, K. N., Gamborg, O. L., Michayluk, M. R., Keller, M. R., and Miller, R. A., The effect of sugars and inorganic salts on cell regeneration and sustained division in plant protoplasts, in *Colloq. Int. C.N.R.S.*, 212, 207, 1973.
71. Kartha, K. K., Michayluk, M. R., Kao, K. N., Gamborg, O. L., and Constabel, F., Callus formation and plant regeneration from mesophyll protoplast of rape plants (*Brassica napus* L. cv. Zephyr), *Plant Sci. Lett.*, 3, 265, 1974.
72. Keller, W. A. and Melchers, G., The effect of high pH and calcium on tobacco leaf protoplast fusion, *Z. Naturforsch.*, 28c, 737, 1973.
73. Kikuta, Y., Saito, W., and Okazawa, Y., Viability and development of potato protoplast culture, in *Protoplasts 1983 — Poster Proceedings*, Potrykus, I., Harms, C. T., Hinner, A., Hütter, R., King, P. J., and Shillito, R. D., Eds., Birkhäuser Verlag, Basel, 1983, 64.
74. Kohlenbach, H. W., Wenzel, G., and Hoffmann, F., Regeneration of *Brassica napus* plantlets in cultures from isolated protoplasts of haploid stem embryos as compared with leaf protoplasts, *Z. Pflanzenphysiol.*, 105, 131, 1982.
75. Kowalczyk, T. P., Mackenzie, I. A., and Cocking, E. C., Plant regeneration from organ explants and protoplasts of medicinal plant *Solanum khasianum* C. B. Clarce var. chatterseeanum Sengupta (syn. *Solanum varianum* Dunal), *Z. Pflanzenphysiol.*, 111, 55, 1983.
76. Li, L. and Kohlenbach, H. W., Somatic embryogenesis in quite a direct way in cultures of mesophyll protoplasts of *Brassica napus* L., *Plant Cell Rep.*, 1, 209, 1982.
77. Lörz, H. and Potrykus, I., Regeneration of plants from mesophyll protoplasts of *Atropa belladonna*, *Experientia*, 35, 313, 1979.
78. Lörz, H., Potrykus, I., and Thomas, E., Somatic embryogenesis from tobacco protoplasts, *Naturwissenschaften*, 64, 439, 1977.
79. Lörz, H., Wernicke, W., and Potrykus, I., Culture and plant regeneration of *Hyoscyamus* protoplasts, *Planta Medica*, 36, 21, 1979.
80. Lu, D. Y., Vasil, V., and Vasil, I. K., Isolation and culture of protoplasts of *Panicum maximum* Jacq. (Guinea Grass) — somatic embryogenesis and plantlet formation, *Z. Pflanzenphysiol.*, 104, 311, 1981.
81. Lu, D. Y., Davey, M. R., and Cocking, E. C., Somatic embryogenesis from mesophyll protoplasts of *Trigonella corniculata* (Leguminosae), *Plant Cell Rep.*, 1, 278, 1982.
82. Lu, D. Y., Pental, D., and Cocking, E. C., Plant regeneration from seedling cotyledon protoplasts, *Z. Pflanzenphysiol.*, 107, 59, 1982.

83. Morgan, A. and Cocking, E. C., Plant regeneration from protoplasts of *Lycopersicon esculentum* Mill., *Z. Pflanzenphysiol.*, 106, 97, 1982.
84. Nagy, J. I. and Maliga, P., Callus induction and plant regeneration from mesophyll protoplasts of *Nicotiana sylvestris, Z. Pflanzenphysiol.*, 78, 453, 1976.
85. Nehls, R., Isolation and regeneration of protoplasts from *Solanum nigrum* L., *Plant Sci. Lett.*, 12, 183, 1978.
86. Nehls, R., Versuche zur Regeneration und Hybridisierung von Protoplasten Höherer Pflanzen und zur Selektion von Fusionsprodukten, Dissertation, University of Kiel, 1981.
87. Nelson, R. S., Creissen, G. P., and Bright, S. W. J., Plant regeneration from protoplasts of *Solanum brevidens, Plant Sci. Lett.*, 30, 355, 1983.
88. Ohyama, K. and Nitsch, J. P., Flowering haploid plants obtained from protoplasts of tobacco leaves, *Plant Cell Physiol.*, 13, 229, 1972.
89. Passiatore, J. E. and Sink, K. C., Plant regeneration from leaf mesophyll protoplasts of selected ornamental *Nicotiana* species, *J. Soc. Hortic. Sci.*, 106, 779, 1981.
90. Power, J. B., Frearson, E. M., George, D., Evans, P. K., Berry, S. F., Hayward, C., and Cocking, E. C., Isolation culture and regeneration of leaf protoplasts in the genus *Petunia, Plant Sci. Lett.*, 7, 51, 1976.
91. Power, J. B., Frearson, E. M., George, D., Evans, P. K., Berry, S. F., Hayward, C., and Cocking, E. C., Isolation, culture and regeneration of leaf protoplasts in the genus *Petunia, Plant Sci. Lett.*, 7, 51, 1976.
92. Raveh, D. and Galun, E., Rapid regeneration of plants from tobacco protoplasts plated at low densities, *Z. Pflanzenphysiol.*, 76, 76, 1975.
93. Santos, A. V. P. Dos, Outka, D. E., Cocking, E. C., and Davey, M. R., Organogenesis and somatic embryogenesis in tissues derived from leaf protoplasts and leaf explants of *Medicago sativa, Z. Pflanzenphysiol.*, 99, 261, 1980.
94. Santos, A. V. P. Dos, Davey, M. R., and Cocking, E. C., Cultural studies of protoplasts and leaf callus of *Trigonella corniculata* and *T. foenum-graecum, Z. Pflanzenphysiol.*, 109, 227, 1982.
95. Saxena, P. K. and Rashid, A., Development of gametophores from isolated protoplasts of the moss *Anoectangium thomsonii* Mitt., *Protoplasma*, 103, 401, 1980.
96. Saxena, P. K., Gill, R., Rashid, A., and Maheshwari, S. C., Plantlet formation from isolated protoplast of *Solanum melongena* L., *Protoplasma*, 106, 355, 1981.
97. Saxena, P. K., Gill, R., Rashid, A., and Maheshwari, S. C., Isolation and culture of protoplasts of *Capsicum annum* L. *Protoplasma*, 103, 357, 1981.
98. Saxena, P. K., Gill, R., Rashid, A., and Maheshwari, S. C., Plantlets from mesophyll protoplasts of *Solanum xanthocarpum, Plant Cell Rep.*, 1, 219, 1982.
99. Schieder, O., Regeneration of haploid and diploid *Datura innoxia* Mill. mesophyll protoplasts to plants, *Z. Pflanzenphysiol.*, 76, 462, 1975.
100. Schieder, O., Attempts in regeneration of mesophyll protoplasts of haploid and diploid wild type lines, and those of chlorophyll deficient strains from different Solanaceae, *Z. Pflanzenphysiol.*, 88, 275, 1977.
101. Schumann, U., Opatrny, A., and Koblitz, H., Plant recovery from long term callus cultures and from suspension culture derived protoplasts of *Solanum phureja, Biochem. Physiol. Pflanzen*, 175, 670, 1980.
102. Scowcroft, W. R. and Larkin, P. J., Isolation, culture and plant regeneration from protoplasts of *Nicotiana debneyi, Aust. J. Plant Physiol.*, 7, 635, 1980.
103. Shahin, E. A. and Shepard, J. F., Cassava mesophyll protoplasts: isolation, proliferation and shoot formation, *Plant Sci. Lett.*, 17, 459, 1980.
104. Shekhawat, N. S. and Galston, A. W., Mesophyll protoplasts of fenngreek (*Trigonella foenumgraecum*): isolation, culture and regeneration, *Plant Cell Rep.*, 2, 119, 1983.
105. Shekhavat, M. S. and Galston, A. W., Isolation, culture and regeneration of moth bean *Vigna aconitifolia* leaf protoplasts, *Plant Sci. Lett.*, 32, 43, 1983.
106. Shepard, J. F., Mutant selection and plant regeneration from potato mesophyll protoplasts, in *Emergent Techniques for the Genetic Improvement of Crops*, Rubenstein, I., Gengenbach, B., and Green, C. E., Eds., University of Minnesota Press, Minneapolis, 1980, 185.
107. Shepard, J. F., Cultivar dependent cultural refinements in potato protoplast regeneration, *Plant Sci. Lett.*, 26, 127, 1982.
108. Shepard, J. F., The regeneration of potato plants from leaf-cell protoplasts, *Sci. Am.*, 246, 112, 1981.
109. Shepard, J. F. and Totten, R. E., Mesophyll cell protoplasts of potato, *Plant Physiol.*, 60, 313, 1977.
110. Shepard, J. F., Bidney, D., and Shahin, E., Potato protoplasts in crop improvement, *Science*, 208, 17, 1980.
111. Sink, K. C. and Power, J. B., The isolation, culture and regeneration of leaf protoplasts of *Petunia parviflora, Plant Sci. Lett.*, 10, 335, 1977.

112. Stumm, I., Meyer, Y., and Abel, W. O., Regeneration of the moss *Physcomitrella patens* (Hedw.) from isolated protoplasts, *Plant Sci. Lett.*, 5, 113, 1975.
113. Takebe, I., Labib, G., and Melchers, G., Regeneration of whole plants from isolated mesophyll protoplasts of tobacco, *Naturwissenschaften*, 58, 318, 1971.
114. Thomas, E., Plant regeneration from shoot culture-derived protoplasts of tetraploid potato (*Solanum tuberosum* cv. Maris Bard), *Plant Sci. Lett.*, 23, 81, 1981.
115. Thomas, E., Hoffmann, F., Potrykus, I., and Wenzel, G., Protoplast regeneration and stem embryogenesis of haploid androgenetic rape, *Mol. Gen. Genet.*, 145, 245, 1976.
116. Thomas, E., Bright, S. W. J., Franklin, J., Lancaster, V. A., Miflin, M. J., and Gibson, R., Variation amongst protoplast-derived potato plants (*Solanum tuberosum* cv. Maris Bard), *Theor. Appl. Genet.*, 62, 65, 1982.
117. Vardi, A., Speigel-Roy, P., and Galun, E., Citrus cell culture: isolation of protoplasts, plating densities, effects of mutagens and regeneration of embryos, *Plant Sci. Lett.*, 4, 231, 1975.
118. Vardi, A., Speigel-Roy, P., and Galun, E., Plant regeneration from *Citrus* protoplasts: variability in methodological requirements among cultivars and species, *Theor. Appl. Genet.*, 62, 171, 1982.
119. Vasil, V. and Vasil, I. K., Regeneration of tobacco and *Petunia* plants from protoplasts and culture of corn protoplasts, *In Vitro*, 10, 83, 1974.
120. Vasil, V. and Vasil, I. K., Isolation and culture of cereal protoplasts. II. Embryogenesis and plantlet formation from protoplasts of *Pennisetum americanum*, *Theor. Appl. Genet.*, 56, 97, 1980.
121. Vasil, I. K., Vasil, V., and Wang, D.-Y., Somatic embryos and plants from cultured protoplasts of *Pennisetum purpureum* Schum. (napier grass), in *Protoplasts 1983 — Poster Proceedings*, Potrykus, I., Harms, C. T., Hinnen, A., Hütter, R., King, P. J., and Shillito, R. D., Eds., Birkhäuser Verlag, Basel, 1983, 58.
122. Vasil, V., Wang, D.-Y., and Vasil, I. K., Plant regeneration from protoplasts of napier grass (*Pennisetum purpureum* Schum.), *Z. Pflanzenphysiol.*, 111, 233, 1983.
123. Wenzel, G. and Schieder, O., Regeneration of isolated protoplasts from nicotinic acid-deficient mutants of the liverwort *Sphaerocarpos donnellii* Aust., *Plant Sci. Lett.*, 1, 421, 1973.
124. Wenzel, G., Schieder, O., Przewozny, T., Sopory, S. K., and Melchers, G., Comparison of single cell culture-derived *Solanum tubersoum* L. plants and a model for their application in breeding programs, *Theor. Appl. Genet.*, 55, 49, 1979.
125. Wernicke, W. and Thomas, E., Studies on morphogenesis from isolated plant protoplasts: shoot formation from mesophyll protoplasts of *Hyoscyamus muticus* and *Nicotiana tabacum*, *Plant Sci. Lett.*, 17, 401, 1980.
126. Wernicke, W., Lörz, H., and Thomas, E., Plant regeneration from leaf protoplast of haploid *Hyoscyamus muticus* L. produced via anther culture, *Plant, Sci. Lett.*, 15, 239, 1979.
127. White, D. W. R., Plant regeneration from mesophyll protoplasts of white clover (*Trifolium repens* L.), *in Protoplasts 1983 — Poster Proceedings*, Potrykus, I., Harms, C. T., Hinnen, A., Hütter, R., King, P. J., and Shillito, R. D., Eds., Birkhäuser Verlag, Basel, 1983, 60.
128. Xiang-hui, Li, Plantlet regeneration from mesophyll protoplasts of *Digitalis lanata Ehrh.*, *Theor. Appl. Genet.*, 60, 345, 1981.
129. Xu, Z.-H. and Davey, M. R., Shoot regeneration from mesophyll protoplasts and leaf explants of *Rehmannia glutinosa*, *Plant Cell Rep.*, 2, 55, 1983.
130. Xu, Z.-H., Davey, M. R., and Cocking, E. C., Organogenesis from root protoplasts of the forage legumes *Medicago sativa* and *Trigonella foenum-graecum*, *Z. Pflanzenphysiol.*, 107, 231, 1982.
131. Xu, Z.-H., Davey, M. R., and Cocking, E. C., Plant regeneration from root protoplasts of *Brassica*, *Plant Sci. Lett.*, 24, 117, 1982.
132. Xuan, L. T. and Menczel, L., Improved protoplast culture and plant regeneration from protoplast-derived callus in *Arabidopsis thaliana*, *Z. Pflanzenphysiol.*, 96, 77, 1980.
133. Zapata, F. Y. and Sink, K. C., Somatic embryogenesis from *Lycopersicon peruvianum* leaf mesophyll protoplasts, *Theor. Appl. Genet.*, 59, 265, 1981.
134. Kao, K. N. and Michayluk, M. R., Nutrient requirements for growth of *Vicia hajastana* cells and protoplasts at a very low population density in liquid media, *Planta*, 126, 105, 1975.
135. Jia, J., Potrykus, I., Lazar, G. B., Saul, M., *Hysocyamus-Nicotiana* fusion hybrids selected via auxotroph complementation and verified by species-specific DNA-hybridization, in *Protoplasts 1983 — Poster Proceedings*, Potrykus, I., Harms, C. T., Hinnen, A., Hütter, R., King, P. J., and Shillito, R. D., Eds., Birkhäuser Verlag, Basel, 1983, 110.
136. Potrykus, I., Harms, C. T., and Lörz, H., Problems in culturing cereal protoplasts, in *Cell Genetics in Higher Plants*, Dudits, D., Farkas, G. L., and Maliga, P., Eds., Publishing House of the Hungarian Academy of Sciences, Budapest, 1976, 129.
137. Nagata, T. and Takebe, I., Plating of isolated tobacco mesophyll protoplasts on agar medium, *Planta*, 99, 12, 1971.
138. Murashige, T. and Skoog, F., A revised medium for rapid growth and bioassays with tobacco tissue cultures, *Physiol. Plant.*, 15, 473, 1962.

139. Gamborg, O. L., Miller, R. A., and Ojima, K., Nutrient requirement of suspension cultures of soybean root cells, *Exp. Cell. Res.*, 50, 151, 1968.

140. Schieder, O., Somatic hybrids between a herbaceous and two tree-*Datura* species, *Z. Pflanzenphysiol.*, 98, 119, 1980.

141. Poirier-Hamon, S., Rao, P. S., and Harada, H., Culture of mesophyll protoplasts and stem segments of *Antirrhinum majus* (snapdragon): growth and organization of embryoids, *J. Exp. Bot.*, 25, 752, 1974.

142. Kohlenbach, H. W., Comparative somatic embryogenesis, in *Frontiers of Plant Tissue Culture 1978*, Thorpe, T. A., Ed., University of Calgary, Canada, 1978, 59.

143. Gamborg, O. L., Davis, B. P., and Stahlhut, R. W., Cell divisions and differentiation in protoplasts from cell cultures of *Glycine* species and leaf tissue of soybean, *Plant Cell. Rep.*, 2, 213, 1983.

144. Maliga, P., Lázár, G., Joó, F., Nagy, A. H., and Meczel, L., Restoration of morphogenic potential in *Nicotiana* by somatic hybridization, *Mol. Gen. Genet.*, 157, 291, 1977.

145. Melchers, G., Sacristan, M. D., and Holder, A. A., Somatic hybrid plants of potato and tomato regenerated from fused protoplasts, *Carlsberg Res. Commun.*, 43, 203, 1978.

146. Gosch, G. and Reinert, J., Cytological identification of colony formation of intergeneric somatic hybrid cells, *Protoplasma*, 96, 23, 1978.

147. Hoffmann, F. and Adachi, R., Arabidobrassica — Chromosomal recombination and morphogenesis in asymmetric intergeneric hybrid cells, *Planta*, 153, 586, 1981.

148. Krumbiegel, G. and Schieder, O., Comparison of somatic and sexual incompatibility between *Datura innoxia* and *Atropa belladonna*, *Planta*, 153, 466, 1981.

149. Gleba, Y. Y., Momot, V. P., Cherep, N. N., and Skarzynskaya, M. V., Intertribal hybrid cell lines of *Atropa belladonna* (x) *Nicotiana chinensis* obtained by cloning individual protoplast fusion products, *Theor. Appl. Genet.*, 62, 75, 1982.

150. Tabaeizadeh, Z., Bergounioux, C., and Perennes, C., Increasing the variability of *Lycopersicon* Mill. by protoplast fusion with *Petunia* L., in *Protoplasts 1983 — Poster Porceedings*, Potrykus, I., Harms, C. T., Hinnen, A., Hütter, R., King, P. J., and Shillito, R. D., Eds., Birkhäuser Verlag, Basel, 1983, 90.

151. Otten. L., De Greve, H., Hernalsteens, J. P., Van Montagu, M., Schieder, O., Straub, J., and Schell, J., Mendelian transmission of genes introduced into plants by the Ti plasmids of *Agrobacterium tumefaciens*, *Mol. Gen. Genet.*, 183, 209, 1981.

Chapter 3

PROTOPLASTS FOR STUDIES OF CELL ORGANELLES

L. C. Fowke, L. R. Griffing, B. G. Mersey, and M. A. Tanchak

TABLE OF CONTENTS

I. INTRODUCTION

Plant protoplasts provide a unique system for studying the structure, chemistry, and function of cell organelles. One of the major advantages of using protoplasts is the relative ease with which they can be lysed by gentle techniques such as osmotic shock or the application of mild detergents and poly-cations. Organelles can be isolated without using the harsh mechanical methods necessary for disrupting plant cells. Isolation of many cell organelles has been achieved using plant protoplasts (see reviews by Quail,[1] Fowke and Gamborg,[2] and Galun[3]).

Plant protoplasts also provide an excellent system to probe the plant plasma membrane which normally is inaccessible due to the presence of a cell wall. Removal of the cell wall permits direct access to this vital cellular component which controls the movement of materials into and out of cells and plays a central role in the process of cell wall formation. By employing plant protoplasts it is also possible to study the inner surface of the plant plasma membrane and associated cell organelles. Large stained fragments of the plasma membrane can be prepared for ultrastructural examination (see below). The intention of this chapter is to illustrate the usefulness of plant protoplasts for studying cell organelles. The discussion will be restricted to cytoskeletal elements and coated vesicles. The application of plant protoplasts for other studies of additional cellular components is highlighted in other chapters of this book, i.e., cell wall, Chapter 6; plasma membrane and tonoplast, Chapter 8.

II. CYTOSKELETAL ELEMENTS

Protoplasts have recently been utilized for studies of cytoskeletal elements of plant cells. Emphasis to date has been focused on microtubules, particularly regarding their relationship to cell wall formation, and information concerning other cytoskeletal components (e.g., contractile elements, intermediate filaments) is rather meager.

A. Microtubules

Microtubules are believed to play an important role in orienting cellulose microfibril deposition during plant cell wall formation.[4-6] Microtubules are located in the cortical cytoplasm of plant cells parallel to newly deposited cellulose microfibrils. Colchicine treatment removes the cortical microtubules and disrupts cellulose microfibril orientation. Cellulose microfibrils are thought to arise from mobile enzyme complexes in the plasma membrane and microtubules may direct the motion of such complexes. The mechanism by which microtubules influence deposition is unknown; however, a number of possibilities have been suggested.[5]

Plant protoplasts provide a unique experimental system in which to study the role of microtubules in plant cell wall formation. Protoplasts produced by enzyme digestion of the cell wall will rapidly regenerate a new wall when cultured in defined media,[2,7] (see Chapter 6). Protoplasts from suspension cultures can synthesize a network of cellulose microfibrils within 1 hr and a complete new wall within a few days. Leaf protoplasts are also capable of regenerating a new wall but only after a lag phase of at least 3 hr.

Cultured plant protoplasts are also interesting because in addition to regenerating a cell wall they usually undergo a marked change in shape as they are converted to cells. Often spherical protoplasts assume a cylindrical shape in culture. Microtubules are thought to indirectly influence the establishment of cell shape by controlling the orientation of cellulose microfibril deposition.

Plant protoplasts can therefore be used to investigate the relationship of cortical microtubules to both the process of cell wall formation and the reestablishment of cell shape. A number of different approaches have been used to examine this relationship. Immunofluorescence techniques are providing a powerful tool for such studies as they permit the examination of the three-dimensional arrangement of microtubules in intact cells. Techniques which have been used successfully for plant cells[8-10] have recently been adapted for use with freshly isolated and cultured higher plant protoplasts in order to follow the microtubule pattern during cell regeneration[11] (Figures 1, 2). Within 24 to 48 hr of culture, protoplasts of *Vicia hajastana* had expanded to form cylindrical cells. The transition from spherical protoplasts to elongate cells was accompanied by a change from randomly arranged microtubules to parallel arrays of transversely oriented microtubules (Figures 1, 2). This experimental approach coupled with the use of polarized light microscopy to follow cellulose orientation and the use of inhibitors affecting both microtubules and cell wall deposition should help to clarify the relationship of microtubules to wall formation and cell shape determination. Inhibitor studies with algal protoplasts have shed some light on this complex relationship. Protoplasts of *Mougeotia* in culture will regenerate filaments composed of cylindrical cells. Experiments with inhibitors which interfere with wall regeneration (Coumarin, Calcofluor) but do not affect cortical microtubules in regenerating *Mougeotia* protoplasts, clearly demonstrate that microtubules alone cannot reestablish the cylindrical cell shape but must be accompanied by ordered cellulose deposition.[12]

Large stained fragments of the protoplast plasma membrane can also be used to provide information concerning the relationship of cortical microtubules to cell wall regeneration. Fragments are prepared by attaching protoplasts to a substrate using polylysine, bursting them osmotically, and then washing and staining the adhering pieces of plasma membrane[13,14] (Figures 3A, 3B). Fragments attached to glass coverslips can be used for immunological studies at the light microscope level while those attached to grids can be examined by electron microscopy. Such preparations are unique in that they permit the visualization of large areas of the inner surface of the plant plasma membrane and thus can be used to examine organelles associated with the plasma membrane.[8,15-17]

Numerous microtubules are firmly attached to plasma membrane fragments from suspension culture protoplasts capable of immediate and rapid cell wall formation (Figure 3) while leaf protoplasts, which exhibit a distinct lag before initiating wall formation, have very few microtubules on their plasma membrane.[16,18] It will be interesting to examine cultured leaf protoplasts during the period of wall deposition to establish whether there is a substantial increase in attached microtubules. Initiation of wall formation by protoplasts of the alga *Mougeotia* seems to be correlated with the attachment of microtubules to the plasma membrane.[19]

Stained plasma membrane fragments have also been used to provide basic information regarding the distribution, length, and chemistry of cortical microtubules. Measurements of microtubules on different membrane fragments revealed average lengths of 2.0 to 5.5 μm with some microtubules exceeding 15 μm in length.[16] Observations of microtubules approaching 25 μm in length on carrot protoplast plasma membrane fragments have contributed to the hypothesis that microtubules may be wound helically around cylindrical plant cells.[20] Light microscope studies of membrane fragments labeled with fluorescent antibodies have similarly provided information regarding microtubule distribution on the plasma membrane as well as preliminary evidence which suggests that plant microtubules have proteins similar to animal microtubule-associated proteins.[14]

B. Contractile Elements

Intracellular motility in both plant and animal cells is believed to result from the interaction of bundles of actin microfilaments with small aggregates of myosin mole-

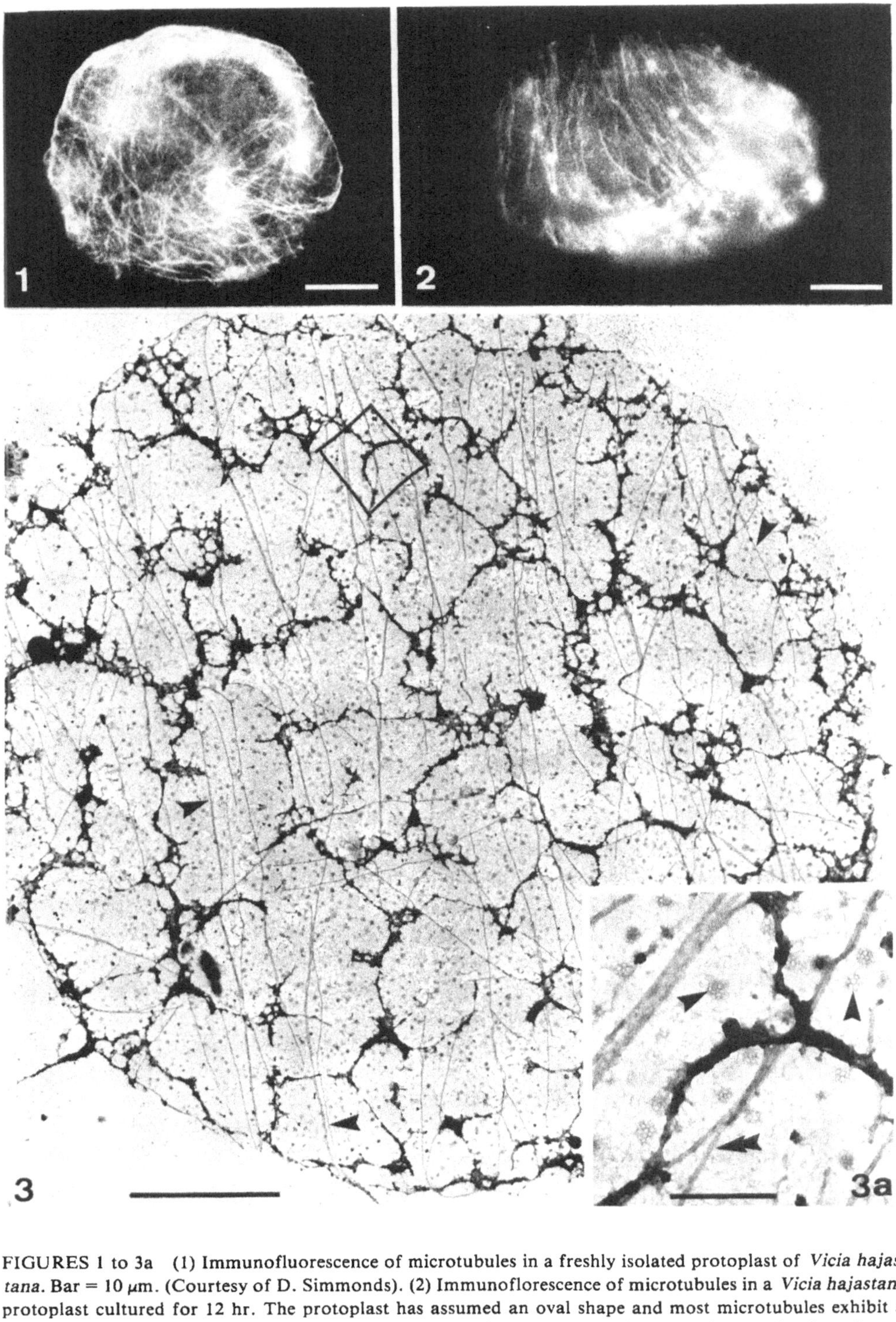

FIGURES 1 to 3a (1) Immunofluorescence of microtubules in a freshly isolated protoplast of *Vicia hajastana*. Bar = 10 μm. (Courtesy of D. Simmonds). (2) Immunoflorescence of microtubules in a *Vicia hajastana* protoplast cultured for 12 hr. The protoplast has assumed an oval shape and most microtubules exhibit a transverse orientation. Bar = 10 μm. (Courtesy of D. Simmonds). (3) Electron micrograph of a tobacco plasma membrane fragment showing numerous microtubules (arrows). The area outlined in black is shown in Figures 3a. Bar = 3 μm. (3a) Enlargement from Figure 3 showing microtubules (double arrow) and coated pits (single arrows) on the plasma membrane. Bar = 0.5 μm.

cules. Very little is known about plant myosin but during the past few years considerable progress has been made in the localization, isolation, and characterization of plant actin, particularly with algal cells.[21] The presence of a rigid cell wall makes plant cells a much more difficult subject than animal cells for *in situ* studies of actin microfilaments or for their isolation. Plant protoplasts offer a number of advantages for studies of actin microfilaments, the major one perhaps being their suitability for organelle isolation. The work of Yamaguchi and Nagai[22] illustrates the potential of protoplasts for microfilament isolation and identification. They prepared protoplasts from leaf cells of the water plant *Vallisneria* which contain bundles of thin filaments and exhibit active cytochalasin B-sensitive cytoplasmic streaming. After mechanically rupturing the protoplasts, bundles of microfilaments were observed in the cytoplasmic suspension (Figure 4). The microfilaments were identified as actin by their affinity for muscle heavy meromyosin (Figure 5).

As mentioned previously, techniques are now available for immunofluorescence studies of microtubules in plant protoplasts. It should, therefore, be possible to apply similar methods to examine microfilaments in both freshly isolated protoplasts and those regenerating cell walls in culture. Results of such studies might shed some light on the proposal that actin microfilaments play a role in cellulose deposition.[5]

Protoplasts are apparently also amenable to microinjection experiments after they have regenerated thin cell walls.[23] Tobacco protoplasts injected with the dye Lucifer Yellow survived the mechanical injury and completed cell division. This technique might be used to introduce either fluorescently labeled actin or actin antibodies into protoplasts to characterize the population of actin microfilaments in living plant cells. This approach has been used with some success with living animal cells as discussed by Jackson.[21] Alternatively, artificial lipid vesicles (liposomes) will fuse with plant protoplasts and thus might provide a vehicle for introducing these molecules into plant cells.[23] This technology is of course only possible using protoplasts rather than intact cells.

Studies of fragments of protoplast plasma membrane may also contribute to our understanding of plant actin. Filaments resembling actin microfilaments have been observed on the inner surface of the plasma membrane of a variety of protoplasts[15,17,18] (Figure 6) but their identity has not been established. Work is in progress to characterize these filaments using fluorescent phallacidin and heavy meromyosin. If the filaments are indeed composed of actin, protoplast membrane fragments could be quite useful for studying the relationship between actin microfilaments and the plant plasma membrane. In animal cells, actin microfilaments are closely associated with the plasma membrane and are involved in a number of processes including cleavage of cells during cytokinesis, ameboid movement, and possible capping.[24] During the capping process, membrane molecules are induced to cluster by antibodies or lectins and subsequently aggregate at the ends of cells. It is interesting that sugar groups on the outer surface of the plant plasma membrane can also be induced to cluster by lectins.[2] Whether actin microfilaments are involved in this process is not known but studies of protoplast plasma membrane fragments might be useful for investigating this possibility.

C. Other Filaments

In addition to microtubules and contractile filaments, animal cells contain a population of tough durable fibers known as intermediate filaments.[24] These filaments are believed to play primarily a structural role in animal cells. Intermediate filaments have not been clearly identified in plant cells; however, recent work with plant protoplasts suggests that a similar type of filament may be present. When protoplasts are burst under isotonic conditions, negatively stained, and examined with the electron microscope, a rather complex cytoskeleton is revealed.[25] A major component of this cyto-

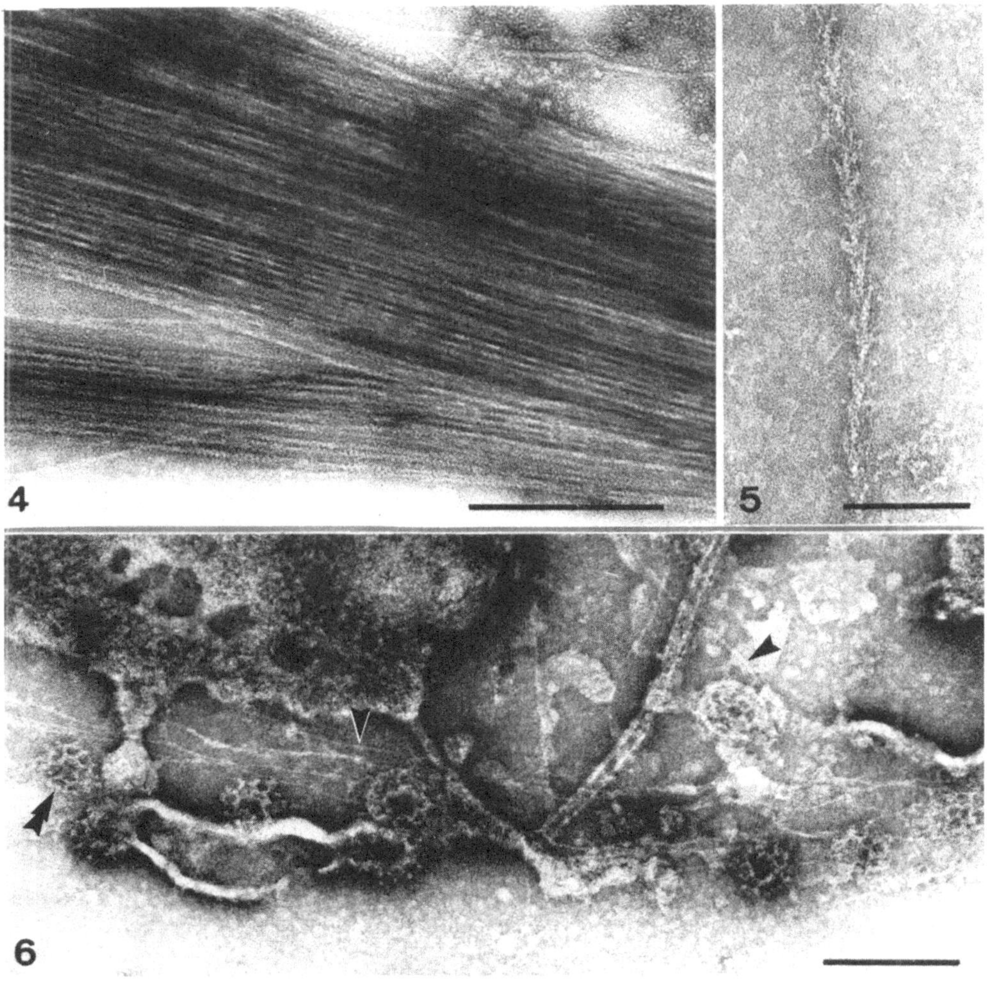

FIGURES 4 to 6 (4) Electron micrograph showing a bundle of negatively stained microfilaments isolated from a protoplast of *Vallisneria*. Bar = 200 nm. (From Yamaguchi, Y. and Nagai, R., *J. Cell Sci.*, 48, 193, 1981. With permission.) (5) Electron micrograph showing a negatively stained microfilament decorated with heavy meromyosin. The microfilament was isolated from a protoplast of *Vallisneria*. Bar = 100 nm. (From Yamaguchi, Y. and Nagai, R., *J. Cell. Sci.*, 48, 193, 1981. With permission.) (6) Electron micrograph showing negatively stained fine filaments (single arrows) on the inner surface of a tobacco protoplast plasma membrane fragment. Note also the coated pits (double arrow). Bar = 200 nm.

skeleton consists of bundles of 7-nm fibrils which are apparently not composed of actin. The fibrils are usually associated with the nucleus and it has been suggested that they may function by anchoring the plant nucleus.[25] Studies of freshly isolated and cultured protoplasts[26] as well as other plant cells (see Powell et al.)[25] by conventional thin sectioning techniques have demonstrated the presence of bundles of filaments of similar size. Whether these filaments are in fact equivalent to animal intermediate filaments remains to be established. Research with protoplasts has contributed information regarding an interesting population of filaments in plant cells and further work will undoubtedly clarify their chemistry and function.

III. COATED VESICLES

Coated vesicles are widespread in plants,[27] but very little information is available concerning their structure, chemistry, or function. In contrast, animal coated vesicles have been well characterized and recent research has clearly demonstrated their participation in endocytosis of specific proteins (see reviews by Goldstein et al.,[28] Pastan and Willingham,[29] Pearse and Bretscher,[30] Brown et al.[31]). It is useful to briefly review the animal literature to provide a framework for further consideration of this cellular component in plants.

A. Animal Coated Vesicles

Morphological, biochemical, and immunological techniques have been utilized to study coated vesicles in a variety of animal cells. Coated vesicles range in diameter from approximately 50 to 150 nm and consist of a membrane vesicle enclosed within a highly patterned coat, which in surface view bears resemblance to a soccer ball.[30]. The basic structural unit of the coat is a "triskelion" which is composed of three heavy chains of clathrin (180 kdalton) and three light chains (33 to 36 kdalton).[32-35] The triskelions are usually arranged to yield a surface pattern of 12 pentagons and a variable number of hexagons.[30] Coated vesicles arise from coated pits, invaginations of the plasma membrane which bear clathrin coat material on their cytoplasmic surface. The clathrin is believed to facilitate the invagination process leading to the release of coated vesicles into the peripheral cytoplasm.[35,36] Coated vesicles can be isolated from animal cells, dissociated to triskelions, and the clathrin coats reassociated in vitro.[32,37,38]

Coated pits and coated vesicles in animal cells function primarily in the process of receptor mediated endocytosis (see reviews by Goldstein et al.,[28] Pastan and Willingham,[29] Brown et al.).[31] A number of proteins (e.g., low density lipoprotein, transferrin, insulin, epidermal growth factor, asialoglycoproteins, α-2-macroglobulin) bind specifically to receptors on the external surface of the plasma membrane and are internalized via coated pits which detach to form free coated vesicles in the cytoplasm. There is not complete agreement as to whether coated vesicles are separate entities or are in fact coated pits connected to the cell surface by thin necks.[39] Serial sectioning work certainly supports the concept of free coated vesicles.[40] Coated vesicles containing internalized protein uncoat rapidly and fuse with each other as well as with a peripherally located membranous component which is often tubular in nature. This compartment referred to as the endosome,[31] receptosome,[41] or CURL[42] has an acidic pH which favors dissociation of the incoming ligand and receptor.[43,44] Many ligands are then delivered to the lysosome of the cell to be degraded while the receptors are returned to the cell surface (reviewed by Brown et al.[31]).

Coated vesicles also seem to form on the Golgi cisternae and it has been suggested that they may be involved in transport of hydrolase enzymes from Golgi to lysosomes.[41] A number of lysosomal enzymes seem to be associated with coated vesicles isolated from rat liver and calf brain.[45] Golgi-associated coated vesicles may also be involved in the transport of glycoproteins to the cell surface.[46]

Finally, it has been suggested that coated vesicles play a role in recycling plasma membrane into the cell. In sea urchin eggs, for example, cortical granule exocytosis is followed by a burst of membrane retrieval via coated vesicles.[47]

B. Plant Coated Vesicles

Basic knowledge regarding the structure, chemistry, and function of plant coated vesicles and coated pits is sparse. They have been described in a wide variety of plant species and seem to resemble quite closely their counterparts in animal cells.[27] Coated vesicles and pits are particularly numerous in cells which are involved in active cell wall formation, e.g., cells undergoing cytokinesis,[14,48,49] young roots and root hairs,[50,51] cotton fibers,[52] and protoplasts.[14] Within such cells they are localized primarily at the plasma membrane and in smaller numbers near dictyosomes.

Very little is known about the function of coated vesicles in plants. The idea that they are exocytotic and are responsible for contributing material to the growing cell wall has received quite wide support (see reviews by Newcomb,[27] Fowke et al.[14]). Others have proposed that coated vesicles are important for protein secretion from plant cells.[50,53] Unfortunately it has not been possible until recently (see below) to determine the direction of movement of coated vesicles and arguments for exocytosis rather than endocytosis have been based entirely on circumstantial evidence.

C. Protoplasts for Studies of Coated Vesicles

Plant protoplasts provide an exceptionally good experimental system to study coated vesicles. The absence of a cell wall should facilitate attempts to determine whether coated vesicles are exocytotic and/or endocytotic and thus contribute to our understanding of their function in plant cells. Cultured plant protoplasts rapidly regenerate a new cell wall[7,14] (see Chapter 6), and during this process, massive exocytosis of matrix polysaccharides and presumably some protein occurs. If coated vesicles contribute materials to the forming wall by exocytosis, cultured protoplasts should provide a favorable system in which to study the process. Alternatively, if coated vesicles are endocytotic, it should be feasible to attach molecules directly to the plasma membrane of protoplasts and follow their internalization.

Protoplasts freshly isolated from suspension cultured cells contain an abundance of coated vesicles and coated pits (Figures 3B, 6, 7). Coated vesicles are characteristically localized at the plasma membrane, near dictyosomes and occasionally close to a tubular membrane system in the cortical cytoplasm.[14] The nature of the tubular membranes is not known and serial sectioning is required to determine their three-dimensional organization. In contrast to protoplasts from suspension cultured cells, freshly isolated leaf protoplasts have very few coated vesicles or coated pits.[18,54]

A particularly useful technique for studying coated vesicles and coated pits on the plasma membrane involves the preparation of large stained fragments of protoplast plasma membrane as described previously. Examination of such membrane fragments reveals the presence of numerous coated structures[15,17] (Figures 3A, 3B, 6). Analysis of stereo pairs suggests that both coated vesicles and coated pits are present but coated pits are far more frequent. Occasionally large flat patches of coat material are observed.[14,17]

The frequency of coated structures on the inner surface of the plasma membrane seems to correlate directly with the cell wall synthesizing ability of the protoplasts. Protoplasts from cultured cells, which initiate wall deposition almost immediately, exhibit a much higher frequency of coated structures than leaf protoplasts, which regenerate a new wall only after a distinct lag phase.[18] Once cell wall deposition has been initiated by cultured leaf protoplasts, the number of coated vesicles and coated pits increases dramatically.[54]

In addition to providing useful information regarding the distribution of coated structures on the plasma membrane, negatively and positively stained fragments can be used to determine basic information regarding the size and general morphology of coated vesicles and coated pits. The general curvature of such structures as well as their coat organization in terms of the relative distribution of hexagons and pentagons has been derived from such preparations.[17]

Recent research efforts have been directed towards the isolation and characterization

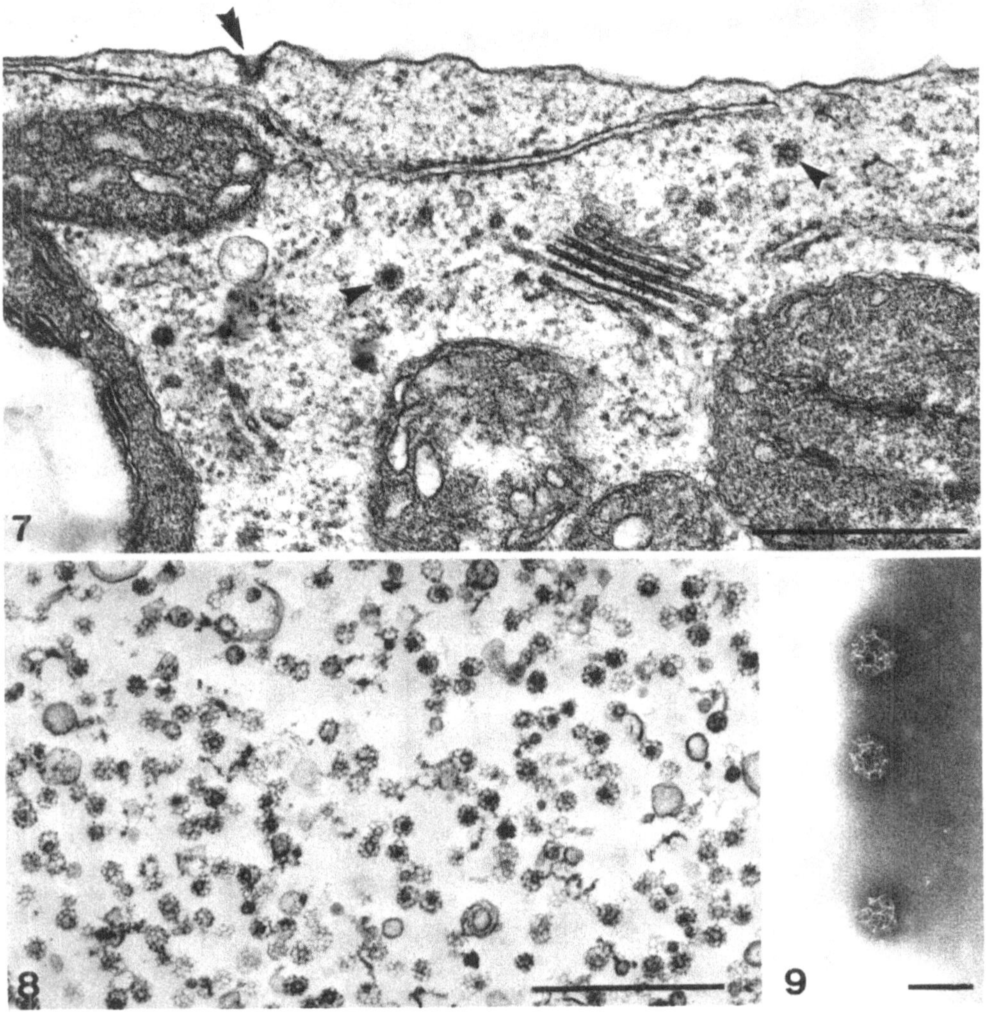

FIGURES 7 to 9 (7) Electron micrograph showing a coated pit (double arrow) at the plasma membrane and coated vesicles (single arrows) in the cytoplasm of a freshly isolated soybean protoplast. Bar = 0.5 μm. (8) Electron micrograph showing vesicles isolated from soybean protoplasts. Bar = 0.5 μm. (9) Electron micrograph showing negatively stained coated vesicles from soybean protoplasts. Bar = 100 nm.

of coated vesicles from plants. Coated vesicles have been successfully isolated from cultured plant cells,[55] soybean hypocotyls,[56] and soybean protoplasts.[57,58] Fractions derived from soybean protoplasts are highly enriched for coated vesicles as indicated by ultrastructural observations (Figures 8, 9); SDS polyacrylamide gel electrophoresis of such fractions demonstrates the presence of a high-molecular-weight polypeptide (approximately 190 kdaltons) which is presumed to be clathrin. The protein composition of plant coated vesicles seems to show some distinct differences from that of animal coated vesicles.

Research with plant protoplasts has provided the only clear demonstration of the direction of movement of coated vesicles in plant cells. Recent experiments with soybean protoplasts indicate that endocytosis of cationized ferritin (CF) can occur via

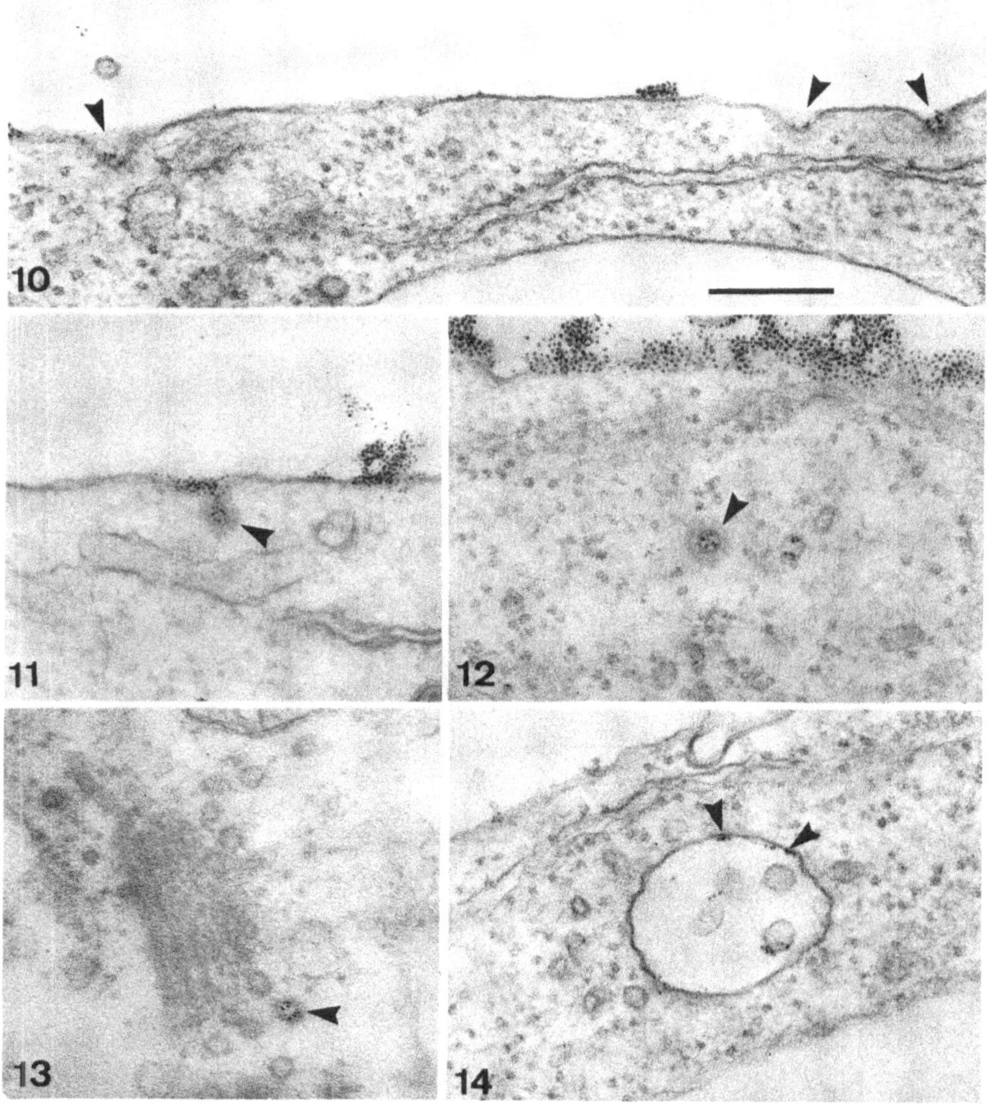

FIGURES 10 to 14 Electron micrographs showing endocytosis of cationized ferritin (CF) by soybean protoplasts. All micrographs same magnification. Bar = 200 nm. (10). CF in coated pits (arrows); (11). CF in a deep coated pit (arrow); (12). CF in coated a vesicle (arrow) in a cytoplasm; (13). CF in a dictyosome vesicle (arrow), (14). CF (arrows) in a multivesicular body. For further details, see Tanchak et al, Reference 60.

coated pits and coated vesicles[59,60] (Figures 10 to 14). The pathway seems to involve binding of CF to the plasma membrane, rapid internalization of CF via coated pits to coated vesicles, uncoating of coated vesicles, and eventual transport of CF to dictyosome cisternae and multivesicular bodies. Occasionally CF was observed in the peripheral tubular membrane compartment. The significance of these observations is presently unknown. It is conceivable that the endocytotic pathway described represents a highly specific uptake mechanism as in many animal cells; however, it is not clear what types of molecules might utilize such a pathway. With intact cells only small molecules would be able to penetrate the cell wall and gain access to the plasma membrane.

Alternatively, the endocytosis via coated membranes may illustrate a mechanism for membrane retrieval. Soybean protoplasts rapidly secrete matrix components[61] presumably via smooth dictyosome vesicles and some method of recycling plasma membrane must exist. In any event, the uptake of CF is unlikely to involve a specific recognition event. Protoplasts generally bear a net negative charge and CF would bind randomly to the surface. Those molecules attached to coated pits would be internalized. The uptake of CF by a coated membrane system in plant protoplasts is certainly intriguing. Further research is required to characterize the process and to determine whether such a mechanism operates in intact plant cells.

IV. SUMMARY

During the past few years protoplasts have provided a valuable tool for cell biologists studying plant cell organelles. This chapter has focused on research with cytoskeletal elements and coated vesicles to illustrate the usefulness of plant protoplasts for such studies. Basic information concerning the chemistry and distribution of microtubules associated with the plant plasma membrane has been obtained from protoplast research. Current work with cultured protoplasts is contributing to our understanding of the role of microtubules in the important processes of cell wall formation and cell shaping. Recent results indicate that protoplasts can also be applied to studies of the structure, distribution, and chemistry of other cytoskeletal elements in plant cells.

Protoplasts derived from rapidly growing cultured plant cells contain numerous coated vesicles and thus are particularly well suited to studies of this interesting cell organelle. Ultrastructural investigations of thin sections of protoplasts and isolated plasma membrane fragments have provided valuable information regarding the distribution and morphology of plant coated vesicles. Protoplasts also offer advantages for the isolation of plant coated vesicles. Fractions highly enriched in coated vesicles have been obtained from soybean protoplasts and biochemical characterization of these organelles is in progress. Using protoplasts, it has recently been possible to demonstrate the internalization of ferritin molecules via coated pits and coated vesicles. The significance of this uptake pathway in plant cells is unclear and requires further investigation.

ACKNOWLEDGMENTS

We wish to acknowledge the excellent technical assistance of Ms. Pat Rennie. Thanks also to Mrs. Beverlee Garnett for help in preparing the manuscript. The research from the Department of Biology, University of Saskatchewan, was supported by the Natural Sciences and Engineering Research Council of Canada.

REFERENCES

1. Quail, P. H., Plant cell fractionation, *Ann. Rev. Plant Physiol.*, 30, 425, 1979.
2. Fowke, L. C. and Gamborg, O. L., Applications of protoplasts to the study of plant cells, *Int. Rev. Cytol.*, 68, 9, 1980.
3. Galun, E., Plant protoplasts as physiological tools, *Ann. Rev. Plant Physiol.*, 32, 237, 1981.
4. Gunning, B. E. S. and Hardham, A. R., Microtubules, *Ann. Rev. Plant Physiol.*, 33, 651, 1982.

5. Heath, I. B. and Seagull, R. W. Oriented cellulose fibrils and the cytoskeleton: a critical comparison of models, in *The Cytoskeleton in Plant Growth and Development*, Lloyd, C. W., Ed., Academic Press, London, 1982, 163.
6. Robinson, D. G. and Quader, H., The microtubule-microfibril syndrome, in *The Cytoskeleton in Plant Growth and Development*, Lloyd, C. W., Ed., Academic Press, London, 1982, 109.
7. Willison, J. H. M. and Klein, A. S., Cell wall regeneration by protoplasts isolated from higher plants, in *Cellulose and Other Natural Polymer Systems: Biogenesis, Structure and Degradation*, Brown, R. M., Ed., Plenum Press, New York, 1982, 61.
8. Lloyd, C. W., Slabas, A. R., Powell, A. J., and Lowe, S. B., Microtubules, protoplasts and plant cell shape, *Planta*, 147, 500, 1980.
9. Wick, S. M., Seagull, R. W., Osborn, M., Weber, K., and Gunning, B. E. S., Immunofluorescence microscopy of organized microtubule arrays in structurally stabilized meristematic plant cells, *J. Cell Biol.*, 89, 685, 1981.
10. Simmonds, D. H., Setterfield, G., Tanchak, M., Brown, D. L., and Rogers, K. A., Microtubule organization in cultured plant cells, in *Plant Tissue Culture 1982*, Fujiwara, A., Ed., Japanese Association of Plant Tissue Culture, Tokyo, 1982, 31.
11. Simmonds, D. H., Setterfield, G., and Brown, D. L., Reorganization of microtubules in protoplasts of *Vicia hajastana* Grossh. during the first 48 hours of culture, in *Protoplasts 1983, Poster Proceedings*, Potrykus, I., Harms, C. T., Hinnen, A., Hutter, R., King, P. J., and Shillito, R. D., Eds., Birkhäuser Verlag, Basel, 1983, 212.
12. Marchant, H. J., The establishment and maintenance of plant cell shape by microtubules, in *The Cytoskeleton in Plant Growth and Development*, Lloyd, C. W., Ed., Academic Press, London, 1982, 295.
13. Marchant, H. J., Microtubules associated with the plasma membrane isolated from protoplasts of the green alga *Mougeotia*, *Exp. Cell Res.*, 115, 25, 1978.
14. Fowke, L. C., Griffing, L. R., Mersey, B. G., and Van der Valk., P., Protoplasts for studies of the plasma membrane and associated cell organelles, in *Protoplasts 1983, Lecture Proceedings*, Potrykus, I., Harms, C. T., Hinnen, A., Hütter, R., King, P. J., and Shillito, R. D., Eds., Birkäuser Verlag, Basel, 1983, 101.
15. Doohan, M. E. and Palevitz, B. A., Microtubules and coated vesicles in guard-cell protoplasts of *Allium cepa* L., *Planta*, 149, 389, 1980.
16. Van der Valk, P., Rennie, P. J., Connolly, J. A., and Fowke, L. C., Distribution of cortical microtubules in tobacco protoplasts. An immunofluorescence microscopic and ultrastructural study, *Protoplasma*, 105, 27, 1980.
17. Van der Valk, P. and Fowke, L. C., Ultrastructural aspects of coated vesicles in tobacco protoplasts, *Can. J. Bot.*, 59, 1307, 1981.
18. Fowke, L. C., Rennie, P. J., and Constabel, F., Organelles associated with the plasma membrane of tobacco leaf protoplasts, *Plant Cell Rep.*, 2, 292, 1983.
19. Marchant, H. J. and Hines, E. R., The role of microtubules and cell-wall deposition in elongation of regenerating protoplasts of *Mougeotia*, *Planta*, 146, 41, 1979.
20. Lloyd, C. W., Helical microtubular arrays in onion root hairs, *Nature (London)*, 305, 311, 1983.
21. Jackson, W. T., Actomyosin, in *The Cytoskeleton in Plant Growth and Development*, Lloyd, C. W., Ed., Academic Press, London, 1982, 3.
22. Yamaguchi, Y. and Nagai, R., Motile apparatus in *Vallisneria* leaf cells. I. Organization of microfilaments, *J. Cell Sci.*, 48, 193, 1981.
23. Steinbiss, H.-H. and Broughton, W. J., Methods and mechanisms of gene uptake in protoplasts, *Int. Rev. Cytol.*, Suppl. 16, 191, 1983.
24. Alberts, B., Bray, D., Lewis, J., Raff, M., Roberts, K., and Watson, J. D., *Molecular Biology of the Cell*, Garland Publishing, New York, 1983.
25. Powell, A. J., Peace, G. W., Slabas, A. R., and Lloyd, C. W., The detergent-resistant cytoskeleton of higher plant protoplasts contains nucleus-associated fibrillar bundles in addition to microtubules, *J. Cell Sci.*, 56, 319, 1982.
26. Olesen, P. and Jensen, C. J., Ultrastructure of intermediate filament bundles associated with the cytoskeleton of protoplasts and cells from maize (*Zea mays* L.) suspensions, in *Protoplasts 1983, Poster Proceedings*, Potrykus, I., Harms, C. T., Hinnen, A., Hütter, R., King, P. J., and Shillito, R. D. Eds., Birkhäuser Verlag, Basel, 1983, 222.
27. Newcomb, E. H., Coated vesicles: their occurrence in different plant cell types, in *Coated Vesicles*, Ockleford, C. J. and Whyte, A., Eds., Cambridge University Press, Cambridge, 1980, 55.
28. Goldstein, J. L., Anderson, R. G. W., and Brown, M. S., Coated pits, coated vesicles, and receptor-mediated endocytosis, *Nature (London)*, 279, 1, 1979.
29. Pastan, I. H. and Willingham, M. C., Receptor-mediated endocytosis of hormones in cultured cells, *Ann. Rev. Physiol.*, 43, 239, 1981.

30. Pearse, B. M. F. and Bretscher, M. S., Membrane recycling by coated vesicles, *Ann. Rev. Biochem.*, 50, 85, 1981.
31. Brown, M. S., Anderson, R. G. W., and Goldstein, J. L., Recycling receptors: the round-trip itinerary of migrant membrane proteins, *Cell*, 32, 663, 1983.
32. Unanue, E. R., Ungewickell, E., and Branton, D., The binding of clathrin triskelions to membranes from coated vesicles, *Cell*, 26, 439, 1981.
33. Ungewickell, E., Biochemical and immunological studies on clathrin light chains and their binding sites on clathrin triskelions, *Eur. Molec. Biol. Org. J.*, 2, 1401, 1983.
34. Winkler, F. K. and Stanley, K. K., Clathrin heavy chain, light chain interactions, *Eur. Molec. Biol. Org. J.*, 2, 1393, 1983.
35. Harrison, S. C. and Kirchhausen, T., Clathrin cages, and coated vesicles, *Cell*, 33, 650, 1983.
36. Heuser, J. and Evans, L., Three-dimensional visualization of coated vesicle formation in fibroblasts, *J. Cell Biol.*, 84, 560, 1980.
37. Nandi, P. K., Prasad, K., Lippoldt, R. E., Alfsen, A., and Edelhock, H., Reversibility of coated vesicle dissociation, *Biochemistry*, 21, 6434, 1982.
38. Kartenbeck, J., Schmid, E., Muller, H., and Franke, W. W., Immunological identification and localization of clathrin and coated vesicles in cultured cells and in tissues, *Exp. Cell Res.*, 133, 191, 1981.
39. Willingham, M. C. and Pastan, I., Formation of receptosomes from plasma membrane coated pits during endocytosis: analysis by serial sections with improved membrane labelling and preservation techniques, *Proc. Natl. Acad. Sci. U.S.A.*, 80, 5617, 1983.
40. Peterson, O. W. and Van Deurs, B., Serial-section analysis of coated pits and vesicles involved in adsorptive pinocytosis in cultured fibroblasts, *J. Cell Biol.*, 96, 277, 1983.
41. Pastan, I. H. and Willingham, M. C., Journey to the center of the cell: role of the receptosome, *Science*, 214, 504, 1981.
42. Geuze, H. J., Slot, J. W., Strous, G. J. A. M., Lodish, H. F., and Schwartz, A. L., Intracellular site of asialoglycoprotein receptor-ligand uncoupling: double-label immunoelectron microscopy during receptor-mediated endocytosis, *Cell*, 32, 277, 1983.
43. Merion, M., Schlesinger, P., Brooks, R. M., Moehring, J. M., Moehring, T. J., and Sly, W. S., Defective acidification of endosomes in Chinese hampster ovary cell mutants "cross-resistnat" to toxins and viruses, *Proc. Natl. Acad. Sci. U.S.A.*, 80, 5315, 1983.
44. Yamashiro, D. J., Fluss, S. P., and Maxfield, F. R., Acidification of endocytic vesicles by an ATP-dependent proton pump, *J. Cell Biol.*, 97, 929, 1983.
45. Campbell, C. H. and Rome, L. H., Coated vesicle transport of lysosomal enzymes, *J. Cell Biol.*, 97, 176a, 1983.
46. Sarasti, J. and Hedman, K., Intracellular vesicles involved in the transport of semlike forest virus membrane proteins to the cell surface, *EMBO J.*, 2, 2001, 1983.
47. Fisher, G. W. and Rebhun, L. I., Sea urchin egg cortical granule exocytosis is followed by a burst of membrane retrieval via uptake into coated vesicles, *Dev. Biol.*, 99, 456, 1983.
48. Franke, W. W. and Herth, W., Morphological evidence for de novo formation of plasma membrane from coated vesicles in exponentially growing cultured plant cells, *Exp. Cell Res.*, 89, 447, 1974.
49. Nakamura, S. and Miki-Hirosige, H., Coated vesicles and cell plate formation in the microspore mother cell, *J. Ultrastruct. Res.*, 80, 302, 1982.
50. Bonnett, H. T., Cortical cell death during lateral root formation, *J. Cell Biol.*, 40, 144, 1969.
51. Robertson, J. G. and Lyttleton, P., Coated and smooth vesicles in the biogenesis of cell walls, plasma membranes, infection threads and peribacteroid membranes in root hairs and nodules of white clover, *J. Cell Sci.*, 58, 63, 1982.
52. Ryser, U., Cotton fiber differentiation: occurrence and distribution of coated and smooth vesicles during primary and secondary wall formation, *Protoplasma*, 98, 223, 1979.
53. Unzelman, J. M. and Healey, P. L., Development, structure and occurrence of secretory trichomes of *Pharbitis*, *Protoplasma*, 80, 285, 1974.
54. Burgess, J., Watts, J. W., Fleming, E. N., and King, J. M., Plasmalemma fine structure in isolated tobacco mesophyll protoplasts, *Planta*, 110, 291, 1973.
55. Mersey, B. G., Fowke, L. C., Constabel, F., and Newcomb, E. H., Preparation of a coated vesicle-enriched fraction from plant cells, *Exp. Cell Res.*, 141, 459, 1982.
56. Widenhoeft, R. E., Schmidt, G. W., and Palevitz, B. A., Comparative studies of plant and animal coated vesicles, *J. Cell Biol.*, 97, 177a, 1983.
57. Mersey, B. G., Griffing, L. R., Rennie, P. J., and Fowke, L. C., Coated vesicles from plant protoplasts, in *Protoplasts 1983*, Poster Proceedings, Potrykus, I., Harms, C. T., Hinnen, A., Hütter, R., King, P. J., and Shillito, R. D., Eds., Birkhäuser Verlag, Basel, 1983, 216.
58. Mersey, B. G., Griffing, L. R., Rennie, P. J., and Fowke, L. C., The isolation of coated vesicles from plant protoplasts, *Planta*, in press.

59. Tanchak, M. A., Griffing, L. R., Mersey, B. G., and Fowke, L. C., Functions of coated vesicles in plant protoplasts: endocytosis of cationized ferritin and transport of peroxidase, *J. Cell Biol.*, 97, 177a, 1983.
60. Tanchak, M. A., Griffing, L. R., Mersey, B. G., and Fowke, L. C., Endocytosis of cationized ferritin by coated vesicles of soybean protoplasts, *Planta*, 162, 481, 1984.
61. Hanke, D. E. and Northcote, D. H., Cell wall formation by soybean callus protoplasts, *J. Cell Sci.*, 14, 29, 1974.

Chapter 4

PROTOPLAST FUSION

F. Constabel and A. J. Cutler*

TABLE OF CONTENTS

* NRCC #24315.

I. HISTORY

At the beginning of the century much discussion centered around the physical properties of cytoplasm, particularly its viscosity. One experimental approach to problems regarding the structure of cytoplasm and cell membrane was protoplast fusion. The formation and subsequent fusion of subprotoplasts within cell walls[1,2] was accomplished by plasmolysis and deplasmolysis treatments, and fusion of protoplasts isolated from the same or different tissues and species was similarly accomplished.[3] Observation of protoplasts subjected to fusion experiments revealed that cells tolerate extreme deformations and distortions, and on occasion membranes agglutinate and coalesce. The application of cell fusion to asexual, i.e., somatic cell hybridization was suggested by Küster as early as 1935.[3] Experiments were hampered by the low number of protoplasts obtainable because these generally had to be isolated by surgical wall removal of plasmolyzed tissues and by inadequate conditions employed in culturing and fusing protoplasts. Modern day fusion was inaugurated by the development of methods to isolate very large numbers of protoplasts by using wall digesting enzymes as proposed by Küster[4] and by the availability of nutrient media which had previously been formulated for the in vitro culture of plant cells and tissues.[5] In addition, powerful agents for the agglutination and fusion of protoplasts soon became known and widely accepted, i.e., Ca^{2+} ions at high pH [6] and polyethylene glycol solutions.[7-9] High frequency fusion was the result, and now somatic cell hybridization became the primary goal of experimentation[10] overshadowing sometimes the interest in fundamental cytological problems. With the discovery of limits in the detection and maintenance of hybrid cells and in the regeneration of hybrid plants (see Chapter 12), investigations, at present, appear to be directed primarily towards problems of compatibility/incompatibility of organelles, specifically problems of organelle segregation.

The technique of protoplast fusion and culture of fusion products has been refined and consolidated in such a manner that several publications provide reliable and step by step procedures.[11-14]

II. SPONTANEOUS FUSION

Cell fusion is a process which is integral to plant development; the most prominent processes are egg fertilization and the differentiation of both water-conducting vessels and articulated laticifers. The breakdown of the cell wall during cell fusion in all these cases has been subject to thorough analysis and description by Sassen.[15] The activity of cellulolytic enzymes was demonstrated during vessel formation in *Hordeum*; chitinolytic activity was found in zygote extracts of *Phycomyces*.

The occurrence of multinucleated cells was detected as soon as enzymatic protoplast isolation techniques were applied to a variety of plant tissues.[16-17] The argument that cell wall degradation would permit dilation of plasmodesmata, fusion, and complete mixing of adjacent cells was well supported by EM documentation and data on the frequency of fusion with plasmolyzed tissues, i.e., cells whose plasmodesmata had been severed.[18] Miller et al.[17] added observations of almost perfect synchrony of the first nuclear division in multinucleates. Unaided fusion of isolated protoplasts has been noted when young tetrads of *Lilium longiflorum* cv. Shungetsu and of *Trillium kamtschaticum* were treated with high concentrations of wall digesting enzymes. Interspecies multinucleated products were obtained at very high frequencies (up to 90%).[19] Power et al.[20] explained the results as due to a ruffled plasma membrane. Observations and analyses of spontaneous fusion appear not to have led to a better understanding of the entrance of the generative nuclei of mature pollen tubes into the egg and primary embryo sac cell of seed plants.

Usually, isolated protoplasts are characterized by smooth surfaces and fusion has to be induced by one of a variety of treatments.

III. AGENTS FOR PROTOPLAST FUSION

A. Virus

A variety of treatments have been employed to fuse homo- and heterospecific protoplasts, even protoplasts and mammalian cells.[21] In developing protoplast fusion protocols it soon became clear that DNA and RNA viruses such as UV irradiated Sendaivirus that is used for animal cell fusion was not effective with plant material.[22] Given the hypothesis that these viruses facilitate agglutination and fusion on account of their hemolytic activity and accomplish fusion by controlled lysis of agglutinated cells, protoplasts, of course, would be an unsuitable substrate.

B. Lipsomes

The use of lipophilic substances for cell fusion has been investigated by Lucy,[23] who found that lysolecithin induced fusion of animal cells. Nagata et al.[24,25] observed that plant protoplasts (which are much more fragile than animal cells) were lysed by this treatment. Martin and MacDonald[26] used positively charged phospholipid vesicles (liposomes) to induce mammalian cell fusion and this technique was applied successfully to *Nicotiana tabacum* leaf protoplasts using the synthetic compound 1,2-*O*-dipentadecylmethyldiene-glycerol-3-phosphoryl-(*N*-ethylamino)-ethandiamine.[24] Up to 30% of protoplasts were fused in these experiments, but both polyethylene glycol and polyvinyl alcohol were more effective, producing up to 45% fusion products. Cationic vesicles would seem to be promising candidates for fusogens since they are lipophilic and should interact strongly with the anionic surface of protoplasts.[27] However, Uchimiya et al.[28] found that although positively charged liposomes (containing stearylamine to impart the charge) caused extensive aggregation of *Nicotiana glutinosa* cell culture protoplasts, very little fusion occurred. In addition, protoplasts were severely damaged by the liposome treatments. In a recent study, Nagata et al.[29] observed that liposomes rather than fusing with protoplasts were taken up by protoplasts via endocytosis. They subsequently demonstrated that genetic material, i.e., encapsulated (TMV) RNA could be efficiently introduced into protoplasts, where its function was expressed.

C. Mineral Salts

Michel[3] experimentally generated homokaryons and heterokaryons with protoplasts derived from parenchyma and epidermis cells by plasmolyzing adjacent protoplasts on a microscope slide with 0.5 *M* $NaNO_3$ for 10 to 20 min. Agglutination and sometimes fusion followed spontaneously. Since enzyme-isolated protoplasts became available solutions of 0.25 *M* $NaNO_3$ have allowed controlled fusion of a variety of cells (oats and maize root,[10]) and the generation of hybrid cells and plants *(Nicotiana glauca + N. langsdorffii*[30]. Binding[31] expanded the number of mineral salts which facilitated protoplast fusion by adding 85% seawater, 0.2 *M* $Ca(NO_3)_2$ at pH 8, and 0.2 *M* $CaCl_2$. In all experiments the frequency of fusion was comparatively low, approximately 1%.

The use of highly basic solutions containing high concentrations of Ca^{2+}[6] increased the frequency of fusion dramatically. Up to 50% of a *Nicotiana* protoplast population was fused when they were packed by centrifugation and incubated at 37°C in a medium containing 0.05 *M* $CaCl_2$ and 0.4 *M* mannitol buffered at pH 10.5. The method led to remarkable success in generating hybrid *Nicotiana tabacum* plants from the fusion products.[32]

D. Water-Soluble Polymers

A universally applicable agent to effect protoplast fusion was found in polyethylene glycol (PEG)[7,8] and was adapted for mammalian cells.[9] PEG was shown to not only have a wide potential for various cell types; it also is easy to handle, and a rather stable agent. Ten years later, the PEG method is still given preference and employed as prescribed in the initial publications as a solution of PEG 1540, 4000, or 6000 at a final concentration of 25 to 30%. The frequency of fusion may vary from 10% up to 40% of protoplasts treated depending on the quality and density of the protoplast preparations, the duration of the treatment (10 to 40 min), pH, and supplemental compounds. The frequency of fusion has sometimes been increased by enrichment of the Ca^{2+} and by an increase of the pH to 9.0 or 10.5.[14] The efficiency of fusion by PEG was greatly enhanced by addition of dimethylsulfoxide.[33,34] The extent of fusion was directly proportional to the concentration of both these compounds.

The universal use of PEG, however, has received some criticism. Its toxicity has been found to affect the viability of fusion products[35] and its affinity for water and its agglutinizing activity often produce severe distortions of the protoplasts.[36] Disturbing, also, was a finding that the activity of commercial grade PEG disappeared when it was purified by reprecipitation and/or dialysis.[37] These authors concluded that although PEG was able to agglutinate erythrocytes, a lipophilic impurity was responsible for fusion. However, a later study of five commercial PEG preparations established that although highly purified PEG was capable of inducing cell fusion, a variety of lipophilic additives enhanced this effect.[38] In at least one commercial sample impurities increased fusion efficiencies. The observation would suggest that variability in the behavior of PEG in different laboratories may be due to variability in the purity of PEG.

Cocking[39] reviewed the various agents used in protoplast fusion and after comparison using *Petunia* material concluded the high pH-Ca^{2+} method with 4% fusion and 440 somatic hybrids per 4×10^6 protoplasts was far superior to the PEG method also with 4% fusion events, but only six somatic hybrids.

Other water-soluble polymers tested successfully include polyvinyl alcohol, dextran, gelatin, and lectin.[40-42]

E. Electric field pulse

Fusion was induced between vacuoles and protoplasts, etiolated and green leaf protoplasts, and between protoplasts of different physiological functions using an electric field pulse technique.[43] The two-step procedure began with application of an alternating, nonuniform electric field to protoplast suspensions. Dielectrophoretic collectors adjusted to 1.5 V and 1 MHz and an electrical conductivity of the suspension medium $<10^{-5}$ sec/cm generated an electrophoresis effect which made the cells attach to each other along the field lines. Injection of an electric field pulse of high intensity (750 to 1000 V/cm) for a short duration (20 to 50 μsec) then led to breakdown of membranes and fusions.[44] In electrofusion Zimmermann and Vienken[45] claim to have found a comparatively superior fusion technique. It avoids the nonphysiological conditions imposed on protoplasts by PEG and/or high pH of the medium. Also, with proper execution of the experiment 60 to 80% of protoplasts were seen to participate in fusion. The fusion frequency, therefore, appears to be significantly higher when facilitated by electric field pulse.[45] Furthermore, the application of electrofusion may be extended to intracellular dielectrophoresis of nuclei and induction of nuclear fusion in previously established heterokaryocytes.[21]

Electrofusion of protoplasts may not have received general acceptance because of lack of the specialized equipment that is required. Also, the technique does not improve the frequency of dinucleated fusion products; as with PEG protoplast fusion often resulted in multinucleated cells.[91] Electrofusion products, so far, appear not to have resulted in somatic hybrid plants.

IV. THE FUSION PROCESS

As fusion has widely been achieved by treating protoplasts with PEG, it is fusion by this method which has most often been studied by electron and light microscopy. Attempts to examine protoplasts in the presence of PEG using TEM have, however, proven difficult, probably because of the difficulty of adequate penetration of the fixative. In all cases, tight adhesion is the first and immediate response by protoplasts to PEG treatment. Adhesion occurs equally well at sites where protoplasts show thin layers of cytoplasm as at those where protoplasts exhibited thicker layers.[45] On the contrary, Power et al.[10] observed that a substantial depth of cytoplasm in both protoplasts is a prerequisite for fusion. PEG-induced fusion of pea leaf and soybean cell culture protoplasts occurred over relatively large areas of the protoplast surface; tight adhesion with Juxtaposition of cell membranes was, however, restricted to small regions, separated by lens-shaped gaps. Membranes, then, were not agglutinated in a continuous pattern as reported for carrot protoplasts by Wallin et al.[8] Both types of membrane contact were observed when tobacco protoplasts were agglutinated with PEG.[47] More importantly, agglutination/adhesion of protoplasts upon treatment with PEG annihilated cell-cell recognition systems which might exist between materials of different plant taxa. For example, adhesion and fusion between carrot and algal protoplasts have been obtained and are well documented.[48,49]

Areas of cytoplasmic continuity due to coalescence of membranes and fusion appears to commence at the edge of the adhering surfaces; most of the cytoplasm of the fusion protoplasts remains separated by the agglutinated membranes at this stage. The fate of these membranes during fusion is uncertain, but some seem to form vesicles which disperse in the cytoplasm as the area of cytoplasmic continuity broadens. Cytoplasmic mixing of participating protoplasts with the membrane of the fusion product appears completed within a few hours.[46,50]

Initially fusion products harbor the total of two or several protoplasts. The membrane is thought to be transformed from two halves to a mosaic of respective parental units; quick rearrangements in the plasmalemma of protoplasts are indicated by movement of label as presented by Williamson et al.[51]

Early observations as presented appear not to have been followed up more recently.

V. THE MECHANISM OF FUSION

Heterospecific protoplast fusion requires the approach, adhesion, and coalescence of two different types of protoplasts. These are physical events and should be described in terms of surface physics. Approach of protoplasts is primarily determined by the long-range electrostatic forces resulting from the zeta potential (net surface charge of particles suspended in a fluid) present at the cell surface. Adhesion occurs when the van der Waals forces acting at the protoplast surfaces begin to overlap. The net van der Waals forces determine the interfacial energy at the protoplast-medium interface. The result of this interaction can be either protoplast fusion or the total failure of adhesion and coalescence. Gerson et al.[52] investigated the interfacial energies of protoplasts from corn roots and *Aureobasidium pullulans* and concluded that the degree of success of fusion when employing PEG media can be predicted by using surface physics. PEG 6000 decreases the zeta potential of lipid monolayers by several hundred millivolts and this action is primarily responsible for tight adhesion.

Attempts to achieve "artificial sexuality" by recharging the membrane surface of one fusion partner and retaining the net negative surface charge of the other and thus allow for directed aggregation and fusion remained unsuccessful.[27] The intrinsic function of PEG is cell adhesion by interpolymer complex formation between polymeric chains of PEG and polysaccharide chains of membrane bound proteins through hydrogen bonds. The polymer also appears to have some detergent-like properties[53] which may be responsible for membrane fusion. These detergent-like properties may be due to lipophilic contaminants[37] as discussed above.

Membrane fluidity is equally important in protoplast fusion. When cells destined for protoplast isolation and fusion are cultured at a low temperature (10°C) their membranes contain larger porportions of phospholipids of low phase transition points, i.e., unsaturated fatty acids, and thus show increased fluidity. Under such conditions and when treated with PEG 6000 protoplasts will progress to fusion more rapidly than those from cells cultured at 25°C.[54-56]

Ca-ions which facilitate PEG mediated fusion appear to interact with the cytoplasmic side of the cell membrane following an initial increase in membrane permeability to exogenous Ca^{2+}. It may also be expected to aggregate negatively charged molecules on the surface of protoplasts. This in turn might aggregate integral membrane proteins. Furthermore, the binding of Ca^{2+} to phosphatidylserine molecules in a membrane yields solid aggregates that allow other phospholipids to form fluid clusters. Aggregation of membrane proteins and the clustering of lipid molecules are thought to be important in membrane fusion.[57]

VI. FUSION PRODUCTS

When cultured in appropriate media[14] fusion products behave like nonfused protoplasts. Within a day or two fusion products may acquire an oval shape indicating cell wall formation. Viable fusion products will show cyclosis and either random dispersal of chloroplasts provided parental material included leaf cells, or chloroplasts will aggregate around nuclei. From this stage on, fusion products cannot any longer be visually recognized among nonfused and equally viable leaf protoplasts, unless a permanent marker through staining has been introduced prior to fusion. Galbraith et al.[58] have demonstrated that dyes can be used to identify heterokaryons that are otherwise indistinguishable from the parental strains. This approach was used by Patnaik et al.,[59] who fused leaf protoplasts of *Nicotiana tabacum* cv. Xanthi with suspension culture protoplasts of *Petunia hybrida* stained with fluorescein isothiocyanate (FITC). Hybrid protoplasts containing both red (chlorophyll) and green (FITC) were isolated manually. The use of intensely fluorescent dyes should allow isolation of heterokaryons by flow cytometry. However, at present most commercial instruments are optimized for use with mammalian cells that are smaller and more mechanically resistant.

Fusion experiments with protoplasts obtained with two different plants result in preparations with hetero- and homospecific fusion products and nonfused protoplasts.[60] In general, only the heterospecific fusion products are of interest for subsequent studies. It is these cells which serve as material for investigations of compatibility and hybridization. In this respect the fate of nuclei has received much attention. Differential staining permitted the recognition of heterospecific nuclei and establishment of the occurrence and viability of heterokaryocytes. In given preparations up to 30% of heterokaryocytes were found to contain one nucleus of each parent only;[61,62] the remainder of the heterospecific fusion products was multinucleated with the two parental nuclei present at various ratios. Heterokaryocytes have been observed to undergo synchronous, sometimes slightly asynchronous, mitoses.[63,64] Also, premature chromosome condensation has been reported.[65] In all cases the taxonomic relationship between cells rendering nuclei to heterokaryocytes appears irrelevant. For example, the fact that upon cell fusion mitotic soybean nuclei stimulated the mitosis of otherwise rather unresponsive nuclei of barley leaf protoplasts[60] would indicate cytoplasmic compatibility and the presence of "universal" compounds inducing the initiation of mitosis.

The heterokaryotic nature of fusion products may be carried on to the progeny; dinucleates may continue to produce dinucleated daughter cells as observed with faba bean and petunia cells.[66] Sometimes, synchronous division of mutinucleated heterokaryocytes will lead to the formation of a small cluster of heterospecific cells. There appears to be no account of callus grown from multinucleated heterokaryocytes and maintaining chimeral character.

Protoplast fusion and formation of heterokaryocytes open the way to somatic cell hybridization. In freshly fused cells two or more nuclei will be found randomly distributed. A day later during synchronous mitosis metaphase chromosome complements of each nucleus may be juxtaposed in one bipolar, occasionally tripolar spindle. Progressive mitosis, then, results in the formation of hybrid daughter nuclei and, after cytokinesis, hybrid cells[61] (see also Chapter 12).

During interphase, fusion of nuclei results from connection between the nuclear envelopes by bridges. Immediately following fusion the heterochromatin of heterospecific nuclei may be readily recognizable.[50] This modus of cell hybridization appears to occur less frequently than that which results from synchronous mitosis. Kao[67] suggests interphase nuclear fusion to be an artifact due to improper use of enzymes during protoplast isolation.

Division of hybrid daughter nuclei revealed that parental genomes retain individuality and do not integrate quickly. Chromosomes of either parent remain grouped together during mitosis and, as evidenced by differential staining, during interphase.[64] Number of chromosomes, size, and DNA content do not appear to infer incompatibility or breakdown of mitoses. Integration usually accompanied by chromosomal aberrations becomes visible only in mitotic figures of cells which have divided a number of times.[62] When cultured for several weeks hybrid cells may show chromosome segregation as documented for *Nicotiana glauca* + soybean hybrids.[68] Zymograms of alcohol dehydrogenase and aspartate amino-transferase of hybrid calli changed over time and eventually resembled soybean zymograms.[69] When protoplasts from such hybrids were "backfused" twice to *N. glauca* leaf protoplasts, a considerable increase in stability of the *N. glauca* chromosomes was observed. "Backfusion", indeed, was suggested to aid in stabilizing somatic hybrid cells.[70]

VII. ANIMAL-PLANT CELL FUSION

Since the first production of interkingdom hybrids[71] there has been much discussion of the unique information obtained from such heterokaryons. For example, Ward[72] suggested that gene transfer via hybridization may be superior to that mediated by gene cloning in some circumstances, in particular if the transfer of large segments of genomes to recipient cells is required. Ward also suggested that the formation of cytoplasmic hybrids may give interesting results and that information on nuclear/cytoplasmic interactions, membrane receptors, the organization of cytoskeletal elements, and the control of gene expression may be accessible. Finally, he suggested that the production of economically important animal products via plant tissue culture techniques may be possible. In most cases, however, one would expect the latter outcome to be more likely after the cloning of specific genes into bacteria. Jones et al.[73] suggested that interkingdom hybrids may be useful for determining the intracellular location of certain products as well as for gene mapping.

Of crucial importance is the question of whether animal genes can be expressed in the environment of a plant cell since, as will be discussed below, the plant character predominates in hybrids. The evidence from gene transfer experiments is inconclusive since mammalian genes have so far not been transferred to plant cells. The bacterial gene for chloramphenicol acetyltransferase was expressed in tobacco after linkage with a plant promoter;[74] however, eukaryotic genes even after such linkage have been expressed poorly, if at all, in plant cells (e.g., References 75 and 76). On a more encouraging note Ward[72] points out that an animal DNA virus has been shown to express in an algal cell[77] and *Drosophila* DNA has been shown to be functional in yeast.[78] Jones et al.[73] also pointed out that it may not be unrealistic to hope for expression of genes of widely different origin in eukaryotic cells if indeed eukaryotes arose according to the endosymbiont theory.

A variety of fusion protocols have been used to obtain interkingdom hybrids. Fusion of *Xenopus* cells and carrot protoplasts was achieved by protease treatment followed by mixing with a high pH (10.4), high calcium (50 m*M*) solution at 30°C.[72] In these experiments fusion efficiencies of 10 to 15% (based on the number of plant protoplasts and heterokaryons) were obtained if a large excess of animal cells was employed (approximately 5:1). Most other work has involved PEG. For example, Jones et al[73] fused HeLa cells and tobacco protoplasts using a 40- to 50-min treatment with 54% w/v PEG 6000 and found that about 0.2% of the protoplasts contained HeLa nuclei. There is one interesting study in which HeLa cells and *Haplopappus gracilis* protoplasts were fused with inactivated Sendai virus.[79] As has been pointed out above, viral fusion has not been successful with plant protoplasts alone.

There have been several studies on the fate of parental cell components after fusion. Mixing of membrane types was shown by scanning electron microscopy in hybrids of avian erythrocytes and tobacco leaf protoplasts.[80] Complete mixing of *Xenopus*-specific membrane components with carrot membrane was demonstrated by immunofluorescent staining and light microscopy.[81,82] Detailed ultrastructural examinations at the EM level have shown the presence of both parental nuclei and complete mixing of cytoplasmic components in erythrocyte/tobacco hybrids[80] and *Xenopus*/carrot hybrids.[83] Plant/animal hybrids tend to resemble the plant partner in appearance and to regenerate a cell wall.[72] This may be due simply to the larger size of the plant partner which results in plant cytoplasm predominating. Heterokaryons between plant protoplasts and amebae tend to resemble the larger ameba partner.[84]

In most cases heterokaryons have been cultured in plant protoplast media, but attempts have been made to produce media of intermediate composition to help maintain the viability of the animal partner. Hadlaczky et al[85] mixed animal and plant media in various compositions and found an optimum mixture which allowed DNA synthesis in both soybean and *Drosophila* cells. There were some heterokaryons which, when cultured in this mixed medium, contained nuclei of both types that retained DNA synthetic activity. Studies weith *Xenopus*/carrot hybrids[72] showed that survival of heterokaryons was not favored in media of intermediate composition but that they responded as did carrot protoplasts. Lima-de-Faria et al.[79] observed that *Haplopappus gracilis* protoplasts maintained viability in media used to culture HeLa cells. Nevertheless, the resulting heterokaryons resembled plant protoplasts.

Plant/animal hybrids have been observed to survive for up to 14 days in culture.[72] Division of the carrot nucleus in *Xenopus*/carrot hybrids occurred occasionally but never division of the animal nucleus. Usually, the *Xenopus* nucleus degenerated fairly quickly. Autoradiographic studies on soybean/*Drosophila* hybrids[85] showed that DNA synthetic activity was sometimes retained by both types of nuclei for up to 48 hr and that in 17.8% of the heterokaryons soybean nuclei had either divided or were in mitosis. Some evidence was presented to suggest that division of *Drosophila* nuclei occasionally occurred. It has been shown that HeLa nuclei in HeLa/tobacco hybrids retained the aiblity to synthesize DNA although they did not divide.[73] Again the plant nucleus in the heterokaryons was frequently capable of division. A couple of studies have suggested that adhesion and possibly fusion between plant and animal nuclei occurs.[86,87]

Formation of stable interkingdom hybrids involving plant cells will probably involve a gradual loss of chromosomes from one partner resulting in the transfer of a limited amount of genetic information to the predominant partner. The production of a stable animal/protozoan hybrid which resembled the animal partner and expressed protozoan antigens after 14 weeks of continuous culture shows that desirable results can be achieved.[88]

VIII. CONCLUSION

Protoplast fusion is foremost a procedure for the generation of fusion products, hybrid cells, and plants. A respectable list of hybrid plants obtained by protoplast fusion (see Chapter 12) would attest to the perfection of the procedure established over the last decennium and practiced today. Present state-of-the-art protoplast fusion is, however, still in need of further improvements. Targets would be techniques which allow precision fusion at high frequency and fusion products with a better potential for subsequent development. PEG and electric field pulse have been widely accepted as fusogenic agents. Preference is given to PEG, as its application is simple, fairly reliable, and despite reservations relatively harmless. As both methods will undergo modifications in the future, one disadvantage common to both should be overcome, i.e., lack of control over the combination of desirable partners. Apart from improvements of selection techniques after fusion which may compensate for deficiencies of the present fusion procedures, novel developments may enable the experimentor to selectively fuse protoplasts at high speed. A combination of cell sorting and fusion would appear as a reasonable solution. The problem of stimulating fusion products to quickly develop into plants while avoiding unwanted variability is similar to that of culturing single protoplasts for plant regeneration. Complex nutrient formulae, nurse cultures, and feeder layers have overcome difficulties remarkably successfully (see Chapter 1), but certainly leave room for new approaches. The conditioning of protoplasts prior to or during fusion for rapid development after fusion by injection of physiological stimuli appears not to have been pursued actively.

Cell fusion is a procedure which will allow further elucidation of the structure of cell membranes. The similarity of the surface of animal cells and plant protoplasts was a striking observation. This similarity became even more evident when both kinds of cells were treated with protease; the rate of fusion increased regardless of the origin of the cells, and interkingdom fusion products showed intermixing of animal and plant membrane components and formation of a hybrid membrane.[72] It remains to be seen whether protoplast fusion will equally enlighten our knowledge of membrane function.

The actual fusion process has received substantial clarification since the introduction of the method using an electric field. Fusion appears to require the emergence of particle-free lipid domains in membranes of adhering protoplasts. Treatment of cells with protease improves such emergence. Breakdown of particle-free lipid domains can be accomplished by electric pulse in the shortest time and under the least possible damage by heat. Protein domains in membranes, while open to breakdown by electric pulse, present a more formidable barrier.[45] In extending this investigation one may ask whether or not walled cells, not protoplasts, can be subject to cell fusion. The answer can be affirmative as fusion may only require wall removal of a small area and orientation of the exposed membrane into the field direction which establishes contact between two cells.[45]

Cell fusion has not, as yet, contributed to a better understanding of natural phenomena like fertilization in higher plants or cell-cell recognition. The PEG as well as the electro-fusion method combine cells in an indiscriminate manner which does not allow possible recognition systems in surfaces to become functional. The fusion of like cells when establishing articulated vessels as occurs in nature has not been simulated. Also, fertilization of egg cells has not been refined to an in vitro method in such a way that the entrance of regenerative nuclei would be better understood; comparable systems with animal cells have been much advanced.[89]

Finally, cell fusion may well be used to contribute to the biology of various phenomena which characterize the physiological state of cells and plants, i.e., to probe cell maturity vs. juvenility, embryonic vs. nonembryonic, mitotic vs. nonmitotic cells. Here fusion is seen as a convenient way of transferring signals for developmental changes from cell to cell. The induction of premature chromosome condensation in resting cells is a case in point which even had diagnostic spin-offs.[90]

Advances in cell fusion with these objectives in mind may well complement the present almost exclusive pursuit of cell fusion for cell hybridization and transformation.

REFERENCES

1. Küster, E., Über die Verschmelzung nackter Protoplasten, *Ber. Dtsch. Bot. Ges.*, 27, 589, 1909.
2. Küster, E., Eine Methode zur Gewinnung abnorm grosser Protoplasten, *Arch. Entwickl. Mech.*, 30, 351, 1910.
3. Michel, W., Über die experimentelle Fusion pflanzlicher Protoplasten, *Arch. Exp. Zellforsch.*, 20, 230, 1937.
4. Küster, E., Über die Gewinnung nackter Protoplasten, *Protoplasma*, 3, 223, 1928.
5. Power, J. B. and Cocking, E. C., Isolation of leaf protoplasts: macromolecule uptake and growth substance response, *J. Exp. Bot.*, 21, 64, 1970.
6. Keller, W. A. and Melchers, G., The effect of high pH and calcium on tobacco leaf protoplast fusion, *Z. Naturforsch.*, 28c, 737, 1973.
7. Kao, K. N. and Michayluk, M. R., A method for high frequency intergeneric fusion of plant protoplasts, *Planta*, 115, 355, 1974.
8. Wallin, A., Glimelius, K. and Eriksson, T., The induction of aggregation and fusion of *Daucus carota* protoplasts by polyethylene glycol, *Z. Pflanzenphysiol.*, 74, 64, 1974.
9. Pontecorvo, G., Production of mammalian somatic cell hybrids by means of polyethylene glycol treatment, *Somatic Cell Genet.*, 1, 397, 1973.
10. Power, J. B., Cummins, S. E., and Cocking, E. C., Fusion of isolated plant protoplasts, *Nature (London)*, 225, 1016, 1970.
11. Thomas, E., *From Single Cells to Plants*, Wykeham, London, 1975.
12. Berry, S. F. and Power, J. B., Fusion of *Petunia* protoplasts, *PMB Newsl.*, 1, 23, 1980.
13. Constabel, F., Fusion of protoplasts by polyethylene glycol (PEG), in *Cell Culture and Somatic Cell Genetics*, Vol 1, Vasil, I. K., Ed., Academic Press, New York, 1984, 414.
14. Kao, K. N., Plant protoplast fusion and isolation of heterokaryocytes, in *Plant Tissue Culture Methods*, #19876 Wetter, L. R. and Constabel, F., Eds., National Research Council of Canada, Ottawa, 1982, 49.
15. Sassen, M. M. A., Breakdown of the plant cell wall during the cell-fusion process, *Acta Bot. Neerland.*, 14, 165, 1965.
16. Hellmann, S. and Reinert, J., Protoplasten aus Zellkulturen von *Daucus carota*, *Protoplasma*, 72, 479, 1971.
17. Miller, R. A., Gamborg, O. L., Keller, W. A., and Kao, K. N., Fusion and division of nuclei in multinucleated soybean protoplasts, *Can. J. Genet. Cytol.*, 13, 347, 1971.
18. Withers, L. A. and Cocking, E. C., Fine-structural studies on spontaneous and induced fusion of higher plant protoplasts, *J. Cell Sci.*, 11, 59, 1972.
19. Ito, M. and Maeda, M., Fusion of meiotic protoplasts in liliaceous plants, *Exp. Cell. Res.*, 80, 453, 1973.
20. Power, J. B., Evans, P. K., and Cocking, E. C., Fusion of plant protoplants, in *Membrane Fusion*, Poste, G. and Nicholson, G. L., Eds., North-Holland, Amsterdam, 1978, 374.

21. Lazar, G. B., Recent developments in plant protoplast fusion and selection technology, in *Protoplasts 1983*, Potrykus, I., Harms, C. T., Hinnan, A., Hütter, R., King, P. J., and Shillito, R. B., Eds., Birkhäuser Verlag, Basel, 1983, 61.
22. Keller, W. A., The Isolation, Fusion and Culture of Plant Protoplasts, Ph.D. thesis, University of Saskatchewan, Saskatoon, Canada, 1972.
23. Lucy, J. A., The fusion of biological membranes, *Nature (London)*, 227, 815, 1970.
24. Nagata, T., Eibl, H., and Melchers, G., Fusion of plant protoplasts induced by positively charged synthetic phospholipid vesicles, *Z. Naturforsch.*, 34c, 460, 1979.
25. Nagata, T., Interaction of liposomes and protoplasts as a model system of protoplast fusion, in *Cell Fusion: Gene Transfer and Transformation*, Beers, R. F. and Basset, E. G., Eds., Raven Press, New York, 1984, 217.
26. Martin, F. J. and MacDonald, R. C., Induction of cell fusion, *J. Cell Biol.*, 70, 506, 1976.
27. Nagata, T. and Melchers, G., Surface charge of protoplasts and their significance in cell-cell interaction, *Planta*, 142, 235, 1978.
28. Uchimiya, H., Kudo, N., Ogawara, T., and Harada, H., Aggregation of plant protoplasts by artificial lipid vesicles, *Plant Physiol.*, 69, 1278, 1982.
29. Nagata, T., Okada, K., Takebe, I., and Matsui, C., Delivery of tobacco mosaic virus RNA into plant protoplasts mediated by reverse-phase evaporation vesicles (liposomes), *Mol. Gen. Genet.*, 184, 161, 1981.
30. Carlson, P. S., Smith, H. H., and Dearing, R. D., Parasexual interspecific plant hybridization, *Proc. Natl. Acad. Sci. U.S.A.*, 69, 2292, 1972.
31. Binding, H., Fusionsversuche mit isolierten Protoplasten von *Petunia hybrida* L., *Z. Pflanzenphysiol.*, 72, 422, 1974.
32. Melchers, G. and Labib, G., Somatic hybridization of plants by fusion of protoplasts, *Mol. Gen. Genet.*, 135, 227, 1974.
33. Norwood, T. H., Zeigler, C. J., and Martin, G. M., Dimethyl sulfoxide enhances polyethylene glycol-mediated somatic cell fusion, *Somatic Cell Genet.*, 2, 263, 1976.
34. Douglas, G. C., Keller, W. A., and Setterfield, G., Somatic hybridization between *Nicotiana rustica* and *N. tabacum*, *Can. J. Bot.*, 59, 220, 1981.
35. Mercer, W. E. and Schlegel, R. A., Phytohemagglutinin enhancement of cell fusion reduces polyethylene glycol cytotoxicity, *Exp. Cell Res.*, 120, 417, 1979.
36. Constabel, F. and Kao, K. N., Agglutination and fusion of plant protoplasts by polyethylene glycol, *Can. J. Bot.*, 52, 1603, 1974.
37. Honda, K., Maeda, Y., Sasakawa, S., Ohno, H., and Tsuchida, E., The components contained in polyethylene glycol of commerical grade (PEG-6000) as cell fusogen, *Biochem. Biophys. Res. Commun.*, 101, 165, 1981.
38. Smith, C. L., Ahkong, Q. F., Fisher, D., and Lucy, J. A., Is purified polyethylene glycol able to induce cell fusion?, *Biochem. Biophys. Acta*, 692, 109, 1982.
39. Cocking, E. C., Concluding remarks, *Plant Cell Cultures, Results and Perspectives*, Sala, F., Parisi, B., Cella, R., and Ciferri, O., Eds., North-Holland, Amsterdam, 1980, 419.
40. Nagata, T., A novel cell-fusion method of protoplasts by polyvinyl alcohol, *Naturwissenschaften*, 65, 263, 1978.
41. Kameya, T., Induction of hybrids through somatic cell fusion with dextran sulfate and gelatin, *Jpn. J. Genet.*, 50, 235, 1975.
42. Glimelius, K., Wallin, A., and Eriksson, T., Concanavalin A improves the polyethylene glycol method for fusing plant protoplasts, *Physiol. Plant.*, 44, 92, 1978.
43. Vienken, J., Ganser, R., Hampp, R., and Zimmermann, U., Electric field-induced fusion of isolated vacuoles and protoplasts of different developmental and metabolic provenience, *Physiol. Plant.* 53, 64, 1981.
44. Zimmermann, U. and Scheurich, P., High frequency fusion of plant protoplasts by electric fields, *Planta*, 151, 26, 1981.
45. Zimmermann, U. and Vienken, J., Electric field-mediated cell-to-cell fusion, in *Cell Fusion: Gene Transfer and Transformation*, Beers, R. F. and Basset, E. G., Eds., Raven Press, New York, 1984, 171.
46. Fowke, L. C., Rennie, P. J., Kirkpatrick, J. W., and Constabel, F., Ultrastructure of fusion products from soybean cell culture and sweet clover leaf protoplasts, *Planta*, 130, 39, 1976.
47. Burgess, J. and Fleming, E. N., Ultrastructural studies of the aggregation and fusion of plant protoplasts, *Planta*, 118, 183, 1974.
48. Fowke, L. C., Gresshoff, P. M., and Marchant, H. J., Transfer of organelles of the alga *Chlamydomonas reinhardii* into carrot cells by protoplast fusion, *Planta*, 144, 341, 1979.
49. Fowke, L. C., Marchant, H. J., and Gresshoff, P. M., Fusion of protoplasts from carrot cell cultures and the green alga *Stigeoclonium*, *Can. J. Bot.*, 59, 1021, 1981.
50. Fowke, L. C., Constabel, F., and Gamborg, O. L., Fine structures of fusion products from soybean cell culture and pea leaf protoplasts, *Planta*, 135, 257, 1977.

51. Williamson, F. A., Fowke, L. C., Constabel, F., and Gamborg, O. L., Labelling of concanavalin A sites on the plasma membrane of soybean protoplasts, *Protoplasma*, 89, 305, 1976.
52. Gerson, D. F., Meadows, M. G., Finkelman, M., and Walden, D. B., The biophysics of protoplast fusion, in *Advances in Protoplast Research*, Kovacs, E. I., Ed., Hungarian Academy of Sciences, Budapest, 1980, 447.
53. Maggio, B., Ahkong, Q. F., and Lucy, J. A., Polyethylene glycol, surface potential and cell fusion, *Biochem. J.*, 158, 647, 1976.
54. Yamada, Y., Hara, Y., Senda, M., Nishihara, M., and Kito, M., Phospholipids of membranes of cultured cells and the products of protoplast fusion, *Phytochemistry*, 18, 423, 1979.
55. Senda, M., Morikawa, H., Katagi, H., Takada, T., and Yamada, Y., Effect of temperature on membrane fluidity and protoplast fusion, *Theor. Appl Genet.*, 57, 33, 1980.
56. Yamada, Y., Hara, Y., Katagi, H., and Senda, M., Effect of low temperature on the membrane fluidity of cultured cells, *Plant Physiol.*, 65, 1099, 1980.
57. Ahkong, Q. F., Fisher, D., Tampion, W., and Lucy, J. A., Mechanisms of cell fusion, *Nature (London)*, 253, 194, 1975.
58. Galbraith, D. W., Mauch, T. J., and Shields, B. A., Analysis of the initial stages of plant protoplast development using 33258 Hoechst: reactivation of the cell cycle, *Physiol. Plant.*, 51, 380, 1981.
59. Patnaik, G., Cocking. E. C., Hamill, J., and Pental, D., A simple procedure for the manual isolation and identification of plant heterokaryons, *Plant Sci. Lett.*, 24, 105, 1982.
60. Kao, K. N., Constabel, F., Michayluk, M. R., and Gamborg, O. L., Plant protoplast fusion and growth of intergeneric hybrid cells, *Planta*, 120, 215, 1974.
61. Constabel, F., Dudits, D., Gamborg, O. L., and Kao, K. N., Nuclear fusion in intergeneric heterokaryons, *Can. J. Bot.*, 53, 2092, 1975.
62. Kao, K. N., Chromosomal behavior in somatic hybrids of soybean — Nicotiana glauca, Mol. Gen. *Genet.*, 150, 225, 1977.
63. Gosch, G. and Reinert, J., Nuclear fusion in intergeneric heterokaryocytes and subsequent mitosis of hybrid nuclei, *Naturwissenschaften*, 63, 534, 1976.
64. Constabel, F., Weber, G., and Kirkpatrick, J. W., Sur la compatibilité des chromosomes dan les hybrides intergénériques de cellules de *Glycine max* x *Vicia hajastana*, *C. R. Acad. Sci. Paris*, 285, 319, 1977.
65. Szabados, L. and Dudits, D., Fusion between interphase and mitotic plant protoplasts, *Exp. Cell Res.*, 127, 442, 1980.
66. Binding, H. and Nehls, R., Somatic cell hybridization of *Vicia faba* + *Petunia hybrida*, *Mol. Gen. Genet.*, 164, 137, 1978.
67. Kao, K. N., Plant protoplast fusion and somatic hybrids, in *Proc. Symp. Plant Tissue Culture*, Science Press, Peking, 1978, 331.
68. Wetter, L. R., Isoenzyme analyses of cultured cells, in *Plant Tissue Culture Methods*, Wetter, L. R. and Constabel, F., Eds., National Research Council of Canada, Ottawa, 1982, 105.
69. Wetter, L. R., Isoenzyme patterns in soybean — *Nicotiana* somatic hybrid cell lines, *Mol. Gen. Genet.*, 150, 231, 1977.
70. Wetter, L. R. and Kao, K. N., Chromosome and isoenzyme studies on cells derived from protoplast fusion of *Nicotiana glauca* with *Glycine max* — *Nicotiana glauca* cell hybrids, *Theor. Appl. Genet.*, 57, 273, 1980.
71. Ahkong, Q. F., Howell, J. I., Lucy, J. A., Safwat, F., Davey, M. R., and Cocking, E. C., Fusion of hen erythrocytes with yeast protoplasts induced by polyethylene glucol, *Nature, (London)*, 255, 66, 1975.
72. Ward, M., Fusion of plant protoplasts with animal cells, in *Cell fusion: Gene transfer and Transformation*, Beers, R. F. and Bassett, E. G., Eds., Raven Press, New York, 1984, 189.
73. Jones, C. W., Mastrangelo, I. A., Smith, H. H., Liu, H. L., and Meck, R. F., Interkingdom fusion between human (HeLa) cells and tobacco hybrid (GGLL) protoplasts, *Science*, 193, 401, 1976.
74. Herrero-Estrella, L., Depicker, A., Van Montagu, M., and Schell, J., Expression of chimaeric genes transferred into plant cells using a Ti-plasmid-derived vector, *Nature (London)*, 303, 209, 1983.
75. Barton, K. A., Binns, A. N., Matzke, A. J. M., and Chilton, M.-D., Regeneration of intact tobacco plants containing full-length copies of genetically engineered T-DNA and transmission of T-DNA to R1 progeny, *Cell*, 32, 1033, 1983.
76. Murai, N., Sutton, D. W., Murray, M. G., Slightom, J. L., Merlo, D. J., Reichert, N. A., Sengupta-Gopalan, C., Stock, C. A., Barber, R. F., Kemp, J. D., and Hall, T. C., Phaseolin gene from bean is expressed after transfer to sunflower via tumor-inducing plasmid vectors, *Science*, 222, 476, 1983.
77. Cairns, E., Doerfler, W., and Schweiger, H.-G., Expression of a DNA animal virus genome in a plant cell, *FEBS Lett.*, 96, 295, 1978.
78. Henikoff, S., Tatchell, K., Hall, B. D., and Nasmyth, K. A., Isolation of a gene from *Drosophila* by complementation in yeast, *Nature (London)*, 289, 33, 1981.

79. Lima-de-Faria, A., Eriksson, T., and Kjellen, L., Fusion of human cells with *Haplopappus* protoplasts by means of Sendai virus, *Hereditas*, 87, 57, 1977.
80. Willis, G. E., Hartmann, J. X., and Lamater, E. D., Electron microscopic study of plant-animal cell fusion, *Protoplasma*, 91, 1, 1977.
81. Ward, M., Davey, M. R., Cocking, E. C., Clothier, R. H., Balls, M., and Lucy, J. A., Double labeling of plant/animal heterokaryons, *Cell Biol. Int. Rep.*, 4, 796, 1980.
82. Ward, M., Davey, M. R., Cocking, E. C., Balls, M., Clothier, R. H., and Lucy, J. A., Animal-specific membrane components visualised on the surface of animal/plant heterokaryons, *Protoplasma*, 104, 75, 1980.
83. Davey, M. R., Clothier, R. H., Balls, M., and Cocking, E. C., An ultrastructural study of the fusion of cultured amphibian cells with higher plant protoplasts, *Protoplasma*, 96, 157, 1978.
84. Rajasekhar, E. W., Chatterjee, S., and Eapen, S., Fusion of plant protoplasts with amoebae induced by polyethylene glycol, *Cytologia*, 45, 149, 1980.
85. Hadlaczky, G. Y., Burg, K., Maroy, P., and Dudits, D., DNA systhesis and division in interkingdom hybrids, *In Vitro*, 16, 647, 1980.
86. Mastrangelo, I. A. and Mitra, J., Chinese hamster ovary chromosomes and antigens in tobacco/hamster heterokaryons, *J. Hered.* 72, 81, 1981.
87. Dudits, D., Rasko, I., Hadlaczky, G., and Lima-de-Faria, A., Fusion of human cells with carrot protoplasts induced by polyethylene glycol, *Hereditas*, 82, 121, 1976.
88. Crane, M. St. J. and Dvorak, J. A., Vertebrate cells express protozoan antigen after hybridization, *Science*, 208, 194, 1980.
89. Shapiro, B. M., Haploid cell fusion, in *Cell Fusion: Gene Transfer and Transformation*, Beers, R. F. and Bassett, E. G., Eds., Raven Press, New York, 1984, 3.
90. Rao, P. N. and Johnson, R. T., Cell fusion and its application to studies on the regulation of the cell cycle, in *Methods in Cell Physiology*, Vol. V, Prescott, D. M., Ed., Academic Press, New York, 1982, 75.
91. Constabel, F., unpublished results.

Chapter 5

PLANT PROTOPLASTS AND THE CELL CYCLE

A. R. Gould and R. J. Daines

TABLE OF CONTENTS

I. INTRODUCTION

It is well appreciated that higher plant protoplasts cannot serve as a botanical analog of animal cells. Protoplasts can be viewed as injured cells which must embark on a program of repair before they are capable of sustained division, and thus cultured plant protoplasts are in a linear rather than a cyclical developmental sequence. Put more simply, protoplasts generally give rise to cells rather than to more protoplasts (subprotoplasts, and the wall-less *Chlamydomonas* line CW-15 accepted and excepted). Consequently, plant protoplasts do not cycle! Certainly, because they are not cells, they cannot be a part of the "cell" cycle. Why then, a chapter on the cell cycle of plant protoplasts, which are not cells, and do not cycle?

The apparent paradox set out above can be resolved only by a clear understanding of the complexity of the process referred to as the cell cycle. Plant cells and plant protoplasts have in common the organelle which provides the signals by which progression through the cell cycle has traditionally been mapped, i.e., the nucleus. The discontinuous synthesis of nuclear DNA between successive mitoses marks out the three most familiar episodes of interphase (G1, S-phase, and G2); but in order to produce viable daughter cells it is necessary to integrate the replication of many different organelles and molecules, of which the nucleus and DNA are only minor, albeit very important, constituents. An obvious difference between cells and protoplasts is the (temporary) absence of a cell wall in the latter; the synthesis or resynthesis of wall components during interphase provides a good example of an important part of the cell cycle which is only loosely linked to the G1-S-G2 sequence. Yet even cell wall biosynthesis has temporal links with the nuclear cycle of DNA replication and partition, as demonstrated by the precise analyses performed by Amino et al.[1] These workers have described specific changes in cell wall composition during certain cell cycle phases. For instance a large increase in the pectin content of cells was observed during cytokinesis. Other laboratories have provided extensive evidence that membrane fractions and microtubule arrays also undergo extensive turnover and reorganization as cells and protoplasts approach the next division. More detailed consideration of the complex nature of the plant cell cycle has appeared in previous reviews,[2,3] and one reason for studying cell cycle kinetics of protoplast populations is to try to understand the intricate program of repair initiated by wall removal and protoplast culture.

Another more recent interest in the cell cycle state of protoplasts has arisen as a by-product of research aimed at unorthodox genetic manipulation (genetic engineering) of plant cells in culture. Unfortunately the rapid technical progress made in plant molecular biology has in general not been matched by similar advances in plant cell biology, and the study of the plant cell cycle has been particularly neglected.[4] With the cycle kinetics of protoplasts only a subset of the neglected field of the plant cell cycle, the volume of literature available for review is limited to the output of only a few laboratories. However, the existing and potential contributions that "protoplast cycle" research can make to both technical and theoretical aspects of plant cell biology (especially in genetic engineering), are of such importance that it is certainly worthwhile to review them. Thus this survey of plant protoplasts and the cell cycle will be short, exclusive of whole cell studies, and slanted towards the future applications of this very specialized field.

II. CYCLE PHASE ESTIMATION WITH PROTOPLASTS

Plant protoplast populations are not always amenable to the well-established methods of cell cycle analysis.[5] For example the Fraction of Labeled Mitoses (FLM) analysis (which for all its shortcomings has been extensively used for both animal and plant cell populations) has yet to be applied to protoplast cultures. Cell cycle phases are commonly expressed either as durations (i.e., hours to complete S-phase) or as frequencies (i.e., fraction of a population engaged in nuclear DNA synthesis).

A. Cycle Phase Duration Analyses

The analysis of phase durations usually requires that cell populations be labeled with radioactive thymidine to identify nuclei in S-phase, followed by sequential sampling to monitor the progress of the labeled cohort towards and through the next mitosis. Protoplast populations have been successfully labeled with tritiated thymidine either before or after enzymatic removal of cell walls,[6] but with protoplasts isolated from leaf tissue, the initial absence of S-phase cells in the population precludes application of this general method of tagging discrete subpopulations.

Comparisons of phase durations in protoplast populations and the cell populations from which they are derived have yet to be performed. This would seem to be a source of very productive research in terms of the optimization of protoplast culture techniques. For example, the difficulties encountered so far with the culture of protoplasts of some monocotyledons could be viewed as a general problem of cell cycle traverse. A comparison of phase durations in the cell population and the protoplast population derived from it would at least identify in which part of interphase a block to further progression was operating. Such an analysis would be best approached via cytological analysis of microautoradiographs prepared from sequential samplings of a protoplast population previously labeled with tritiated thymidine. Another approach would be to monitor DNA precursor uptake into the trichloracetic acid-insoluble fraction during the culture of recently isolated protoplasts. As an example, this method has been convincingly applied by Zelcer and Galun[7] in studies on biosynthesis of macromolecules and the effects of coumarin in freshly isolated tobacco mesophyll protoplasts.

However, there are certain pitfalls in the use of a method which does not distinguish between nuclear DNA synthesis and the replication of the cytoplasmic (mitochondrial and chloroplast) genomes. First, the obvious control needed in experiments which measure rates of precursor uptake into protoplasts is a comparative estimate of precursor incorporation into cells from which the protoplasts are isolated. Second, changes in rates of thymidine incorporation are not reliable indicators of changes in numbers of cells or protoplasts engaged in nuclear DNA synthesis or of changes in rate of progress of cells through S-phase: changes in pool size or the rate of mitochondriogenesis could easily account for small changes in the rate of thymidine incorporation. Third, the inability of a cell or protoplast population to incorporate thymidine does not indicate that there is a block in the cycle. Numerous studies have demonstrated that the duration of the G1-phase is greatly extended in cultured cells as compared to meristems of the same species.[3] Thus even a delay of 30 hr or more before the initiation of DNA synthesis would not be unusual for a population of G1 cells, and G2 cells could take considerably longer to traverse mitosis and G1 before they began to incorporate thymidine in the next round of nuclear replication. A general failure to address any of the above requirements for satisfactory kinetic analysis of protoplast populations[8] is unfortunately representative of the low level of understanding of cycle kinetics in the plant tissue culture community. However, all of these problems can be resolved by supplementing the precursor incorporation studies with simple autoradiographic techniques which unequivocally demonstrate presence or absence of nuclear DNA synthesis.[5,9]

B. Cycle Distribution Analyses

Several methods are available which estimate what fraction of a population resides in a particular cycle phase; all of them require the use of dyes which bind specifically and quantitatively to nuclear DNA. Then measurement of the amount of dye bound per nucleus, either by absorption or emission (fluorescent) microspectrophotometry, allows assignment of individual cells to one of the three interphase cycle compartments.[5,9] The Feulgen staining protocol is probably the most well known of the cytological methods which can be used to quantify nuclear DNA on a single-cell basis. Feulgen-stained protoplasts have been used to estimate cell cycle distributions by absorption microspectorphotometry in a variety of situations, and automated grain counting in autoradiograph preparations can provide additional characterization of cycle phases beyond simple assignment to G1, S, G2, or mitosis.[10] The smallest amount of DNA that can be reliably detected by Feulgen microspectrophotometry is probably of the order of 1 pg per nucleus, and this a major limitation of the Feulgen method, especially when measurements need to be made on single chromosomes. Difficulties with reproducibility, absolute measurements, and other factors,[11] have made Feulgen microspectrophotometry a little-used technique.

The availability of DNA-specific fluorochromes significantly increases our ability to measure small amounts of nuclear DNA. The Hoechst dyes 33258 or 33342, propidium iodide, ethidium bromide, or mithramycin, all allow measurement of subpicogram amounts of DNA by fluorescent microspectrophotometry. Laloue et al.[12] present a typical staining protocol (using the bis-benzimidazole derivative Hoechst 33258) which is very rapid, and unlike the Feulgen technique, does not require a hydrolysis step. Quantitative fluorescent staining of nuclear DNA has been successfully used to trace the progression of protoplasts towards mitosis,[13] and in general the use of emission microspectrophotometry has proved much more reliable and sensitive for such work than the older Feulgen based techniques. Even so there remain some difficulties with the new fluorescent methods (for example, photobleaching of the dye-DNA complex), and it still takes a long time to generate statistically significant numbers of measurements, by scanning single nuclei one by one with the microspectrophotometer.

C. Flow Cytometric Estimations

Technical advances in the design and operation of flow cytometers promise such great improvements in efficiency for both analysis and manipulation of protoplast populations that a separate consideration of the technology, in the context of cell cycle research, is warranted. In its simplest conception, the flow cytometer passes a stream of droplets containing dyed cells, through a laser beam tuned to elicit a fluorescent signal from the dye. The fluorescent emission is recorded by a detector focused on the stream of droplets. The detector has such a rapid response time, and the droplet stream can be made to flow at such high velocities, that many thousands of measurements can be made each second. This technology has been applied to plant protoplasts labeled with vital stains such as fluorescein isothiocyanate (FITC) by Redenbaugh et al.[14] with the added sophistication that the droplet stream can be deflected by changing the polarity of a pair of charged plates through which the stream falls. This is the essential mechanism by which flow sorting is achieved: the signals from the fluorescence detector are evaluated by a computer, and specific classes of cells (or in this case, protoplasts) can be separated from the main population by rapid, computer-initiated polarity switches of the deflector plates. The most obvious application of such technology in protoplast research is in high speed sorting of homo- and heterokaryons after fusion of differently stained protoplasts. The most common approach has been to stain one protoplast population to fluoresce yellow (with FITC) and another to fluoresce red (with RITC, rhodamine isothiocyanate); the detector can then distinguish heterokaryons resulting from protoplast fusion as giving both yellow and red signals, and this combination can be made to trigger a deflection out of the main droplet stream, and ultimate collection in a separate container.

Flow cytometry and sorting have made great contributions to the study of the cell-cycle in animal systems, and so there has been a very understandable interest in transplanting this technology to the plant sciences. Two of the most attractive possibilities would be rapid analysis of cell cycle distributions and sorting live protoplasts in terms of their cell cycle phase (i.e., selection synchrony). Both of these objectives rely on the fluorescent dyes (mentioned above) which bind stoichiometrically to nuclear DNA, and these goals can be routinely achieved with animal cell populations. However, significant technical difficulties accrue when plant protoplasts are used in such work. Although cells isolated from *Asparagus* fronds and stained with DNA-specific dyes have been successfully analyzed in a flow system,[33] in general plant cells in suspension culture are too clumped to be useful in flow cytometry applications, and some workers have been forced to use nuclei isolated from plant tissue to obtain flow analyses of cycle distribution.[15] The problem with any method which isolates nuclei or protoplasts from a cell population and then uses the isolated population for a cycle distribution analysis, is to know how accurately the cycle distribution of the isolate reflects the distribution in the original cell population. In some situations this is not a valid criticism because the cycle status of the protoplasts is the parameter under study, but it has been shown that certain purification methods enrich protoplast populations with one cycle phase at the expense of the others.[3]

To date, all of the flow cytometric analyses of cycle distribution with plant protoplasts have used staining protocols which require fixation of protoplasts prior to staining. As the only authors to report on DNA staining in nonfixed protoplasts, Puite and Ten Broeke[16] present data which force the disappointing but unavoidable conclusion that present staining methods do not allow quantitation of nuclear DNA content in live protoplasts. This is the major technical difficulty which at present prevents flow sorting of protoplasts in terms of cycle phase, and work in our laboratory confirms that DNA-specific dyes, which will vitally stain the nuclei of animal cells, are ineffective with live plant protoplasts. Even with animal cells the kinetics of DNA staining with many vital dyes are far from simple: the final intensity of fluorescence obtained apparently depends on the complex interaction of several rate constants (for transport in and out across the plasmalemma, in and out across the nuclear membrane, etc.), and a transport system that may rely on microtubules. Thus considering the "injured" state of protoplasts which have recently suffered wall removal, and their unusual and artificially maintained osmotic relations with the culture environment, the failure of dyes such as Hoechst 33258 to stain nuclei of live protoplasts is not so surprising. Prior staining of live cells with "33258" (which gives good nuclear fluorescence) and subsequent isolation of protoplasts is ineffective: the dye seems to be lost from the protoplasts, because occasional lysis of individual protoplasts in the presence of the dye immediately causes bright nuclear fluorescence.[33]

This staining problem with live protoplasts, and incompatabilities which arise because commercial versions of the flow cytometer apparatus have been designed with animal cells in mind, should not daunt intending "botanical" users of this sophisticated machinery. The potential of the flow approach has been long felt in the animal sciences, and the applications possible with totipotent plant cells and protoplasts are limited only by imagination, and what will probably be a temporary dearth of plant-oriented users.

III. CYCLE PHASE MANIPULATION OF PROTOPLAST POPULATIONS

Plant cell and protoplast culture techniques are now sufficiently well advanced to allow sensible alternatives to the initial, crude attempts to genetically transform plant cells. With the benefit of hindsight, these initial attempts could be characterized as the application of DNA of dubious biochemical and genetic integrity, to plant cells of dubious genetic stability and unknown competence. Cell cycle considerations could and should play a major role in at least two aspects of research on genetic modification of plant cells.

A. Competence for Genetic Transformation

Microbial geneticists were the first to realize that the ability to take up and integrate foreign DNA into the host genome is not inherited, but is rather an expression of the metabolic state of the bacterial cell, a state which can be induced by very specific nutritional constraints imposed by the conditions of culture. Although similar conclusions can be drawn from the successful genetic transformation of mammalian cells in culture,[17] the concept of competence is rarely considered by plant cell biologists.

The phenomenon of competence must arise as the result of efficient completion of a multi-step process, which begins with binding of exogenous DNA to the host cell and is completed when some new function (encoded in the foreign DNA) is expressed by the transformed cell. Internalization of foreign DNA, its transport through the cytoplasm, its entry into the nucleus, and structural integration with host DNA are only a few of the requirements for successful transformation (see Chapter 13). Partial or total failure to complete any one of the intermediate steps in the transformation program will result in reduction or abolition of competence. Dynamic cyclic alterations in the structure and physiology of cells occur as they progress from one division to the next, and these alterations are very likely to be factors linked to the expression of competence, via effects on the efficiency of specific intermediate steps in the transformation process. Thus from both theoretical considerations and from results obtained with animal systems,[18] the involvement of specific cell cycle phases in the expression of competence for genetic transformation is strongly indicated.

Plant protoplasts are often the material of choice in genetic transformation studies, because the absence of the cell wall should presumably remove one barrier to DNA entry. Thus the first obligatory step towards successful transformation of protoplasts is the binding of the transforming vector to the protoplast membrane. Studies with tobacco mosaic virus (TMV)[19] and with isolated DNA[20] suggest that even in this very first interaction between host and vector, the cell cycle phase of the host can exert quantitative effects. It has been postulated that large differences in the amount of TMV or purified DNA bound to protoplasts of different cycle phase are due to a cyclical fluctuation of the electrostatic charge on the protoplast membrane.[10,19,20] Protoplasts in S-phase appear to be relatively inefficient binders of DNA and TMV when binding is mediated by polycations such as poly-L-ornithine, and measurements of the charge on the protoplast membrane suggest that S-phase protoplasts have less negative charge than protoplasts of other cell cycle stage.[6,19] In terms of competence this could mean that S-phase protoplasts are less likely to be transformed than protoplasts elsewhere in the cycle, but it should be remembered that the efficiency of transformation will be the result of many factors, of which amount of vector initially bound is but one.

The nuclear chromatin undergoes elaborate reorganization during interphase and a strong case can be made for a relationship between transformability and some transient activity or conformation of the euchromatin. Constitutive heterochromatin is very highly condensed for nearly all of the cell cycle and is usually considered to be genetically inert; for this reason, heterochromatin would be an unattractive candidate as a site for insertion of foreign genetic material. However, if gene insertion into heterochromatin was an objective, logic suggests that the time to make a transforming vector available to the host genome would be late in S-phase, this often being the time when heterochromatin is temporarily decondensed as heterochromatic sequences are replicated. This argument could be applied to the euchromatic fraction of the genome, and then the whole of S-phase would be characterized as the "competent phase of the cell cycle". Additionally, the fact that particular fractions of a eucaryotic genome are replicated at particular times during S-phase might allow some control over the location in which a transforming vector inserts a foreign gene into the host genome (i.e., by using protoplasts at some specific stage within S-phase).

All of this speculation assumes that replication forks in the DNA of S-phase nuclei are somehow susceptible to an illegitimate recombination event which leads to stable integration of a new gene and transformation of the host genome. Certainly, the involvement of S-phase in transformation competence of both animal cells and plant protoplasts in culture have been considered by several authors.[18,20] However, Nagata et al.[21] consider that mitosis is more likely to be the phase most competent for genetic transformation. These authors base their conclusions on results of studies on the infection of tobacco protoplasts with TMV RNA encapsulated in liposomes (see Chapter 9). The absence of a nuclear membrane during the mitotic phases of the cell cycle prompts these workers to suggest that vector-access to the host genome will be facilitated if transformation is attempted with protoplast populations that have been accumulated in mitosis. Clearly, it is possible that the nature of the transforming vector will dictate which phase of the cell cycle will be the most susceptible to transformation. If transformation is defined as the stable integration of foreign DNA into the nuclear genome of the host, then the highly condensed chromatin of mitotic cells would seem an unlikely target for a transforming vector. However, if transformation is defined merely as infection of plant cells or protoplasts with a cytoplasmically replicating RNA virus, then the state of the nucleus in terms of cell cycle stage may be irrelevant as a factor in competence.

If a transforming vector remains undegraded and available to the host cell for a period of time which is longer than the duration of the cell cycle, then the preceding discussion is superfluous: as each protoplast passes through the hypothetical competent phase of the cycle, the transforming vector will be present and integration could occur. However the powerful degradative effects of plant nucleases are well known, and it has been shown that a high level of nuclease activity is present throughout the division cycle of some cultured cells.[22] Also if exponentially dividing asynchronous suspension cultures are used for the isolation of protoplasts, all stages of the cycle should be represented, and thus some fraction of the population should be in the "competent" phase. However, cycle phase manipulation of protoplast populations still offers an opportunity to increase the efficiency with which transformation of protoplasts can be achieved.

B. Isolation of Chromosomes

In the previous section, protoplasts were considered as hosts for insertion of vectors carrying foreign genes, and cycle phase manipulation was suggested as a means for increasing "transformability" or competence of protoplasts. Conversely, protoplasts may also represent a valuable source of transforming genes, and cycle phase manipulation plays a key role here, too, because the approach relies on whole chromosomes or fragments of chromosomes as the transforming vectors (see Chapter 12).

The accumulation of cells in mitosis by the use of various inhibitors is sometimes erroneously referred to as cell cycle synchronization. Mitotic accumulation is only a subset of cell cycle synchrony and does not guarantee that release from inhibition will result in synchronous cell cycle traverse. This semantic quibble aside, very useful technical developments in mitotic accumulation of plant cells in culture are now available as prerequisites for the bulk isolation of plant chromosomes. The three most productive approaches are application of a single inhibition which collects cells in or just before mitosis, cycle arrest of cells in S-phase and then release of the accumulated cohort to progress towards mitosis in a synchronous wave, or a combination of both of these strategies. When a suitable fraction of the cell population has been arrested in mitosis, protoplasts are made and chromosomes isolated by lysis. The early work on plant chromosome isolation by Malmberg and Griesbach[23] relied on fluorodeoxyuridine and colchicine to produce cell populations with relatively high mitotic indices. Meiotic chromosomes were also isolated from naturally synchronous meiocytes; the efficiencies of chromosome isolation in all cases seemed to be directly related to the degree of synchrony or mitotic accumulation achieved. Dudits' group has also used both single and double block approaches depending on species;[24,25] with suspension cultures of poppy *Papaver somniferum* adequate mitotic accumulation (30 to 40%) could be obtained with treatments of colchicine alone. In contrast, wheat cell suspensions required a pretreatment with hydroxyurea before colchicine could induce similarly high mitotic indices. The paper by Hadlaczky et al.[25] contains particularly convincing photomicrographs of wheat and poppy cell populations with high mitotic indices, and thousands of isolated wheat chromosomes in a single field.

The great advantage of using isolated chromosomes as transforming vectors would be the ability to transfer numbers of linked genes at the same time. This advantage does to some extent presume that specific chromosome types can be separated from the rest of the karotype. In an exciting technical advance, de Laat and Blaas[26] have recently reported on the flow sorting of individual chromosomes of the species *Haplopappus gracilis*, a member of the Compositae which has only four chromosomes in the diploid set. The karyotype of *H. gracilis* is particularly suitable for discrimination by a flow cytometer/sorter, because the acrocentric and the metacentric pair differ significantly in length and DNA content. Using a commercially available flow sorter and isolated chromosomes stained with ethidium bromide, de Laat and Blaas[26] claim resolution that would enable chromosomes differing in DNA content by only 10% to be separated. As with the paper mentioned above from the Dudits group, the inclusion of micrographs of whole fields of chromosomes of one kind (i.e., either acrocentric or metacentric) in the de Laat and Blaas publication leaves no doubt of the power and precision of the flow sorting technique. Bulk isolations of human X chromosomes have already been made by flow sorting,[27] and further developments along these lines with plant material could lead to such sophisticated possibilities as bulk isolation of individual chromosomes of crop species and subsequent construction of chromosome-specific genomic libraries.

C. Cell Cycle Effects During and After Protoplast Fusion

As mentioned above in the section on competence it has been suggested that cell cycle related variations exist in the charge on the membrane of plant protoplasts. These charge fluctuations were deduced from electrophoretic studies on single protoplasts and from the separation of S-phase protoplasts from other cycle stages by a method which depends on electrostatic differences.[19] In related experiments[28] results suggest that fusion of protoplasts is not completely random; deviations from randomness of fusion were calculated from counts of binucleates of various nuclear constitution (i.e., G1/G1, G1/G2, G1/S, S/G2, S/S, G2/G2), compared to expected frequencies of each class depending on the abundance of each phase before fusion. As with the work on competence for transformation,[20] it was concluded that charge on the protoplast membrane was a factor in the observed deviations from absolute randomness of fusion. Thus cell cycle status of the plant cell population used to make protoplasts could have significant effects on the frequencies of specific types of binucleates obtained; this will not be a function only of abundance (of each cycle stage) if membrane charge plays a significant role in fusion efficiency and also varies with position in the cell cycle.

From the foregoing discussion it seems that cycle phase can have effects on protoplast interaction during fusion, but the interaction of plant nuclei in a common cytoplasm after fusion provides even stronger evidence that cycle phase must be taken into consideration in genetic manipulation studies. In 1970 Johnson and Rao[29] reported on the phenomenon of Premature Chromosome Condensation (PCC) following fusion of interphase with mitotic mammalian cells in culture. It took 10 years before similar work with plant protoplasts came to light.[30] This lag was due mainly to deficiencies in synchronization and mitotic accumulation methods available for plant tissue cultures and to the need for very rapid isolation techniques which would allow mitotic cells to be made into protoplasts before dividing cells entered the G1 phase. PCC results from the formation of binucleate cells (or plant protoplasts) which are heterophasic, that is the nuclei in the binucleate are in different cycle stages. In general, a mitotic nucleus in a heterophasic combination exerts a dominant effect on its partner nucleus. If the partner is in the G1 phase, its chromatin undergoes condensation as though for mitosis, but as no intervening S-phase has occurred the chromosomal threads are single stranded (i.e., there is only one chromatid). If the partner nucleus is in S-phase the induction of premature condensation by the mitotic nucleus has much more destructive effects on the integrity of the interphase chromosomes: the S-phase chromatin suffers "pulverization" to small fragments. Other cell cycle effects in heterophasic fusions have been observed; for example in mitotic/G2 combinations, the G2 nucleus enters mitosis much more rapidly than it would normally, and the timing of the onset and cessation of DNA synthesis can be altered. Clearly, however, the striking consensus from research on PCC is that there is a strong signal from mitotic nuclei, which initiates premature condensation and is compatible with nuclei from intergeneric and even interkingdom cell and protoplast fusions. Dudits et al.[31] propose that the fragmentation of S-phase nuclei by PCC is a serendipitous event that could lead to integration of small pieces of one genome into another. Certainly work on animal cell fusions would support this contention,[32] and once again the major impact of a cell cycle approach seems to be in the area of genetic manipulation.

In summary, cell cycle studies with plant protoplasts, although not presently in vogue among plant scientists, obviously have much to offer. In the field of genetic manipulation of plant cells, cell cycle considerations have a major role to play in terms of host cell competence and the possible use of isolated chromosomes as genetic vectors. In addition, the PCC approach to an understanding of cell cycle control in plants is now a possibility due to recent developments in protoplast fusion.

REFERENCES

1. Amino, S., Takeuchi, Y., and Komamine, A., Changes in cell wall constituents during the cell cycle in a synchronous culture of *Catharanthus roseus, Physiol. Plant.*, 60, 326, 1984.
2. Dyer, A. F., The visible events of mitotic cell division, in *Cell Division in Higher Plants,* Yeoman, M. M., Ed., Academic Press, London, 1976, chap. 2.
3. Gould, A. R., Control of the cell cycle in cultured plant cells, *CRC Crit. Rev. Plant Sci.*, 1, 315, 1984.
4. King, P. J., Cell proliferation and growth in suspension cultures, *Int. Rev. Cytol,* 11A, (Suppl.) 25, 1980.
5. Gould, A. R., Cell cycle analysis by conventional methods, in *Cell Culture and Somatic Cell Genetics of Plants,* Vol. 1, Vasil, I. K., Ed., Academic Press, New York, 1984, 753.

6. Ashmore, S. E., Kinetic and Cytological Analyses of Plant Tissue Cultures, Ph.D. thesis, Australian National University, Canberra, 1981.
7. Zelcer, A. and Galun, E., Culture of newly isolated tobacco protoplasts: cell division and precursor incorporation following a transient exposure to coumarin, *Plant Sci. Lett.*, 18, 185, 1980.
8. Kaur-Sawhney, R., Flores, H. E., and Galston, A. W., Polyamine induced DNA synthesis and mitosis in oat leaf protoplasts, *Plant Physiol.*, 65, 368, 1980.
9. Gould, A. R., Staining and nuclear cytology of cultured cells, in *Cell Culture and Somatic Cell Genetics of Plants*, Vol. 1, Vasil, I. K., Ed., Academic Press, New York, 1984, 698.
10. Gould, A. R., Combined microspectrophotometry and automated quantitative autoradiography applied to the analysis of the plant cell cycle, *J. Cell Sci.*, 39, 235, 1979.
11. Miksche, J. P., Dhillon, S. S., Berlyn, G. P., and Landauer, K. J., Nonspecific light loss and intrinsic DNA variation problems associated with Feulgen DNA cytophotometry, *J. Histochem. Cytochem.*, 27, 1377, 1979.
12. Laloue, M., Courtois, D., and Manigualt, P., Convenient and rapid fluorescent staining of plant cell nuclei with "33258" Hoechst, *Plant Sci. Lett.*, 17, 175, 1980.
13. Galbraith, D. W., Mauch, T. J., and Shields, B. A., Analysis of the initial stages of plant protoplast development using 33258 Hoechst: reactivation of the cell cycle, *Physiol. Plant.*, 51, 380, 1981.
14. Redenbaugh, K., Ruzin, S., Bartholomew, J., and Bassham, J. A., Characterization and separation of plant protoplasts via flow cytometry and cell sorting, *Z. Pflanzenphysiol.*, 107, 65, 1982.
15. Galbraith, D. W., Harkins, K. R., Maddox, J. M., Ayres, N. M., Sharma, D. P., and Firoozabady, D. P., Rapid flow cytometric analysis of the cell cycle in intact plant tissues, *Science*, 220, 1049, 1983.
16. Puite, K. J. and Ten Broeke, W. R. R., DNA staining of fixed and non-fixed plant protoplasts for flow cytometry with Hoechst 33342, *Plant Sci. Lett.*, 32, 79, 1983.
17. Wigler, M., Sweet, R., Sim, G. K., Wold, B., Pellicer, A., Lacy, E., Maniatis, T., Silverstein, S., and Axel, R., Transformation of mammalian cells with genes from procaryotes and eucaryotes, *Cell*, 16, 777, 1979.
18. Sompayrac, L. M. and Danna, K. J., Efficient infection of monkey cells with DNA of simian virus 40, *Proc. Natl. Acad. Sci. U.S.A.*, 78, 7575, 1981.
19. Gould, A. R., Ashmore, S. E., and Gibbs, A. J., Cell cycle related changes in the quantity of TMV virions bound to protoplasts of *Nicotiana sylvestris*, *Protoplasma*, 108, 211, 1981.
20. Gould, A. R., Ashmore, S. E., Interaction of purified DNA with protoplasts of different cell cycle stage: the concept of a competent phase for plant cell transformation, *Theor. Appl. Genet.*, 64, 7, 1982.
21. Nagata, T., Okada, K., and Takebe, I., Mitotic protoplasts and their infection with Tobacco Mosaic Virus RNA encapsulated in liposomes, *Plant Cell Rep.*, 1, 250, 1982.
22. Cress, D. E., Uptake of plasmid DNA by protoplasts from synchronized soybean cell suspension cultures, *Z. Pflanzenphysiol.*, 105, 467, 1981.
23. Malmberg, R. L. and Griesbach, R. J., The isolation of mitotic and meiotic chromosomes from plant protoplasts, *Plant Sci. Lett.*, 17, 141, 1980.
24. Szabados, L., Hadlaczky, Gy., and Dudits, D., Uptake of isolated plant chromosomes by plant protoplasts, *Plants*, 151, 141, 1981.
25. Hadlaczky, Gy., Bisztray, Gy., Praznovsky, T., and Dudits, D., Mass isolation of plant chromosomes and nuclei, *Planta*, 157, 278, 1983.
26. deLaat, A. M. M. and Blaas, J., Flow cytometric characterization and sorting of plant chromosomes, *Theor. Appl. Genet.*, 67, 463, 1984.
27. Davis, K. E., Young, D. B., Elles, R. G., Hill, M. E., and Williamson, R., Cloning of a representative genomic library of the human X chromosome after sorting by flow cytometry, *Nature (London)*, 293, 374, 1981.
28. Ashmore, S. E. and Gould, A. R., Protoplasts fusion and the cell cycle, *Plant Cell Rep.*, 1, 225, 1982.
29. Johnson, R. T. and Rao, P. N., Mammalian cell fusion: induction of premature chromosome condensation in interphase nuclei, *Nature (London)*, 226, 717, 1970.
30. Szabados, L. and Dudits, D., Fusion between interphase and mitotic plant protoplasts. Induction of premature chromosome condensation, *Exp. Cell Res.*, 127, 442, 1980.
31. Dudits, D., Szabados, L., and Hadlaczky, Gy., Premature chromosome condensation in plant cells and its potential use in genetic manipulation, in *Premature Chromosome Condensation*, Academic Press, New York, 1982, 359.
32. Tsukamoto, K., Klein, R., and Hatanaka, M., Insertion of muntjac gene segment into hamster cells by cell fusion, *J. Cell. Physiol.*, 104, 225, 1980.
33. Daines, R. J., unpublished results.

Chapter 6

ISOLATED PROTOPLASTS AS LABORATORY TOOLS IN THE STUDY OF CELL WALL DEPOSITION

J. H. M. Willison

TABLE OF CONTENTS

I. INTRODUCTION

It is a characteristic of plant cells that they are bounded by a cell wall which they have synthesized and secreted by their individual activity.[1,2] When this wall is stripped away, the protoplast is left in an isolated and unfamiliar situation. In culture, viable protoplasts usually progress toward reestablishing their familiar environment by re-creating cell walls.[3-6] This process of forming a protected environment for the cell is important for the subsequent development of the cell, as made evident by the fact that cytokinesis either does not occur or is abnormal in the absence of a sufficiently substantial cell wall,[7-10] and that protoplasts which are unable to regenerate substantial walls do not form callus.[8] In order to recreate their protected environments, cells must synthesize and secrete suitable materials, which must assemble on the cell surface. Yet, before their isolation as protoplasts, many cells from which protoplasts are prepared may not have been actively secreting cell walls. Even if they were cell-wall secretors, they were doing so in a walled environment very different from that of the isolated protoplast. That certain isolated protoplasts have been shown to generate novel intracellular "stress" proteins[11-13] and to produce extracellular materials characteristic of wounding[14,15] indicates that the process of cell wall regeneration should not be viewed as a continuance of an existing process of wall secretion, nor as a reestablishment of a previously completed developmental program, but as a response to protoplast isolation. The precise response can be expected to be influenced by a complex of genetic, developmental, and environmental factors.

It is clear from the foregoing comments that isolated protoplasts have certain limitations as experimental systems which must be recognized, yet they also appear to offer unique advantages in the investigation of cell wall synthesis and secretion and in the investigation of the developmental roles of the cell wall, advantages which for the most part have not been fully exploited. The surface of the protoplast is naked; therefore processes such as cellulose synthesis which occur in intimate association with the plasma membrane can be observed more readily in the electron microscope.[16-18] This same nakedness makes it possible to apply chemical probes, such as lectins,[5] or immunoglobulin-marker complexes, which would not penetrate cell walls. Recent advances in methods of producing protoplast ghosts (see Chapter 3) open the possibility of examining, in novel ways, interrelationships between elements of the cytoskeleton and the developing wall, particularly if, as seems possible,[102] the ghost retains some of its metabolic capacity. Prat[19] noted that isolated protoplasts must sequentially reestablish, with some synchrony, the processes of biosynthesis, secretion, and assembly of the cell wall, which might permit these processes to be separately examined to some extent. That a wound response appears to have been elicited in certain isolated protoplasts suggests that protoplasts might become useful single-cell experimental systems for the study of wounding. From a developmental perspective, the removal of the cell wall is a challenge to the cell wall synthetic capacity of the cell, and it may be possible (once the "protoplast response" per se is better understood) to use protoplasts as tools in the study of the restriction of developmental potential in particular cell lineages. There can be little doubt that the most powerful experimental capacity of isolated protoplasts is the extent to which they can be manipulated genetically (through fusion, transformation, cytoplast formation, and so on, see Chapters 10 to 13), but I find it difficult to foresee how this capacity can be used to help in understanding cell wall biogenesis, at least in the immediate future. In the shorter term, I perceive that the greatest value will come from attempts to understand initially inexplicable responses of isolated protoplasts, leading the thinking observer to ask questions which could not previously have been formulated. Why, for example, should leaf epidermal protoplasts, unlike their mesophyll counterparts, produce extracellular fibers in the presence of calcium?[20] Is there an experimental value to the observations that isolated protoplasts, unlike the cells from which they originate, are internally electropositive?[8,21]

II. MICROFIBRILS AND CELLULOSE

A. Cellulose Content

Deposition of microfibrillar material on the surfaces of isolated protoplasts has been examined and reviewed extensively at the cytological level[4-6] and will be treated only briefly here. In initial studies, this microfibrillar material was assumed to be cellulosic and the findings were assumed to have relevance to understanding cellulose biogenesis. It is only relatively recently that adequate confirmation of the assumption that cellulose is synthesized by isolated protoplasts has been obtained.[14,22-28] This confirmation, however, has contained a strong cautionary note. In tobacco protoplasts, the protoplast wall contains only 4 to 7% cellulose,[24-27] which differs from that of the mother tissue with respect to molecular weight distribution.[25] Leaf mosophyll cell walls from which protoplasts were isolated in these studies contained 50 to 60% cellulose;[24,26,27] tobacco tissue cultures, also used as protoplast sources, had walls containing 35 to 50% cellulose, varying according to source and culture conditions.[24-27] Others, working with protoplasts isolated from various sources, have similarly produced results which suggest that cellulose is present in relatively small amounts.[14,23,29] This cellulose does not demonstrate normal crystallinity,[22,28] and it seems probable that these microfibrils contain an unusually thin core of crystalline cellulose, having dimensions in the "subelementary fibril" range, which is coated with noncellulosic glycans.[24] It is not until the protoplasts have regenerated sufficient wall to permit normal cell division (i.e., with a cell plate) that cellulose synthesis begins to return to normal levels.[23,26]

B. Cytology

Many electron microscopic methods have been used in demonstrating that most protoplasts isolated from higher plants deposit microfibrillar material at their surfaces.[16-19,22,30,31] Usually, all (or most) parts of protoplast surfaces appear to be equally capable of participating in microfibril synthesis once the process has been initiated, and microfibril deposition is initiated randomly and spreads randomly over the surface.[16,18,30-33] It seems probable that tobacco epidermal protoplasts are exceptions to this generalization,[20] producing fibers from a localized zone of the cell surface (Figure 1), and it is possible that some protoplasts which generate walls slowly (e.g., mature leaves of onion and leek)[19,30] initiate envelope synthesis in localized zones from which the envelope spreads to cover the surface. It is difficult, however, to make observations which allow us to draw unequivocal conclusions since one either sees a large area of protoplast surface at insufficient resolution, or small parts of the surface at sufficient resolution. Even scanning electron microscopy may render patches of microfibrils invisible.[33,34]

Most contemporary hypotheses of microfibril biogenesis suggest that complexes of proteins, associated with microfibril tips and bound to the plasma membrane, assemble microfibrils at their termini.[1,35-38] There have been many attempts to find such structures in higher plant protoplasts, but with remarkably little success. The freeze-fracture particle arrays which Robenek and Peveling[39,40] suggested were involved in microfibril biogenesis are probably induced by plasmolysis and have no functional role.[41] The surface projections more cautiously put forward as microfibril-synthesizing candidates by Burgess and Linstead[17] are now seen to be unlikely to be involved since their apparent association with microfibril tips is artifactual.[33,34] No evidence for globular "terminal complexes"[36] has been found at the visible ends of microfibrils on deep-etched outer surfaces of the plasma membrane.[6,16,32] Impressions of terminal complexes in

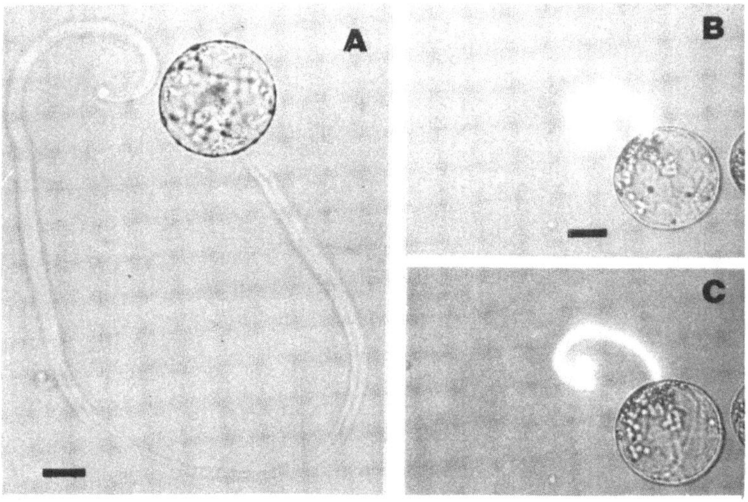

FIGURE 1. Extracellular fibers emanating from tobacco epidermal protoplasts. (A) After 2 days in the incubation medium; bright-field; bar 25 μm. (B) and (C) After 15 hr in the incubation medium, then 30 min (B) and 65 min (C) after addition of Calcofluor to the incubation medium; fluorescence and bright-field combined; bar 25 μm. (From Fannin, F. F. and Shaw, J. G., *Planta*, 159, 282, 1983. With permission.)

freeze-fractured plasma membrane E fracture faces were not found in a study of mesophyll protoplasts,[34] but have been reported in regenerating protoplasts derived from a carrot suspension culture.[42] "Rosettes" of intramembrane particles, found on P fracture faces in many tissues and proposed as being widely associated with cellulose synthesis,[35,42,43] appear to be absent both from tobacco mesophyll protoplasts[36] and from carrot suspension culture protoplasts[42] during active wall regeneration. These largely negative findings are particularly striking in view of the fact that similar techniques appear, relatively readily, to have produced promising results for the isolated protoplasts (wound-induced aplanospores) of *Boergesenia* and *Valonia*,[42,44,45] which suggest that cellulose microfibrils are deposited around these algal protoplasts by means comparable with *Oocystis* autospores, in which "terminal complexes" are clear.[46] The reason for this contrast in results probably lies with the different sizes of cellulose crystallites produced. By contrast with higher plant protoplasts, *Boergesenia* and *Valonia* produce relatively large crystallites of highly crystalline cellulose.

It is probably too early to reject the "terminal-complex mediated cellulose-synthesis" hypothesis for higher plant protoplasts, as Grout[32] did in favor of a hypothesis involving precipitation of cellulose within the plasma membrane from preformed polymers. It does seem, however, that higher plant protoplasts provide an argument for proposing that different types of cellulose might be assembled by different means. In this respect it would be interesting to see details of membrane architecture in the highly differentiated membrane which must be presumed to exist around tobacco epidermal protoplasts (Figure 1).

Kinetics of deposition of wall material vary according to protoplast source and culture conditions.[4,6] The following are examples of genetically and developmentally determined effects: actively growing cells, such as from exponential phase tissue cultures[14,18,47] or from meristem tissue,[48] begin to regenerate wall immediately; protoplasts from certain differentiated cells, such as leaf mesophyll of *Nicotiana*,[16,32,49] *An-*

tirrhinum,[49] and *Allium*,[30] have a lag phase of 8 to 45 hr before wall formation begins, older mesophyll having a longer lag phase;[30] protoplasts from stationary phase cultures[47] and differentiated tissues of certain species (e.g., mesophyll of grasses[8]) appear to be intrinsically deficient in their capacity to regenerate walls. The following are examples of the influence of culture media: higher concentrations of osmotica tend to inhibit wall formation;[50,51] in the absence of added sucrose, the inorganic salts medium (RO.6) supports little wall regeneration,[52] but when ammonium is included (medium WO.6), even in the absence of sucrose, wall regeneration proceeds normally;[7] hormones are usually components of culture media and in one careful study, at suitable concentrations, have been demonstrated to increase the efficiency of wall formation (i.e., percent protoplasts regenerating walls),[53] but well-controlled studies of wall regeneration kinetics (using for example, Galbraith's simple quantitative wall-assay method[54] and the multiple-drop-array testing technique[55]) have not been attempted. With respect to this last point, it must be realized that changes in wall composition, as well as regeneration rate, might be expected in relation to changes in culture media (see later).

C. Experimental Inhibition

Direct inhibition of cell wall regeneration has been attempted with some success. Dichlorobenzonitrile (DCB), a specific inhibitor of cellulose synthesis,[56] clearly inhibits the deposition of cellulose and of microfibrillar material[6,57,58] for an indefinite period. This inhibition is rapidly reversed upon removal of DCB; eliminating the lag phase in protoplasts which normally reveal one[57,58] and supporting the concept that the lag is a period of developmental redetermination. Because DCB had no effect on DNA nor protein systhesis, Galbraith and Shields[58] suggest that this inhibition might be "directly associated with the cellulose synthesizing complex at the plasma membrane." DCB does not inhibit ethylene production in protoplasts,[59] even though ethylene, too, might be a plasma membrane product.[60] Other inhibitors of wall regeneration by isolated protoplasts, such as coumarin, are both less effective and less specific than DCB.[56,61,62]

Indirect inhibition of cellulose synthesis and microfibril formation have been obtained. Inhibition of glucan synthetase I (presumptive cellulose synthetase) by actinomycin-D and cycloheximide appears to have been demonstrated in protoplasts from carrot suspension cultures,[63] in a system which, unexpectedly, has a lag phase for cellulose synthesis. The result was interpreted[63] to indicate that transcription and subsequent translation are required for cellulose synthesis in this system. The description of the regeneration of a thin wall around anucleate microplasts in one case[64] but not in another[42] must indicate, however, that postisolation transcription is not an absolute necessity in all systems. Cell wall regeneration by *Polyphysa* protoplasts is inhibited when the protoplasts are cultured in the presence of sufficient Concanavalin A (Con-A),[65] presumably as a result of Con-A binding to individual glycosylated membrane-surface sites (or cross-linking of such sites) which must be available for the synthesis of Calcofluor-positive material. α-Glucosidase was also an effective inhibitor.[65] However, some contrary results have been obtained by Williamson and colleagues[66,67] working with soybean and leek protoplasts. This difference may be attributable to the wide difference in taxa, if not, then to experimental protocol, since long-term exposure to Con-A was not reported for the higher plant protoplasts. Soybean protoplasts were briefly exposed to Con-A and microfibrils were generated during subsequent culturing in Con-A free medium;[66] leek protoplasts[67] sloughed off Con-A after it had been supplied to fresh isolates, and would not bind Con-A after a relatively brief incubation (suggesting that the binding sites were protected by extracellular material). In view of the evidence that, in the alga *Oocystis*, fairly rapid turnover of cellulose-synthesizing "terminal complexes" appears to occur,[68] more investigation of possible Con-A inhi-

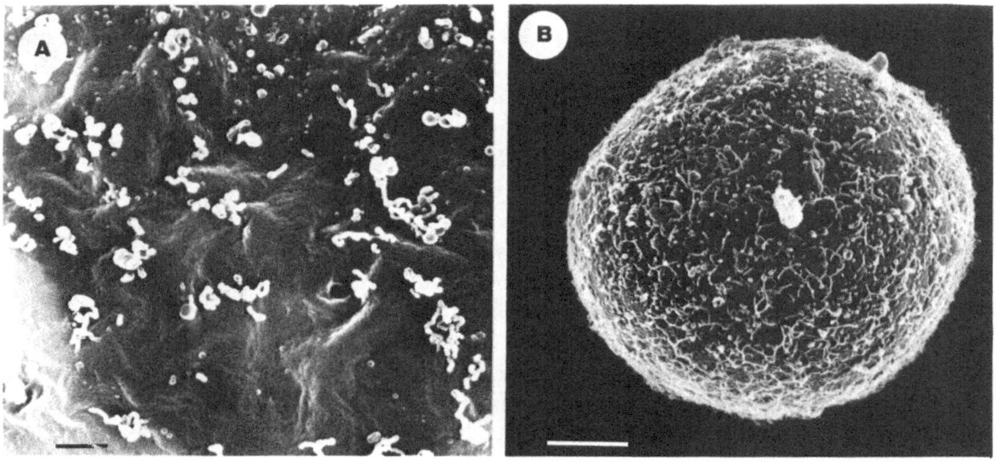

FIGURE 2. (A) Surface of a *Physcomitrella* protoplast incubated 4 hr in a medium containing congo red. Short twisted fibers have arisen at various places on the surface. Bar 1 μm. (B) *Physcomitrella* protoplast incubated 2 hr in congo red medium, followed by 2 hr in normal medium. The relatively long and straight fibrils and their frequency are typical of a 2-hr incubation in normal medium alone. Bar 5 μm. (From Burgess, J., *Micron*, 13, 185, 1982. With permission.)

bition of cellulose deposition seems deserved. A number of agents have been shown to disrupt microfibril deposition at isolated protoplast surfaces by interfering with the process at a postsynthetic stage, particularly the dyes Calcofluor and congo red[62,69] (Figure 2). These dyes appear to prevent crystallization by binding to polymer molecules before crystallization can occur,[70,71] resulting in the production of excess amorphous glucan by *Physcomitrella* protoplasts[62] and of loss of crystallinity in the extracellular material surrounding tobacco mesophyll protoplasts.[69] These results demonstrate that polymerization and crystallization are separable processes in regenerating protoplasts, which is particularly interesting because the low cellulose content of regenerated walls raises the possibility that a noncellulosic crystalline glucan is affected. These vital dyes should continue to be valuable experimental tools in cell wall regeneration studies. Do they, for example, have any effect on the binding of microtubules to protoplast ghosts, or on the morphology of the cytoplasmic surface of the plasma membrane (see Chapter 3 for typical micrographs)?

D. Membrane Potential

When tobacco mesophyll protoplasts are cultured in the presence of cellulase, no wall is regenerated and huge uninucleate protoplasts sometimes arise after several weeks in culture.[10] Similar enlargement has been reported for corn and oat isolates which were incompetent at wall regeneration.[8] When protoplasts are initially isolated there is a marked change in the plasma membrane potential, from the normal internal −75 mV to about +10 mV.[8,72] The enlarged wall-less corn and oat protoplasts retain this electropositivity,[8] but protoplasts which regenerate walls slowly regain typical electronegativity, particularly when turgor is reestablished.[8,72] Plasma membrane in tobacco leaf tissue has been reported to depolarize in response to exposure to cell wall degrading enzymes, even when the enzymes were denatured.[72] In view of the reported relationship between membrane potential and β-glucan synthetase (both 1,3 and 1,4) activity,[73] the findings are interesting, but more work is required before the precise significance becomes clear. During isolation, protoplasts probably suffer both from osmotic shock[13,72,74] and from enzymatic damage to the plasma membrane;[72,74] and the reinsertion of glucan synthetases into the plasma membrane[63] may be only one of many membrane features requiring repair. The complex pattern of changes in electrophoretic mobility (reflecting net surface charge) which has been reported to occur during the period of recovery from the process of isolation[75] is understandable in this light, and further study using this technique in combination with the wider range of cell wall synthesis inhibitors now available seems deserved.

E. Degree of Polymerization

Not only do protoplasts produce little cellulose during the early stages of wall regeneration,[23,77] but this cellulose displays an unusually low degree of polymerization (DP).[23,25] The DP ranges of celluloses in tissues from which protoplasts have been isolated typically fall in the range <200 to 4000, with 30 to 50% >1000;[23,25] 65 to 85% of the cellulose in the walls initially regenerated by isolated protoplasts has DP <500,[23,25] most of the remainder being of relatively high DP. In pulse-chase experiments,[25] it was found that only some of the radioisotopic label could be chased into the higher DP cellulose, indicating that, at most, only a proportion of the low DP material consists of growing chains and that separate pools of celluloses having distinctive molecular weights exist. It is interesting to note in this context, that Burgess and Linstead[76] have described short fibrils, around 60 nm in length (DP ~120, if cellulosic), in wall material extracted from briefly sonicated protoplasts cultured for relatively short periods. While it is possible that these short fibrils arose as a result of sonication (see Kolpak and Blackwell,[77] for example), microfibrils produced later by these protoplasts were considerably longer even after sonication.[76]

III. NONCELLULOSIC MATERIALS

A. Polysaccharides

Just as in electron micrographs the microfibrils of pectocellulosic walls stand out as distinct and relatively uniform components against an amorphous background, so in relative terms, the chemically and physically uniform cellulose contrasts with the complex heterogeneous mixture of covalently linked polysaccharides and proteins which make up the remainder of the wall. Primary cell walls (those degraded in preparing isolated protoplasts) vary in composition and linkage both according to source and state of development.[78-80] As walls develop, not only are materials added, but routine saccharide transfer (turnover) occurs between pools of extracellular polysaccharide, at least in the case of 1,3-glucan and cellulose in cotton fibers.[78,81]

Although the walls regenerated around isolated protoplasts have been less studied than those in tissue, it is becoming clear that the situation is at least as complicated in isolated protoplasts as in tissues. Not only might developmental regulation and response to isolation-shock operate during regeneration, but the cell wall also contributes anomalously to the regulation of its own development. In tissues, and to a lesser extent in tissue cultures, newly secreted extracellular material is confined by the surrounding walls and cells, but around isolated protoplasts the confinement is greatly reduced, resulting in loss to the culture medium of some of the materials which would otherwise appear in the wall around freshly isolated protoplasts,[6,82-84] while less is lost around those which have regenerated substantial quantities of wall.[82-84] We must presume that the capacity of the regenerating wall to capture secreted products is progressive. Relatively little attention has been paid to walls regenerated when protoplasts are cultured on solid media, but it is clear that the loss of secreted material is much reduced over that which occurs to liquid media: the wall is more compact and there is less budding,[50,85,86] secreted β-glycerophosphatases are retained rather than lost,[86] and the polysaccharides of the regenerated wall have neutral sugar and uronic acid compositions similar to those of agar-cultured cells, unlike the relatively deficient wall of protoplasts cultured in liquid media.[83]

Partial analyses of the chemical composition of polysaccharidic components of walls regenerated by higher plant protoplasts isolated from the following sources have been analyzed: soybean cell cultures,[14,82] *Vinca* cell culture,[28,29,83] carrot cell culture,[84] and tobacco mesophyll.[24,26,27] Furthermore, cell wall hydroxyproline-containing glycoprotein has been examined in the cell wall of *Vinca* protoplasts.[87] Only cautious generalizations are possible. Glucans commonly predominate in the regenerated walls,[24,27,29,83,84] and the levels are higher than in the walls of the parent tissue;[27,29,84] in some cases, a large proportion of this glucan consists of β1,3-linked products (callose and pachyman) unique to the protoplast wall,[14,28] and some of this material may also appear in the culture medium.[14] Polysaccharides lost to the culture medium are variable; pectic fractions rich in arabinose and galactose commonly predominate,[14,24,27,28,82,84] but acidic polyuronides may also be found.[24,28,82,84] It appears that the mixture of molecules appearing both in the regenerated envelope and in the culture medium varies from source to source. It is less clear that the culture medium affects the composition of the wall; Takeuchi and Komamine[83] found virtually no effect of varying concentrations of mannitol, 2,4-D, and BAP on the glucan envelope of *Vinca* protoplasts, but Blaschek and Franz[27] suspect that an osmotic pressure effect is operative in tobacco in view of changes in the cell wall composition of tobacco suspension culture induced by the addition of 0.7 *M* mannitol. At this point it should be noted (1) that there have been reports of influences of both cytokinin[88] and auxin[89] on matrix polysaccharides in tissue systems and that Takeuchi and Komamine[83] did not analyze the culture medium for secreted products and (2) that the culture medium used by Blaschek and co-workers[24] has been considered to be poor for supporting wall regeneration.[53]

Studies using radioisotopic tracers[14,24,29,82,84] have confirmed that products analyzed in wall residues are postisolation products. Glucose[14,24,29,82,84] and sucrose[24] are suitable precursors for the entire spectrum of polysaccharides. Label from myo-inositol appears in uronic acids and pentoses only;[24,84] mannitol is not utilized.[24] Label was transferred from uridinediphosphoglucose (UDPG) to the β1,3-glucan synthesized by soybean protoplasts, but very little to cellulose.[14] This last observation clearly implies that UDPG cannot be utilized for cellulose synthesis when supplied to the outside of the plasma membrane, but does not imply that secreted 1,3-glucan cannot be turned over for subsequent cellulose synthesis, according to the Meier model,[81] since the label was supplied to soybean protoplasts for only a relatively short period and small amounts did appear in cellulose.

That glucans appear predominantly in the regenerated wall may be related to the different cellular origins of glucan and nonglucan polysaccharides, glucans being a plasma membrane product while the heteropolymers are of endomembrane origin and therefore not anchored to the cell when released.[28,82] The significance of the unusual production of large quantities of β1,3-glucan is unclear. The fact that this is a wound product has been noted,[14] as has the fact that 1,3-glucans are found in cell plates which are early stages in wall ontogeny.[28] Alternatively, perhaps the surface of the isolated protoplast is an unsuitable environment for efficient interconversion of 1,3-glucan to cellulose, resulting in its accumulation.

B. Suberization

Using thin-layer and gas chromatography, Rao and co-workers[15] have demonstrated that tomato fruit protoplasts regenerate an envelope which includes the polymeric

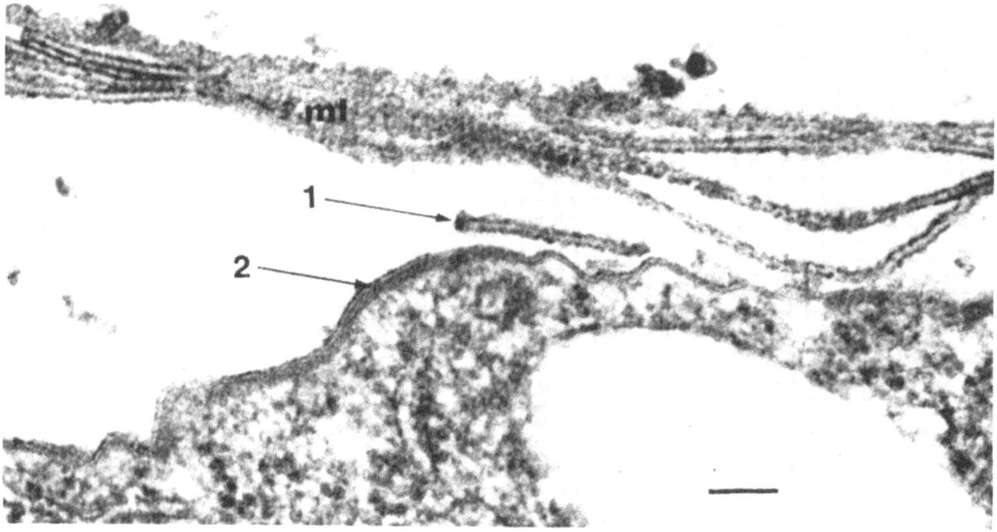

FIGURE 3. Wall regeneration by tomato fruit protoplast after 4 days in culture. The multilamellated wall (m*l*) appears to be derived from individual plates (1 and 2) which arise from the plasma membrane. Bar = 100 nm. (From Willison, J. H. M., in *Encyclopedia of Plant Physiology New Series,* Vol. 13B, Springer-Verlag, Berlin, 1981, 513. With permission.)

lipid, suberin, a characteristic wound product.[90] Tomato fruit protoplasts, unlike those from any other source so far investigated, regenerate walls which include a multilamellated component (Figure 3).[91-93] Fine-structural study of the deposition of this material suggests that it arises as plates from the plasma membrane surface (Figure 3), and that the plates fuse extracellularly to produce continuous lamellae.[93] Waxes, which might be expected in association with suberin, have not yet been analyzed.

IV. CELL WALL REGENERATION AND MORPHOGENESIS

The reader should refer to chapters by Fowke and by Kohlenbach in this volume for discussion of the developmental roles of the regenerated cell wall. Of particular contemporary interest are the influence of the cytoskeleton on the physical organization of the regenerated wall,[6,94-97] and the related morphogenetic role of the wall in the subsequent development of the regenerated cell, both in its relatively normal manifestations[98-100] and in its aberrant manifestations.[7-10,82,101] Any such discussion, however, must take account of the important differences between cell wall synthesis and secretion around experimentally isolated protoplasts and that which occurs in the relatively benign environment of the exoplasmic space surrounding a normal cell.

REFERENCES

1. Willison, J. H. M., Secretion of cell wall material in higher plants, in *Encyclopedia of Plant Physiology New Series*, Vol. 13B, Springer-Verlag, Berlin, 1981, 513.
2. Morre, D. J. and Mollenhauer, H. H., Interactions among cytoplasm, endomembranes, and the cell surface, in *Encyclopedia of Plant Physiology New Series*, Vol. 3, Springer-Verlag, Berlin, 1976, 288.
3. Cocking, E. C., Plant cell protoplasts — isolation and development, *Ann. Rev. Plant Physiol.*, 23, 29, 1972.
4. Willison, J. H. M., Synthesis of cell walls by higher plant protoplasts, in *Microbial and Plant Protoplasts*, Peberdy, J. F., Rose, A. H., Rogers, H. J., and Cocking, E. C., Eds., Academic Press, London, 1976, 283.
5. Fowke, L. C. and Gamborg, O. L., Applications of protoplasts to the study of plant cells, *Int. Rev. Cytol.*, 68, 9, 1980.
6. Willison, J. H. M. and Klein, A. S., Cell-wall regeneration by protoplasts isolated from higher plants, in *Cellulose and Other Natural Polymer Systems, Biogenesis, Structure and Degradation*, Brown, R. M., Jr., Ed., Plenum Press, New York, 1982, 61.
7. Meyer, Y. and Abel, W. O., Budding and cleavage division of tobacco mesophyll protoplasts in relation to pseudo-wall and wall formation, *Planta*, 125, 1, 1975.
8. Kinnersley, A. M., Racusen, R. H., and Galston, A. W., A comparison of regenerated cell walls in tobacco and cereal protoplasts, *Planta*, 139, 155, 1978.
9. Meyer, Y. and Herth, W., Interaction of cell-wall formation and cell division in higher plant cells, in *Cellulose and Other Natural Polymer Systems, Biogenesis, Structure and Degradation*, Brown, R. M., Jr., Ed., Plenum Press, New York, 1982, 149.
10. Schilde-Rentschler, L., Role of the cell wall in the ability of tobacco protoplasts to form callus, *Planta*, 135, 177, 1977.
11. Fleck, J., Durr, A., Lett, M. C., and Hirth, L., Changes in protein synthesis during the initial stage of life of tobacco protoplasts, *Planta*, 145, 279, 1979.
12. Fleck, J., Durr, A., Fritsch, C., Lett, M. C., and Hirth, L., Comparison of proteins synthesized *in vivo* and *in vitro* by mRNA from isolated protoplasts, *Planta*, 148, 453, 1980.
13. Fleck, J., Durr, A., Fritsch, C., Vernet, T., and Hirth, L., Osmotic-shock 'stress proteins' in protoplasts of *Nicotiana sylvestris*, *Plant Sci. Lett.*, 26, 159, 1982.
14. Klein, A. S., Montezinos, D., and Delmer, D. P., Cellulose and 1,3-glucan synthesis during the early stages of wall regeneration in soybean protoplasts, *Planta*, 152, 105, 1981.
15. Rao, G. S. R. L., Willison, J. H. M., and Ratnayake, W. M. N., Suberin production by isolated tomato fruit protoplasts, *Plant Physiol.*, 75, 716, 1984.
16. Willison, J. H. M. and Cocking, E. C., Microfibril synthesis at the surface of tobacco mesophyll protoplasts, a freeze-etch study, *Protoplasma*, 84, 147, 1975.
17. Burgess, J. and Linstead, P. J., Scanning electron microscopy of cell wall formation around isolated plant protoplasts, *Planta*, 131, 173, 1976.
18. Williamson, F. A., Fowke, L. C., Weber, G., Constabel, F., and Gamborg, O., Microfibril deposition on cultured protoplasts of *Vicia hajastana*, *Protoplasma*, 91, 213, 1977.
19. Prat, R., Contribution à l'étude des protoplastes végétaux. II. Ultrastructure du protoplaste isolé et régénération de sa paroi, *J. Microsc. (Paris)*, 18, 65, 1973.
20. Fannin, F. F. and Shaw, J. G., Production of extracellular fibers by tobacco leaf epidermal protoplasts, *Planta*, 159, 282, 1983.
21. Racusen, R. H., Kinnersley, A. M., and Galston, A. W., Osmotically induced changes in electrical properties of plant protoplast membranes, *Science*, 198, 405, 1977.
22. Herth, W. and Meyer, Y., Ultrastructural and chemical analysis of the wall fibrils synthesized by tobacco mesophyll protoplasts, *Biol. Cell.*, 30, 33, 1977.
23. Asamizu, T., Tanaka, K., Takebe, L., and Nishi, A., Change in molecular size of cellulose during regeneration of cell wall on carrot protoplasts, *Physiol. Plant.*, 40, 215, 1977.
24. Blaschek, W., Haass, D., Koehler, H., and Franz, G., Cell wall regeneration by *Nicotiana tabacum* protoplasts: chemical and biochemical aspects, *Plant Sci. Lett.*, 22, 47, 1981.
25. Blaschek, W., Koehler, H., Semler, U., and Franz, G., Molecular weight distribution of cellulose in primary cell walls, *Planta*, 154, 550, 1982.
26. Franz, G., Blaschek, W., Haass, D., and Koehler, H., Biosynthesis of cellulose: studies with tobacco protoplasts and cultured cells, *J. Appl. Polym. Sci. Appl. Polym. Symp.*, 37, 145, 1983.
27. Blaschek, W. and Franz, G., Influence of growth conditions on the composition of cell wall polysaccharides from cultured tobacco cells, *Plant Cell Rep.*, 2, 257, 1983.
28. Takeuchi, Y. and Komamine, A., Glucans in the cell walls regenerated from *Vinca rosea* protoplasts, *Plant Cell Physiol.*, 22, 1585, 1981.
29. Takeuchi, Y. and Komamine, A., Composition of the cell wall formed by protoplasts isolated from cell suspension cultures of *Vinca rosea*, *Planta*, 140, 227, 1978.

30. Prat, R. and Williamson, F. A., Chronologie de la sécrétion de parois par les protoplastes végétaux, *Soc. Bot. Fr. (Colloq. Secret. Veg.)*, 123, 33, 1976.
31. Burgess, J. and Fleming, E. N., Ultrastructural observations of cell wall regeneration around isolated tobacco protoplasts, *J. Cell Sci.*, 14, 439, 1974.
32. Grout, B. W. W., Cellulose microfibril deposition at the plasmalemma surface of regenerating tobacco mesophyll protoplasts: a deep etch study, *Planta*, 123, 275, 1975.
33. Burgess, J., Linstead, P. J., and Harnden, J. M., The interpretation of scanning electron micrographs, *Micron*, 8, 181, 1977.
34. Willison, J. H. M. and Grout, B. W. W., Further observations on cell wall formation around isolated protoplasts of tobacco and tomato, *Planta*, 140, 53, 1978.
35. Mueller, S. and Brown, R. M., Jr., Evidence for an intramembrane component associated with a cellulose microfibril synthesizing complex in higher plants, *J. Cell Biol.*, 84, 351, 1980.
36. Willison, J. H. M., The morphology of supposed cellulose-synthesizing structures in higher plants, *J. Appl. Polym. Sci. Appl. Polym. Symp.*, 37, 91, 1983.
37. Heath, I. B. and Seagull, R. W., Oriented cellulose fibrils and the cytoskeleton: a critical comparison of models, in *The Cytoskeleton in Plant Growth and Development*, Lloyd, C. W., Ed., Academic Press, New York, 1982, 163.
38. Delmer, D. P., Biosynthesis of cellulose, *Adv. Carbohydr. Chem. Biochem.*, 41, 105, 1983.
39. Robenek, H. and Peveling, E., Veränderungen den Plasmalemmas während der Zellwandregeneration an isolierten Protoplasten aus dem Sprosskallus von *Skimmia japonica* Thunb., *Planta*, 127, 281, 1975.
40. Robenek, H. and Peveling, E., Ultrastructure of the cell wall regeneration of isolated protoplasts of *Skimmia japonica* Thunb., *Planta*, 136, 135, 1977.
41. Wilkinson, M. J. and Northcote, D. H., Plasma membrane ultrastructure during plant protoplast plasmolysis, isolation and wall regeneration: a freeze-fracture study, *J. Cell Sci.*, 42, 401, 1980.
42. Brown, R. M., Jr., Haigler, C. H., Suttie, J., White, A. R., Roberts, E., Smith, C., Itoh, T., and Cooper, K., The biosynthesis and degradation of cellulose, *J. Appl. Polym. Sci. Appl. Polym. Symp.* 37, 33, 1983.
43. Mueller, S. C., Cellulose-microfibril assembly and orientation in higher plant cells with particular reference to seedlings of *Zea mays*, in *Cellulose and Other Natural Polymer Systems, Biogenesis, Structure and Degradation*, Brown, R. M., Jr., Ed., Plenum Press, New York, 1982, 87.
44. Mizuta, S. and Wada, S., Microfibrillar structure of growing cell wall in a coenocytic green alga *Boergesenia forbesii*, *Bot. Mag. (Tokyo)*, 94, 343, 1981.
45. Itoh, T., O'Neil, R., and Brown, R. M., Jr., The assembly of cellulose microfibrils in selected Siphonocladalean algae, *J. Cell Biol.*, 97, 416a, 1983.
46. Brown, R. M., Jr. and Montezinos, D., Cellulose microfibrils: visualization of biosynthetic and orienting complexes in association with the plasma membrane, *Proc. Natl. Acad. Sci. U.S.A.*, 73, 143, 1976.
47. Reinert, J. and Hellmann, S., Aspects of nuclear division and cell wall formation in protoplasts of different origin, *Colloq. Int. C.N.R.S.*, 212, 273, 1973.
48. Gamborg, O. L., Shyluk, J., and Kartha, K. K., Factors affecting the isolation and callus formation in protoplasts from the shoot apices of *Pisum sativum* L., *Plant Sci. Lett.*, 4, 285, 1975.
49. Burgess, J., Linstead, P. J., and Bonsall, V. E., Observations on the time course of wall development at the surface of isolated protoplasts, *Planta*, 139, 85, 1978.
50. Wallin, A. and Eriksson, T., Protoplast cultures from cell suspensions of *Daucus carota*, *Physiol. Plant.*, 28, 33, 1973.
51. Pearce, R. S. and Cocking, E. C., Behaviour in culture of isolated protoplasts from "Paul's Scarlet" rose suspension culture cells, *Protoplasma*, 77, 165, 1973.
52. Meyer, Y. and Abel, W. O., Importance of the wall for cell division and in the activity of the cytoplasm in cultured tobacco protoplasts, *Planta*, 123, 33, 1975.
53. Uchimiya, H. and Murashige, T., Influence of the nutrient medium on the recovery of dividing cells from tobacco protoplasts, *Plant Physiol.*, 57, 424, 1976.
54. Galbraith, D. W., Microfluorimetric quantitation of cellulose synthesis by plant protoplasts using Calcofluor White, *Physiol. Plant.*, 53, 111, 1981.
55. Harms, C. T., Lörz, H., and Potrykus, I., Multiple-drop-array (MDA) technique for the large-scale testing of culture media variations in hanging microdrop cultures of single cell systems. II. Determination of phytochrome combinations for optimal division response in *Nicotiana tabacum* protoplast cultures. *Plant Sci. Lett.*, 14, 237, 1979.
56. Montezinos, D. and Delmer, D. P., Characterization of inhibitors of cellulose synthesis in cotton fibers, *Planta*, 148, 305, 1980.
57. Meyer, Y. and Herth, W., Chemical inhibition of cell wall formation and cytokinesis, but not of nuclear division, in protoplasts of *Nicotiana tabacum* cultivated *in vitro*, *Planta*, 142, 253, 1978.

58. Galbraith, D. W. and Shields, B. A., The effects of inhibitors of cell wall synthesis on tobacco protoplast development, *Physiol. Plant.*, 55, 25, 1982.
59. Pilet, P. E., Facciotti, D., and Bucher, D., Ethylene production and cell wall formation in mesophyll protoplasts, *Experientia*, 38, 93, 1982.
60. Anderson, J. D., Lieberman, M., and Stewart, R. N., Ethylene production by isolated protoplasts, *Plant Physiol.*, 63, 931, 1979.
61. Burgess, J. and Linstead, P. J., Coumarin inhibition of microfibril formation at the surface of cultured protoplasts, *Planta*, 133, 267, 1977.
62. Burgess, J. and Linstead, P. J., Congo red inhibition of wall regeneration by *Physcomitrella* protoplasts: structural aspects, *Micron*, 13, 185, 1982.
63. Langebartels, C., Seitz, U., and Seitz, H. U., β-glucan synthetase activities in regenerating protoplasts from carrot suspension cultures, *Plant Sci. Lett.*, 22, 327, 1981.
64. Bilkey, P. C., Davey, M. R., and Cocking, E. C., Isolation, origin and properties of enucleate plant microplasts, *Protoplasma*, 110, 147, 1982.
65. Zimmer, B. and Werz, G., Concanavalin A affects polysaccharidic wall formation and mitotic activity in *Polyphysa (Acetabularia) cliftonii* protoplasts, *Exp. Cell Res.*, 126, 299, 1980.
66. Williamson, F. A., Fowke, L. C., Constabel, F. C., and Gamborg, O. L., Labelling of Concanavalin A sites on the plasma membrane of soybean protoplasts, *Protoplasma*, 89, 305, 1976.
67. Williamson, F. A., Concanavalin A binding sites on the plasma membrane of leek stem protoplasts, *Planta*, 144, 209, 1979.
68. Robinson, D. G. and Quader, H., Structure, synthesis, and orientation of microfibrils. IX. A freeze-fracture investigation of the *Oocystis* plasma membrane after inhibitor treatments, *Eur. J. Cell Biol.*, 25, 278, 1981.
69. Hahne, G., Herth, W., and Hoffmann, F., Wall formation and cell division in fluorescence-labelled plant protoplasts, *Protoplasma*, 115, 217, 1983.
70. Herth, W., Calcofluor white and congo red inhibit chitin microfibril assembly of *Poterioochromonas*: evidence for a gap between polymerization and microfibril formation, *J. Cell Biol.*, 87, 442, 1980.
71. Haigler, C. H., Brown, R. M., Jr., and Benziman, M., Calcofluor white ST alters the *in vivo* assembly of cellulose microfibrils, *Science*, 210, 903, 1980.
72. Morris, P., Linstead, P., and Thain, J. F., Comparative studies on leaf tissue and isolated protoplasts. III. Effects of wall-degrading enzymes and osmotic stress, *J. Exp. Bot.*, 32, 801, 1981.
73. Bacic, A. and Delmer, D. P., Stimulation of membrane-associated polysaccharide synthetases by a membrane potential in developing cotton fibers, *Planta*, 152, 346, 1981.
74. Taylor, A. R. D. and Hall, J. L., Some physiological properties of protoplasts isolated from maize and tobacco tissues, *J. Exp. Bot.*, 27, 383, 1976.
75. Coutts, R. H. A. and Grout, B. W. W., Early stages in cell wall regeneration of tobacco mesophyll protoplasts: an electrophoretic study, *Plant Sci. Lett.*, 4, 103, 1975.
76. Burgess, J. and Linstead, P. J., Structure and association of wall fibrils produced by regenerating tobacco protoplasts, *Planta*, 146, 203, 1979.
77. Kolpak, F. J. and Blackwell, J., Deformation of cotton and bacterial cellulose, *Text. Res. J.*, 45, 568, 1975.
78. Huwyler, H. R., Franz, G., and Meier, H., Changes in the composition of cotton fibre cell walls during development, *Planta*, 146, 635, 1979.
79. Meinert, M. C. and Delmer, D. P., Changes in biochemical composition of the cell wall of the cotton fiber during development, *Plant Physiol.* 59, 1088, 1977.
80. Darvill, A., McNeil, M., Albersheim, P. A., and Delmer, D. P., The primary cell walls of flowering plants, in *The Biochemistry of Plants*, Vol. 1, Stumpf, P. K. and Conn, E. E., Eds., Academic Press, New York, 1980, 91.
81. Meier, H., Buchs, L., Buchala, A. J., and Homewood, T., (1 → 3)-β-D-Glucan (callose) is a probable intermediate in biosynthesis of cellulose of cotton fibres, *Nature (London)*, 289, 821, 1981.
82. Hanke, D. E. and Northcote, D. H., Cell wall formation by soybean callus protoplasts, *J. Cell Sci.*, 14, 29, 1974.
83. Takeuchi, Y. and Komamine, A., Effects of culture conditions on cell division and composition of regenerated cell walls in *Vinca rosea* protoplasts, *Plant Cell Physiol.*, 23, 249, 1982.
84. Asamizu, T. and Nishi, A., Regenerated cell wall of carrot protoplasts isolated from suspension-culture cells, *Physiol. Plant.*, 48, 207, 1980.
85. Frearson, E. M., Power, J. B., and Cocking, E. C., The isolation, culture, and regeneration of *Petunia* leaf protoplasts, *Dev. Biol.*, 33, 130, 1973.
86. Roland, J. C. and Prat, R., Les protoplastes at quelques problemes concernant le role et l'elaboration des parois, *Colloq. Int. C.N.R.S.*, 212, 243, 1973.
87. Tanaka, M. and Uchida T., Heterogeneity of hydroxyproline-containing glycoproteins in protoplasts from a *Vinca rosea* suspension culture, *Plant Cell Physiol.*, 20, 1295, 1979.

88. Halmer, P. and Thorpe, T. A., Kinetin induced changes in cell wall composition of tobacco callus, *Phytochemistry,* 15, 1585, 1976.
89. Nishitani, K., Shibaoka, H., and Masuda, Y., Growth and cell wall changes in azuki bean epicotyls. II. Changes in wall polysaccharides during auxin-induced growth of excised segments, *Plant Cell Physiol.,* 20, 463, 1979.
90. Dean, B. B. and Kolattukudy, P. E. Synthesis of suberin during wound-healing in jade leaves, tomato fruit and bean pods, *Plant Physiol.,* 58, 411, 1976.
91. Pojnar, E., Willison, J. H. M., and Cocking, E. C., Cell wall regeneration by isolated tomato fruit protoplasts, *Protoplasma,* 64, 460, 1967.
92. Willison, J. H. M. and Cocking, E. C., The production of microfibrils at the surface of isolated tomato-fruit protoplasts, *Protoplasma,* 75, 397, 1972.
93. Willison, J. H. M., Fine structural changes occurring during the culture of isolated tomato fruit protoplasts, *Colloq. Int. C.N.R.S.,* 212, 215, 1973.
94. Fowke, L. C., Griffing, L. R., Mersey, B. G., and Van der Valk, P., Protoplasts for studies of the plasma membrane and associated cell organelles, in *Protoplasts 1983,* Potrykus, I., Ed., Birkhäuser Verlag, Basel, 1983, 101.
95. Simmonds, D. H., Setterfield, G., and Brown, D. L., Reorganization of microtubules in protoplasts of *Vicia hajastana,* Grossh. during the first 48 hours of culturing, *Proc. 6th Int. Protoplast Symp.,* Potrykus, I., Ed., Birkhäuser Verlag, Basel, 1983, 212.
96. Lloyd., C. W., Slabas, A. R., Powell, A. J., and Lowe, S. B., Microtubules, protoplasts and plant cell shape, *Planta,* 147, 500, 1980.
97. Zimmer, B. and Werz, G., Cytoskeletal elements and their involvement in *Polyphysa (Acetabularia)* protoplast differentiation, *Exp. Cell Res.,* 131, 105, 1981.
98. Burgess, J. and Linstead, P. J., Studies on the growth and development of protoplasts of the moss, *Physcomitrella patens,* and its control by light *Planta,* 151, 331, 1981.
99. Burgess, J. and Linstead, P. J., Cell-wall differentiation during growth of electrically polarised protoplasts of *Physcomitrella, Planta,* 156, 241, 1982.
100. Marchant, H. J. and Hines, E. R., The role of microtubules and cell-wall deposition in elongation of regenerating protoplasts of *Mougeotia, Planta,* 146, 41, 1979.
101. Herth, W. and Meyer, Y., Cytology of budding and cleavage in tobacco mesophyll protoplasts cultivated in saline medium, *Planta,* 142, 11, 1978.
102. Fowke, L. C., personal communication.

Chapter 7

CYTODIFFERENTIATION

Hans Willy Kohlenbach

TABLE OF CONTENTS

I. INTRODUCTION

Protoplast isolation and culture is a very efficient, and perhaps absolutely necessary methodology for crop improvement by genetic engineering with cells of higher plants. In addition, isolated protoplasts offer novel opportunities for fundamental research. In developmental physiology, protoplasts can be subject to investigations of the formation of the cell wall, the transition from quiescent noncycling G_o-cells to cycling ones, differentiation at the cellular level, as well as morphogenetic processes leading to organ primordia and nonzygotic embryos. Depending on the goal of the research, protoplasts or cells and clusters regenerated from them are important, unique material.

When dealing with cytodifferentiation in cultures arising from isolated protoplasts, it seems appropriate to confine descriptions to the early stages of development, including microcolonies. These developmental stages usually are still under the influence of osmotic stabilizers, e.g., mannitol. Apart from this factor, i.e., treatment of protoplasts with osmotica, which makes derivatives of protoplasts and same age derivatives of cells two different kinds of material, further and more fundamental differences between protoplasts and comparable cells can be detected. Isolated protoplasts differ from isolated cells of the same source essentially by various physiological activities, e.g., RNA metabolism, protein synthesis, photosynthetic rates.[1,2] This is mainly due to the following factors involved in the isolation procedure: treatment with enzymes, osmotic stress, the replacement of wall pressure by osmotic pressure with osmotic stabilizers, and the removal of the cell wall. Differences resulting from different isolation procedures are expected to diminish with time after isolation and finally disappear.

After isolation, protoplasts rather quickly begin to regenerate their cell walls, provided this process is not prevented by the use of a salt mixture for osmotic stabilization[3] or by an inhibitor.[3,4] Protoplasts of noncycling cells can be induced to divide by employing media with auxins and cytokinins, once a new cell wall has formed. In mesophyll protoplasts, biochemical processes associated with the regeneration of the cell wall will overlap with processes linked to the "change of program" of the regenerating cells. This "change of program" makes G_o-cells, specialized for a specific function, e.g., photosynthesis, become mitotic cells, open to various kinds of differentiation. It has been demonstrated[5] that the protein pattern of isolated mesophyll protoplasts does not differ from that of the source tissue, but that the pattern of protein synthesis is changed, apparently as a result of the trauma and stress suffered by the cells during protoplast isolation and as a result of a change in gene activation. It thus seems that for the study of the transition of G_o-cells to dividing cells, mechanically isolated cells[6] rather than isolated protoplasts would be the material of choice, since the former obviously suffer less stress. One advantage of protoplasts, however, is that in principle they permit an examination of fundamental phenomena "of cytodifferentiation under aspects of somatic genetics." It is possible "to influence the process of differentiation by introducing foreign genetic material via incorporation or fusion experiments."[7]

In this article, which is confined to reactions of protoplasts and their derivatives during the early phases of culturing, "cytodifferentiation" denotes not only irreversible, but also some transient, morphogenetic phenomena.

II. CYTODIFFERENTIATION OF ISOLATED PROTOPLASTS INDEPENDENT OF CELL DIVISION

A. Changes in Cell Size and Shape

When a protoplast has become a regenerated cell due to formation of a new cell wall, its spherical shape usually is lost. In most cases the cells become somewhat elongated and oval. In two cases remarkable cell elongation has been observed. Spherical protoplasts isolated from a suspension culture of tobacco (*Nicotiana tabacum* cv. Bright Yellow 2 Go) showing a diameter of 30 to 40 μm, elongated to cylindrical cells of 300 to 400 μm in length.[8] The basal medium contained 10 g/ℓ sucrose and 0.4 *M* mannitol and it was supplemented with NAA and BAP. Maximum elongation occurred in a culture with 0.1 mg/ℓ NAA and 1.0 mg/ℓ BAP. Increasing concentrations of NAA from 0.1 to 5.0 mg/ℓ decreased the elongation growth but increased the ratio of dividing cells without extraordinary elongation. Under optimal conditions elongation began 2 days after starting the culture and from the fourth to the seventh day continued linearly with an elongation rate of about 80 μm/day. The expansion in cell width was negligible. After 8 days and after nearly all cells had participated, elongation ceased. During the period of linear elongation, only a few cell divisions occurred. Since the cells elongated nearly synchronously and without undergoing cell division, the system seems to be well suited for analysis of unidirectional cell elongation and for investigation of microtubule arrangement in relation to growth. In this context it should be mentioned that, also in cell suspension cultures of tobacco, elongation and formation of filamentous cells may occur as a result of hormone[9] or a long-term low temperature treatment.[10] The omission of 2,4-D as the sole auxin or the addition of FUdR in the presence of 1.0 mg/ℓ 2,4-D to cultures of spheroidal cell clusters of carrot arrests cell division but allows cell elongation.[11] In tobacco cell cultures (*Nicotiana tabacum* cv. Xanthi line TX1), 0.5 m*M* DL-α-difluoromethyl ornithine inhibits cell division and leads to an enormous cell enlargement.[12] The question arises whether these responses might be given by the corresponding protoplasts also. In *Brassica rapa* macrocells representing about 100 single cells are produced by the cell division inhibiting herbicide, 2,6-dichlorobenzonitrile.[4] It should be mentioned further that the parallel arrays of cortical microtubules present in cultured cells of *Vicia hajastana* (which elongate without dividing when in media without auxin but with 0.025 mg/ℓ gibberellic acid) are disorganized in protoplasts isolated from them[13] (see also Chapter 3, Figures 1, 2).

The second example of specific cell elongation in cultures of isolated protoplasts has been observed in only three experiments until now, and the factors responsible for this reaction are still unknown. However, the phenomenon seems so remarkable that it will be presented. Isolated haploid stem embryo protoplasts of *Brassica napus* (strain H1) occasionally form long thin tubes which have a certain similarity to pollen tubes or tubes of other germinating spores (Figure 1). They develop from circular areas of the cells. In these tubes a thin peripheral cytoplasmic layer surrounds an elongated central vacuole. Often a greater portion of cytoplasm is located at the tip. The highest number of such tubes was observed in cultures which had been under the influence of rifampicin for 4 days and which had been subjected to a low temperature treatment for 6 days. As lower numbers of tubes were also found in cultures which had had only cold or rifampicin treatment or neither, the factors responsible for these structures remain unknown.[14]

B. The Formation of Tracheary Elements and of Alkaloid Cells

Protoplasts from cotyledons and primary leaves of *Zinnia elegans* L. cv. Dahlienblütige Riesen may directly differentiate into tracheary elements, "directly" referring to differentiation not preceded by cell division. Since these protoplasts are isolated from specialized mesophyll cells, it means further that their differentiation occurs without any relation to a cell cycle.[7] These findings confirm the results already obtained with mechanically isolated mesophyll cells from mature leaves of *Zinnia elegans*.[15] The results disprove the long held theory that at least one cell cycle is an absolute prerequisite for cytodifferentiation. The fact that enzymatically isolated mesophyll protoplasts and

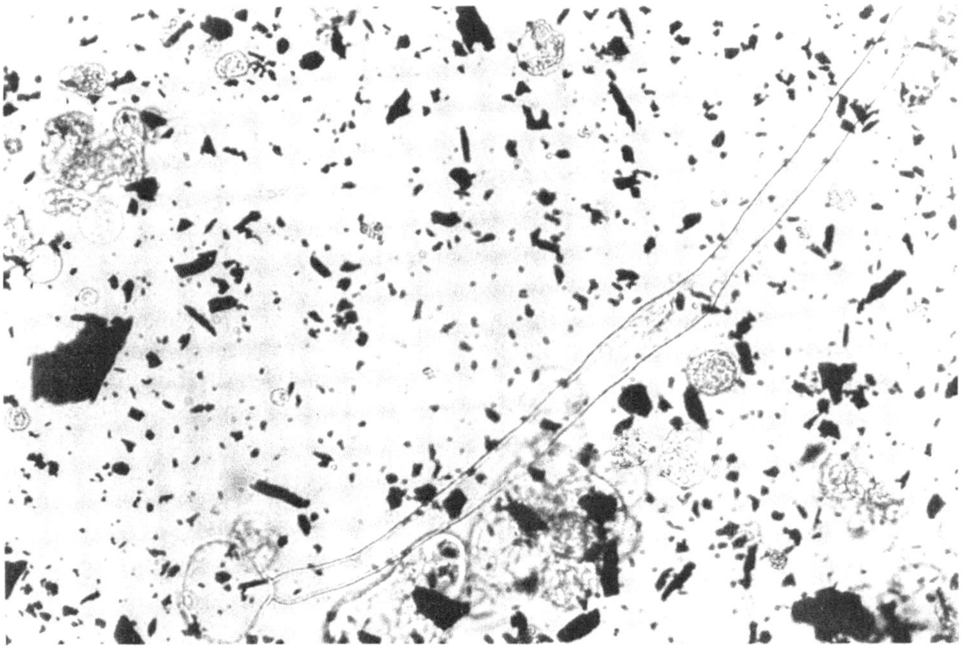

FIGURE 1. *Brassica napus*. Stem embryo protoplast forming a long, thin tube. The black particles are activated charcoal. (Körber and Kohlenbach)

mechanically isolated mesophyll cells would behave so similarly could by no means have been expected from the beginning, if one considers the differences between isolated protoplasts and cells as discussed in the introduction.

The protocol used with *Zinnia* is as follows. Protoplasts were isolated from leaves of 12-day-old seedlings in axenic culture. The following two enzyme solutions were used:

For cotyledons	1.0% Cellulase Onozuka R10
	0.5% Macerozyme R10
For primary leaves	0.5% Cellulase Onozuka R10
	0.2% Macerozyme R10
	0.1% Pectinase (Serva)

The enzymes were dissolved in culture medium with 0.45 *M* glucosemonohydrate and without phytohormones, pH adjusted to 5.5. Incubation: 17 hr; protoplast density 10^5/mℓ; 1.2 mℓ suspension per petri dish.

The phytohormones NAA and BAP were tested in various concentrations; 2 mg/ℓ NAA and 1 mg/ℓ BAP were found to produce the best yield, i.e., 158 tracheary elements from protoplasts of primary leaves in one petri dish. Tracheary elements of spherical, cylindrical, and more complicated forms, and with spiral, reticulate, and pitted wall thickenings were found. Intermediate forms, which possessed simultaneously still intact green chloroplasts and a nearly complete secondary wall thickening, could also be observed (Figure 2).

The potential for direct transformation of isolated mesophyll protoplasts into cells of another type is obviously not restricted to *Zinnia*. In cultures of mesophyll protoplasts of *Macleaya*, occasionally single cells can be detected which contain yellow col-

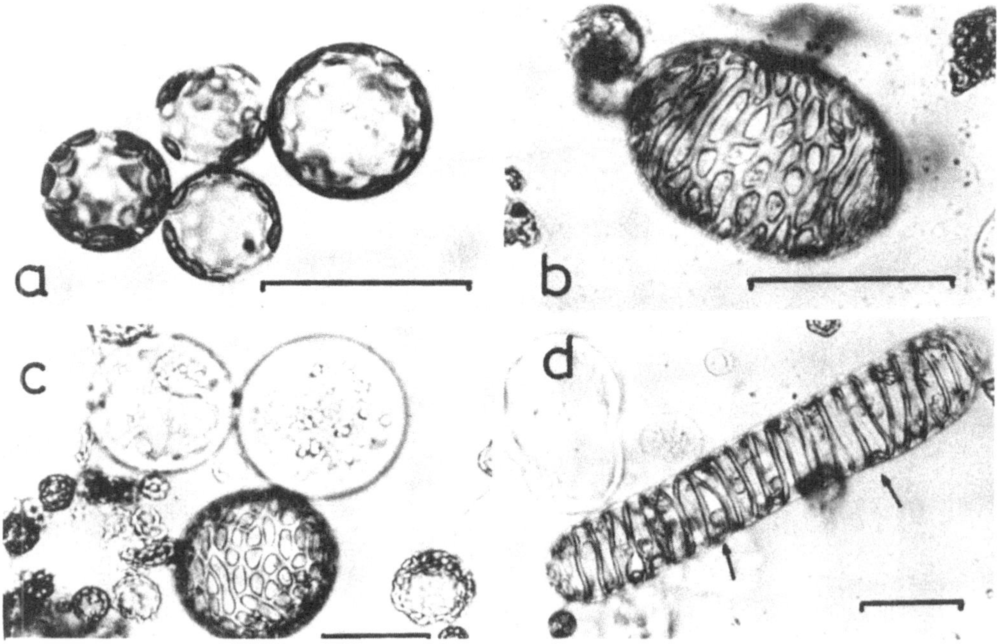

FIGURE 2. *Zinnia elegans.* (a) Freshly isolated protoplasts of primary leaves. (b to d) Tracheary elements in a 7-day-old culture of primary leave protoplasts, (b) and (c) with pitted wall patternings, (d) with residual chloroplasts (see arrows) and spiral wall texture. Scale 50 μm, (from Kohlenbach, H. W. and Schöpke, C., *Naturwissenschaften,* 68, 576, 1981. With permission.)

ored vacuoles in addition to chloroplasts. They must be regarded as an intermediate stage in the transformation of mesophyll protoplasts to alkaloid cells without any preceding cell division, because alkaloid cells regenerated from protoplasts of alkaloid cells already present in the leaf would not possess chloroplasts. Also, the number of alkaloid cells detected is dependent on the medium used.[16]

The occurrence of tracheary elements in cultures of haploid mesophyll protoplasts of *Nicotiana* has been reported.[17] In addition to *Zinnia* the differentiation of protoplasts into tracheary elements was more thoroughly investigated with haploid stem embryo protoplasts of *Brassica napus* (strain H1).[18] As with *Zinnia,* no cell division is necessary between isolation and differentiation. Unlike *Zinnia,* however, a cell division might have preceded the isolation, since in the experiments with *Brassica napus* the protoplasts had been isolated from material which also contained proliferating tissues. Although for this reason the results obtained with protoplasts from *Brassica* stem embryos cannot be considered as an example of direct cytodifferentiation without any relation to a cell cycle, this material is well suited for studies of the formation of tracheary elements from protoplasts.

The *Brassica* protoplasts were isolated using 0.2% PATE (pectic acid transeliminase, Hoechst), 0.3% Driselase, and 0.3% Cellulase Onozuka R10, in 0.4 *M* mannitol, at pH 7.6. Incubation: 16 to 17 hr; protoplast density 2×10^5/mℓ. The culture media were supplemented with 1.0 mg/ℓ 2,4-D and 0.5 mg/ℓ zeatin riboside.

In cultures of *Brassica* stem embryo protoplasts, tracheary elements with spiral, reticulate, and pitted secondary wall patternings occur. The first elements can be found 4 days after isolation. Their yield varies (Table 1 and 2). There obviously is an influence of the cytokinin concentration and the nature of the osmotic stabilizer; glucose favors

Table 1
NUMBER OF TRACHEARY ELEMENTS IN CULTURES OF ISOLATED PROTOPLASTS

Experiment A

No. of petri dish	Number of tracheary elements/m*l*
1	600
2	340
3	440
4	520

Note: First occurrence: 6 days after isolation
Scored: 24 days after isolation
Medium with 0.05% activated charcoal

From Kolenbach, H. W., Korber, M., and Li, L. C., *Z. Pflanzenphysiol.*, 107, 367, 1982. With permission.

Table 2
NUMBER OF TRACHEARY ELEMENTS IN CULTURES OF ISOLATED PROTOPLASTS

Experiment B

No. of petri dish	Total number of tracheary elements/m*l*	Number of globular tracheary elements/ m*l*	Globular tracheary elements in % of total number
1	160	76	47.5
2	140	70	50.
3	85	22	25.9
4	120	52	43.3
5	100	46	46.

Note: First occurrence: 13 days after isolation
Scored: 22 days after isolation
Medium with 0.05% activated charcoal.

From Kolenbach, H. W., Korber, M., and Li, L. C., *Z. Pflanzenphysiol.*, 107, 367, 1982. With permission.

the formation of tracheary elements more than mannitol (Table 3). The high percentage of spherical tracheary elements, up to 50%, is remarkable (Table 2). In these tracheary elements the previously spherical shape of the protoplasts is preserved. At least in these cases, regeneration of the wall must have been closely connected with the formation of the secondary wall pattern, for usually regeneration of the cell wall leads to the loss of the globular shape. Figure 3 also demonstrates that early stages of wall regeneration can be closely connected with the process of tracheary element formation; it shows a budding stage integrated into a tracheary element. The protoplasts used for these investigations have been isolated from a haploid stem embryo system. This is certainly of no significance for differentiation, but it might facilitate genetic manipulations.

Table 3
NUMBER OF TRACHEARY ELEMENTS IN DIFFERENT MEDIA PER PETRI DISH

Osmotic stabilizer	0.4*M*	Glucose	0.4	Mannitol
Concentration of cytokinins in mg/*l*	0.2	0.5	0.2	0.05
BAP	14	7	2	1
Kinetin	16	7	1	3
Zeatin riboside	11	6	2	1

Note: Scored: 13 days after isolation
Media: 8 days in N 9
5 days in B 5 + 0.05 mg/*l* GA_3 with different cytokinins and different osmotic stabilizers
Activated charcoal was not added in this experiment.

From Kohlenbach, H. W., Korber, M., and Li, L. C., *Z. Pflanzenphysiol.*, 107, 367, 1982. With permission.

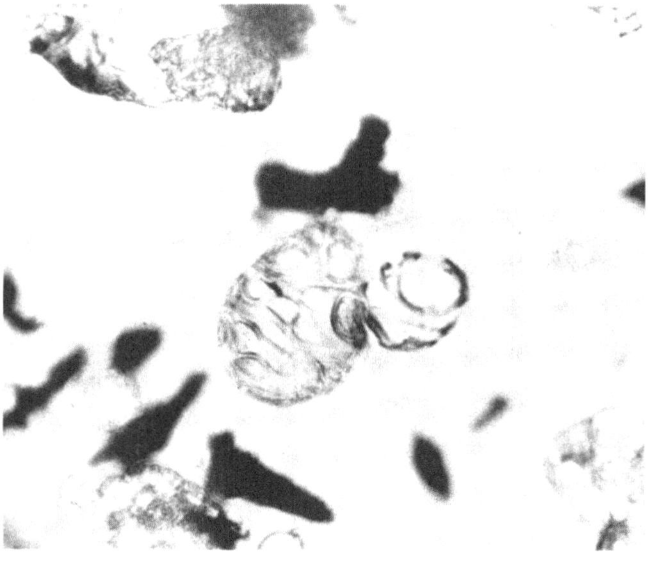

FIGURE 3. *Brassica napus.* A tracheary element derived from a stem embryo protoplast with "budding". The black particles are activated charcoal. (From Kohlenbach, H. W., Körber, M., and Li, L. C., *Z. Pflanzenphysiol.*, 107, 367, 1982. With permission.)

III. CYTODIFFERENTIATION IN CELL CLUSTERS AND MICROCOLONIES DERIVED FROM PROTOPLASTS BY CELL DIVISION GROWTH

The growth pattern of cell clusters and microcolonies obviously depends mainly on phytohormones. Their combination determines whether the clusters and microcolonies are loose and contain elongating vacuolated cells or more compact and spheroidal with densely cytoplasmic cells. In cultures of mesophyll protoplasts of *Brassica napus* (strain H1), for instance, the former type of development is obtained with 0.5 mg/ℓ NAA, 0.25 mg/ℓ 2,4-D, and 0.1 mg/ℓ BAP in the culture medium, whereas the latter occurs when 1.0 mg/ℓ 2,4-D and 0.5 mg/ℓ zeatin riboside are used. Activated charcoal is of influence, too. With *Brassica napus* and *Solanum tuberosum* it favors the formation of microcolonies.[19,20] However, detailed reports of systematic investigations of growth patterns appear to be lacking.

In addition to these phenomena, various kinds of differentiation can be observed in cell clusters and microcolonies: the formation of hair cells,[21] and tracheary elements,[16,22] the synthesis and accumulation of anthocyanin,[23] the development of alkaloid cells, and the formation of organ primordia and nonzygotic embryos.

A. The Formation of Alkaloid Cells in *Macleaya*

In small cell clusters derived from mesophyll protoplasts of *Macleaya cordata* (Willd.) R.Br. and *Macleaya microcarpa* (Maxim.) Fedde[16] alkaloid cells can be observed. In these cells, which are characterized by a yellow vacuole, alkaloids are accumulated as was shown for callus and suspension cultures.[24]

The protoplasts were isolated from leaves taken from plants in axenic culture propagated by shoot tips. The enzyme solutions for the isolation either contained 0.2% PATE (pectic acid transeliminase, Hoechst), 0.5% Cellulase Onozuka R10 in combination with 0.5 *M* glucose, and 0.02% Tween 80®, at pH 7.6, incubation time 16 hr, or 0.1% Pectolyase Y 23, 2.0% Cellulase Onozuka R10 in combination with 0.5 *M* glucose, and 0.02% Tween 80®, at pH 5.5, incubation time 2 hr. A modified 8p medium[25] or a modified XM medium[26] were used as culture media. The density of cultured protoplasts was 2 to 5 × 10^4/mℓ. The former medium was supplemented with 0.2 mg/ℓ 2,4-D, 0.5 mg/ℓ zeatin, and 1.0 mg/ℓ NAA, whereas the latter contained 1.0 mg/ℓ 2,4-D and 0.15 mg/ℓ BAP.

Cytodifferentiation occurred readily in both media. In medium XM-M6, up to 70% of the cell clusters contained alkaloid cells. Glucose as sole carbohydrate enhanced the culture of *Macleaya* mesophyll protoplasts. The differentiation of cells began 2 to 3 weeks after initiation of the culture with a faint yellow coloring near the plastids. Thereafter larger, colored vesicles developed which then fused with the vacuole. Older alkaloid cells turned red or brown. Reddish brown crystalline substances, obviously due to a saturation within the vacuole, may be formed. It is remarkable that cell clusters can be observed which consist mostly or exclusively of alkaloid cells. This behavior is quite different from that of suspension cultures of *Macleaya* callus where 10% alkaloid cells is the maximum.[24]

B. The Formation of Organ Primordia

Callus grown from mesophyll protoplasts of rapeseed (strain H1) has a high rhizogenic capacity. Under various hormone combinations, calluses of only a few millimeters in diameter produce numerous radially arranged root primordia covered by a fur of root hairs. Very early in culture, small cell clusters can become single root primordia (Figure 4). This process is enhanced by adding activated charcoal to the liquid medium, final concentration 0.05%.[19]

In *Hyoscyamus muticus* L. and *Nicotiana tabacum* L. cv. atropurpurea, shoot buds are formed at microcolonies derived from mesophyll protoplasts. They develop about 2 weeks after rapidly growing microcolonies have been transferred from media with high levels of phytohormones (5 to 20 μ*M* 4-CPA or NAA and 1 to 5 μ*M* BAP or zeatin) to media with low auxin concentrations. The formation of primordia com-

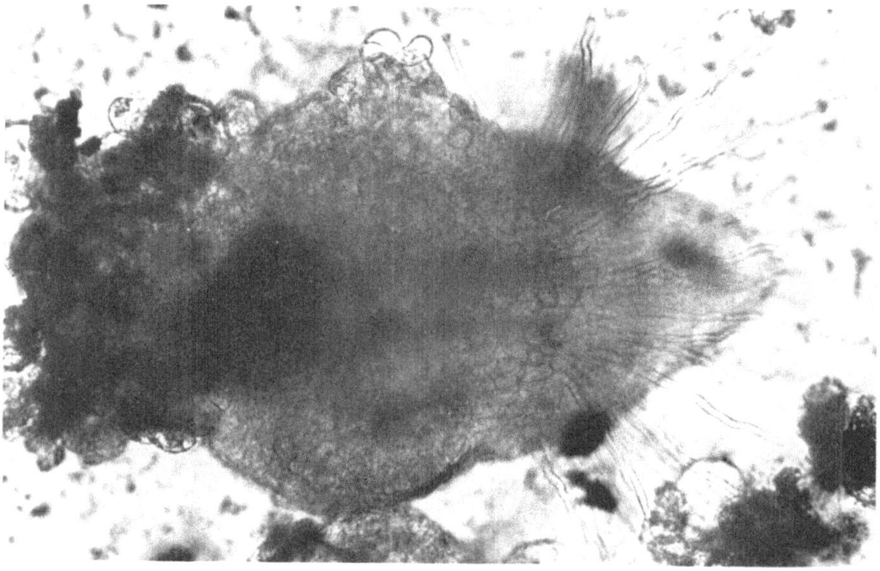

FIGURE 4. *Brassica napus*. Root primordium already with root hairs formed by a small cell cluster derived from a haploid leaf protoplast. The black particles are activated charcoal. (From Kohlenbach, H. W., Wenzel, G., and Hoffmann, F., *Z. Pflanzenphysiol.*, 105, 131, 1982. With permission.)

mences soon after the initiation of divisions. The authors[27] question how many divisions are required before expression of morphogenetic ability from an already differentiated cell can occur. Is it possible to get direct expression of morphogenesis?

C. The Formation of Nonzygotic Embryos

As an example of somatic embryogenesis in cultures derived from isolated protoplasts, a system featuring mesophyll protoplasts of *Brassica napus* will be presented with some detail.[28] Somatic embryos and plantlets were produced with microcolonies in quite a direct way using a four-step procedure (Table 4).

Protoplasts were isolated from leaves from axenic plants which had been regenerated from an androgenic stem embryo system of *Brassica napus* cv. Tower and cv. Loras. The protoplasts were isolated by an incubation of thin leaf sections with a solution of 0.025% PATE and 0.125% Cellulase Onozuka R10 containing 0.55 *M* mannitol, at pH 7.6 for 14 hr. Basal media were a modified Nitsch medium (M-I); MS-13 after (19) (M-II and M-III), and Murashige and Skoog medium (M-IV).

In M-I growth by division of the mesophyll protoplasts resulted in light green cell clusters within 3 weeks. About 2 weeks after replacing medium M-I with M-II, lentiform to spherical microcolonies of more than 0.5 mm in diameter developed. Many of these possessed a central compact globular structure which consisted of small densely packed cells and which must be regarded as a proembryo because of its appearance and further development. After 2 to 3 weeks in M-II, up to 350 microcolonies per petri dish were individually transferred to M-III. Proembryos became oval-shaped and bipolar, one pole being pigmented and forming the cotyledon and plumula region, the other one being colorless forming the root primordium (Figure 5). At present the highest frequency of microcolonies transferred into M-III developing into embryos is 27%, and this result was obtained employing a medium containing 0.01 mg/ℓ 2,4-D and 1.0

Table 4
CONDITIONS FOR THE FORMATION OF EMBRYOS FROM ISOLATED MESOPHYLL PROTOPLASTS OF *BRASSICA NAPUS* CV. "TOWER"

	Induction of Division Growth of Protoplasts
M I	Mod. Nitsch med. supplemented with (in mg/ℓ) 0.5 2,4-D; 0.5 NAA; 0.5 BAP 0.55 *M* mannitol 3 weeks (in the dark for 2 weeks, then illuminated after adding 0.5 mℓ M-I but with 0.4 *M* mannitol)
	Formation of Proembryos in Microcolonies
M II	MS medium supplemented with (in mg/ℓ) 0.2 2,4-D; 3.0 Kinetin; 500 inositol; 500 glutamine 0.4 *M* mannitol 2 — 3 weeks
	Development of Embryos
M III	MS medium supplemented with (in mg/ℓ) 500 inositol; 500 glutamine and reduced hormones (e.g., 0.01 2,4-D 1.0 BAP) 0.4 *M* mannitol 4 — 6 weeks
	Development of rooted Plants
M IV	MS medium without hormones (solid medium) At least 4 weeks

From Li, L. C. and Kohlenbach, H. W., *Plant Cell Rep.*, 1, 209, 1982. With permission.

mg/ℓ BAP. The development of individual embryos could be observed when single microcolonies were cultured in hanging-droplets of 0.03 mℓ M-III (Figure 6). In most cases, the development of proembryos to embryos was accompanied by some proliferation of the peripheral cells of the microcolonies surrounding the proembryos. The proliferation of this pale unorganized tissue was, however, limited. After 3 to 4 weeks its cells degenerated and died and the embryos became free of surrounding tissue (Figure 5).

It is most interesting and important to note that in the example given the embryos developed more or less directly, i.e., without an intervening callus and with a major proportion of cells being integrated into the embryo. Each embryo can be traced back to a single protoplast. A similar behavior has been reported for mesophyll protoplasts of the following species: *Nicotiana tabacum* L.,[29] *Nicotiana sylvestris* L.,[30] *Lycopersicon peruvianum* (L.) Mil.,[31] *Ranunculus sceleratus* L.,[32] and *Medicago sativa* L.[33]

In addition to these observations, two further reports of early development of embryos in cultures of mesophyll protoplasts[34,35] and some reports of embryogenesis with protoplasts of other cell types must be mentioned. For these investigations protoplasts were isolated from roots of *Daucus carota* L.[36] and *Medicago sativa* L.;[34,37] cotyledons of *Medicago sativa* L.;[34] cell suspension cultures of *Atropa belladonna* L.,[38] *Hyoscy-*

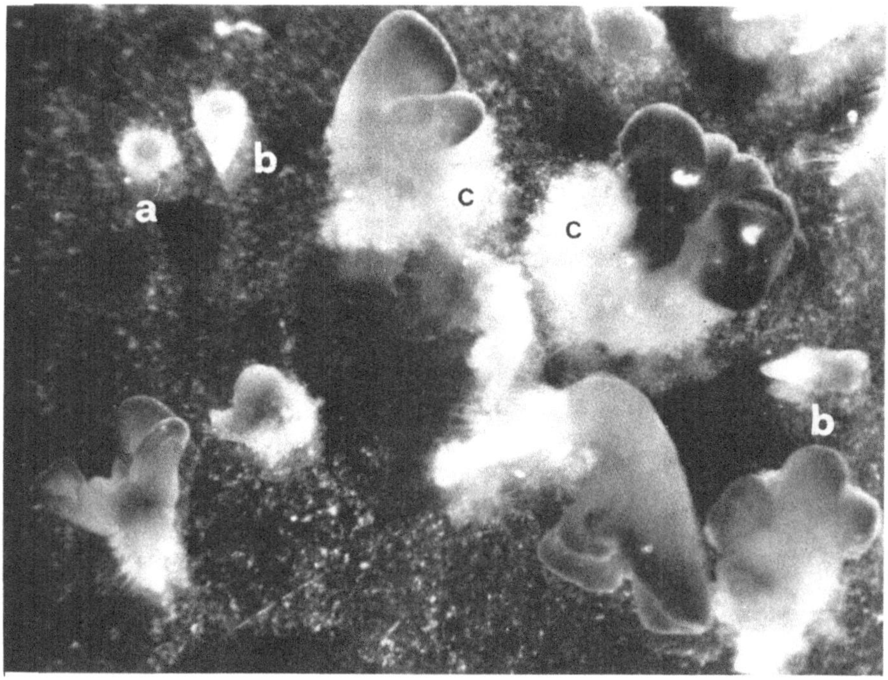

FIGURE 5. *Brassica napus*. Somatic embryos from mesophyll protoplasts at different stages of development after 40 days in M III. (a) Microcolony with a compact structure (proembryo) within it, (b) Early stages of embryo formation with differentiating roots, (c) Unorganized cells in degeneration. (From Li, L. C. and Kohlenbach, H. W., *Plant Cell Rep.*, 1, 209, 1982. With permission.)

FIGURE 6. *Brassica napus*. Somatic embryo developing to a plantlet. The embryo arose from a microcolony after this had been transferred to a hanging droplet. The microcolony had derived from a mesophyll protoplast. (Li and Kohlenbach)

amus muticus L.[39] ovular and nucellar calluses of *Citrus*;[40,41] embryogenic suspension cultures of *Daucus carota* L.,[42] *Solanum melongena* L.,[43] *Pennisetum americanum* (L.) K. Schum.,[44] *Pennisetum purpureum* Schum.;[45] and somatic embryos of *Daucus carota* L.[46]

In the case of root protoplasts of *Medicago*, somatic embryos can appear as early as 30 to 35 days after isolation.[34] In the same study, dealing not only with root but also with cotyledon and leaf protoplasts of *Medicago* seedlings, it is demonstrated (at least for this material) that the procedure of diluting the osmotic stabilizer is of great influence on the further development of the protoplasts; the method of dilution appears to determine whether the microcolonies can form embryos or only roots.

There is no convincing observation of "completely direct" development of somatic embryos from noninduced protoplasts, where "completely direct" refers to all cells derived from one protoplast being completely integrated into an embryo. Still, mesophyll protoplasts offer an opportunity to analyze somatic embryogenesis *ab initio*.

IV. SUMMARY

Isolated protoplasts can be used for fundamental research on the dedifferentiation of cells, the transition from quiescent noncycling cells to cycling ones, differentiation at the cellular level either independent of cell division or after cell division in cell clusters and microcolonies, and on the formation of organ primordia and nonzygotic embryos. The formation of tracheary elements in *Zinnia* and in *Brassica*, of alkaloid cells in *Macleaya* and of nonzygotic embryos in *Brassica* is presented with details. One advantage of protoplasts in comparison with isolated cells is that they may permit examination of fundamental phenomena of differentiation which are under the influence of foreign genetic material introduced via incorporation or fusion.

REFERENCES

1. Kulikowski, R. R. and Mascarenhas, J. B., RNA synthesis in whole cells and protoplasts — a comparison, *Plant Physiol.*, 61, 575, 1978.
2. Burgess, J., Plant cells without walls?, *Nature (London)*, 275, 588, 1978.
3. Meyer, Y. and Aspart, L., The first mitotic cycle of mesophyll protoplasts, in *Protoplasts 1983, Lecture Proc., 6th Int. Protoplast Symp.*, Potrykus, I., Harms, C. T., Hinnen, A., Hütter, R., King, P. J., and Shillito, R. D., Eds., Birkhäuser Verlag, Basel, 1983; *Exp. Suppl.*, 46, 93, 1983.
4. Torriani, U. and Potrykus, I., Attempts to develop a plant analogue to the oocyte system, in *Protoplasts 1983, Poster Proc., 6th Int. Protoplast Symp.*, Potrykus, I., Harms, C. T., Hinnen, A., Hütter, R., King, P. J., and Shillito, R. D., Eds., Birkhäuser Verlag, Basel, 1983; *Exp. Suppl.*, 45, 268, 1983.
5. Fleck, J., Durr, A., and Hirth, L., Gene expression in freshly isolated protoplasts from *Nicotiana sylvestris*, in *Protoplasts 1983, Poster Proc., 6th Int. Protoplast Symp.*, Potrykus, I., Harms, C. T., Hinnen, A., Hütter, R., King, P. J., and Shillito, R. D., Eds., Birkhäuser Verlag, Basel, 1983; *Exp. Suppl.*, 45, 240, 1983.
6. Kohlenbach, H. W., Culture of isolated mesophyll cells, in *Cell Culture and Somatic Cell Genetics of Plants*, Vol. 1, Vasil, I. K., Ed., Academic Press, New York, 1984, 204.
7. Kohlenbach, H. W. and Schöpke, C., Cytodifferentiation to tracheary elements from isolated mesophyll protoplasts of *Zinnia elegans*, *Naturwissenschaften*, 68, 576, 1981.
8. Hasezawa, S. and Syono, K., Hormonal control of elongation of tobacco cells derived from protoplasts, *Plant Cell Physiol.*, 24, 127, 1983.
9. Opatrny, Z. and Opatrna, J., The specifity of the effect of 2,4-D and NAA of the growth, micromorphology, and occurrence of starch in long-term *Nicotiana tabacum* L. cell strains, *Biol. Plantarum (Praha)*, 18, 359, 1976.

10. Opatrny, Z. and Opatrna, J., The cytokinine-like effect of a lowered temperature on the micromorphology of *Nicotiana tabacum* L. cell strains, *Biol. Plantarum (Praha)*, 17, 314, 1976.
11. Lloyd, C. W., Lowe, S. B., and Peace, G. W., The mode of action of 2,4-D in counteracting the elongation of carrot cells grown in culture, *J. Cell. Sci.*, 45, 257, 1980.
12. Berlin, J. and Forche, E., DL-α-difluoromethyl ornithine causes enlargement of cultured tobacco cells, *Z. Pflanzenphysiol.*, 101, 277, 1981.
13. Simmonds, D. H., Setterfield, G., and Brown, D. L., Reorganization of microtubules in protoplasts of *Vicia hajastana*, Grossh. during the first 48 hours of culturing, in *Protoplasts 1983, Poster Proc., 6th Int. Protoplast Symp.*, Potrykus, I., Harms, C. T., Hinnen, A., Hütter, R., King, P. J., and Shillito, R. D., Eds., Birkhäuser Verlag, Basel, 1983; *Exp. Suppl.*, 45, 212, 1983.
14. Körber, M. and Kohlenbach, H. W., unpublished.
15. Kohlenbach, H. W. and Schmidt, B., Cytodifferenzierung in Form einer direkten Umwandlung isolierter Mesophyllzellen zu Tracheiden, *Z. Pflanzenphysiol.*, 75, 369, 1975.
16. Lang, H. and Kohlenbach, H. W., Differentiation of alkaloid cells in cultures of *Macleaya* mesophyll protoplasts, *Planta Medica*, 46, 78, 1982.
17. Potrykus, I., Lörz, H., and Harms, C. T., On some selected problems and results concerning culture and genetic modification of higher plant protoplasts, in *Plant Tissue Culture and its Biotechnological Application, Proc. 1st Int. Congr. on Medicinal Plant Research, Section B, Munich*, Barz, W., Reinhard, E., and Zenk, M. H., Eds., Springer-Verlag, New York, 1977, 323.
18. Kohlenbach, H. W., Körber, M., and Li, L. C., Cytodifferentiation of protoplasts isolated from a stem embryo system of *Brassica napus* to tracheary elements, *Z. Pflanzenphysiol.*, 107, 367, 1982.
19. Kohlenbach, H. W., Wenzel, G., and Hoffmann, F., Regeneration of *Brassica napus* plantlets in cultures from isolated protoplasts of haploid stem embryos as compared with leaf protoplasts, *Z. Pflanzenphysiol.*, 105, 131, 1982.
20. Carlberg, I., Glimelius, K., and Eriksson, T., Improved culture ability of shoot culture derived protoplasts of *Solanum tuberosum* L. by use of activated charcoal, in *Protoplasts 1983, Poster Proc., 6th Int. Symp.*, Potrykus, I., Harms, C. T., Hinnen, A., Hütter, R., King, P. J., and Shillito, R. D., Eds., Birkhäuser Verlag, Basel, 1983; *Exp. Suppl.*, 45, 258, 1983.
21. Krumbiegel, G. and Schieder, O., Selection of somatic hybrids after fusion of protoplasts from *Datura innoxia* Mill. and *Atropa belladonna* L., *Planta*, 145, 371, 1979.
22. Koblitz, H., Isolierung und Kultivierung von Protoplasten aus Calluskulturen von *Catharanthus roseus*, *Biochem. Physiol. Pflanzen*, 167, 489, 1975.
23. Hoffmann, F., Phytohormone — Werkzeuge pflanzlicher Zellkulturtechniken, *Chemie Unserer Zeit*, 11, 108, 1977.
24. Böhm, H., Regulation of alkaloid production in plant cell cultures, in *Frontiers of Plant Tissue Culture 1978, 4th Int. Congr. Plant Tissue and Cell Culture*, Thorpe, T. A., Ed., University of Calgary, 1978, 201.
25. Kao, K. N. and Michayluk, M. R., Nutritional requirements for growth of *Vicia hajastana* cells and protoplasts at a very low population density in liquid media, *Planta*, 126, 105, 1975.
26. Xuan, L. T. and Menczel, L., Improved protoplast culture and plant regeneration from protoplast-derived callus in *Arabidopsis thaliana*, *Z. Pflanzenphysiol.*, 96, 77, 1980.
27. Wernicke, W. and Thomas E., Studies on morphogenesis from isolated plant protoplasts: shoot formation from mesophyll protoplasts of *Hyoscyamus muticus* and *Nicotiana tabacum*, *Plant Sci. Lett.*, 17, 401, 1980.
28. Li, L. C. and Kohlenbach, H. W., Somatic embryogenesis in quite a direct way in cultures of mesophyll protoplasts of *Brassica napus* L., *Plant Cell Rep.*, 1, 209, 1982.
29. Lörz, H., Potrykus, I., and Thomas E., Somatic embryogenesis from tobacco protoplasts, *Naturwissenschaften*, 64, 439, 1977.
30. Facciotti, D. and Pilet, P. E., Plants and embryoids from haploid *Nicotiana sylvestris* protoplasts, *Plant Sci. Lett.*, 15, 1, 1979.
31. Zapata, F. J. and Sink, K. C., Somatic embryogenesis from *Lycopersicon peruvianum* leaf mesophyll protoplasts, *Theor. Appl. Genet.*, 59, 265, 1981.
32. Dorion, N., Chupeau, Y., and Bourgin, J. P., Isolation, culture and regeneration into plants of *Ranunculus sceleratus* L. leaf protoplasts, *Plant Sci. Lett.*, 5, 325, 1975.
33. Kao, K. N. and Michayluk, M. R., Plant regeneration from mesophyll protoplasts of alfalfa, *Z. Pflanzenphysiol.*, 96, 135, 1980.
34. Lu, D. Y., Davey, M. R., and Cocking, E. C., A comparison of the cultural behaviour of protoplasts from leaves, cotyledons and roots of *Medicago sativa*, *Plant Sci. Lett.*, 31, 87, 1983.
35. Lu, D. Y., Davey, M. R., and Cocking, E. C., Somatic embryogenesis from mesophyll protoplasts of *Trigonella corniculata* (Leguminosae), *Plant Cell Rep.*, 1, 278, 1982.
36. Kameya, T. and Uchimiya, H., Embryoids derived from isolated protoplasts of carrot, *Planta*, 103, 356, 1972.

37. Xu, Z. H., Davey, M. R., and Cocking, E. C., Organogenesis from root protoplasts of the forage legumes *Medicago sativa* and *Trigonella foenum-graecum*, *Z. Pflanzenphysiol.*, 107, 231, 1982.
38. Gosch, G., Bajaj, Y. P. S., and Reinert, J., Isolation, culture and induction of embryogenesis in protoplasts from cell-suspensions of *Atropa belladonna*, *Protoplasma*, 86, 405, 1975.
39. Lörz, H., Wernicke, W., and Potrykus, I., Culture and plant regeneration of *Hyoscyamus* protoplasts, *Planta Medica*, 36, 21, 1979.
40. Vardi, A., Spiegel-Roy, P., and Galun, E., Citrus cell culture: isolation of protoplasts, plating densities, effects of mutagens and regeneration of embryos, *Plant Sci. Lett.*, 4, 231, 1975.
41. Vardi, A., Spiegel-Roy, P., and Galun, E., Plant regeneration from *Citrus* protoplasts: variability in methodological requirements among cultivars and species, *Theor. Appl. Genet.*, 62, 171, 1982.
42. Grambow, H. J., Kao, K. N., Miller, R. A., and Gamborg, O. L., Cell division and plant development from protoplasts of carrot cell suspension cultures, *Planta*, 103, 348, 1972.
43. Gleddie, S. C., Keller, W. A. and Setterfield, G., Somatic embryogenesis and plant regeneration from protoplasts of eggplant (*Solanum melongena* L.), in *Protoplasts 1983, Poster Proc., 6th Int. Protoplast Symp.*, Potrykus, I., Harms, C. T., Hinnen, A., Hütter, R., King, P. J., and Shillito, R. D., Eds., Birkhäuser Verlag, Basel, 1983; *Exp. Suppl.*, 45, 66, 1983.
44. Vasil, V. and Vasil, I. K., Isolation and culture of cereal protoplasts, II. Embryogenesis and plantlet formation from protoplasts of *Pennisetum americanum*, *Theor. Appl. Genet.*, 56, 97, 1980.
45. Vasil, V., Wang, D. Y., and Vasil, I. K., Plant regeneration from protoplasts of Napier grass (*Pennisetum purpureum* Schum.), *Z. Pflanzenphysiol.*, 111, 233, 1983.
46. Nomura, K., Nitta, T., Fujimura, T., and Komamine, A., Isolation of protoplasts from somatic embryos of carrot, *Plant Cell, Tissue Organ Cult.*, 1, 211, 1982.

Chapter 8

THE USE OF PROTOPLASTS FOR STUDIES ON MEMBRANE TRANSPORT IN PLANTS

Robert T. Leonard and Lisa Rayder

TABLE OF CONTENTS

I. INTRODUCTION

All plant cells expend metabolic energy to absorb and retain specific mineral nutrient ions and organic solutes. Most of these substances, while utilized in the cytoplasm, are stored in the large vacuole which is characteristic of plant cells. Hence, transport of solutes through the plasma membrane and the tonoplast membrane are important cellular activities. The mechanism of energy coupling to solute transport at these membranes is not fully understood and is the subject of much ongoing research.[1]

Most of the information on membrane transport in plants has come from experiments utilizing excised tissues usually (but not exclusively) obtained from roots. However, results from experiments with cells organized into a tissue can be difficult to interpret in terms of cellular mechanisms.[2] For many studies on membrane transport in plant cells, it would be desirable to work with a relatively homogeneous population of individual cells which are not organized into a tissue. Also, the presence of a thick and resistant cell wall poses several additional technical problems. For example, the cell wall provides a formidable physical barrier restricting experimental access to the living protoplast. Removal of the cell wall to expose the protoplast would allow several research techniques currently used by animal cell biologists to be employed by the plant cell biologist.

Isolated protoplasts offer much promise as a model system for studies on membrane transport in plants. This is true for the reasons already mentioned and because protoplasts represent an excellent starting point for the purification of intact vacuoles. The ability to obtain vacuoles in high yield and purity is important for experiments on solute transport at the tonoplast membrane.

In this article, we discuss the methods used for, and results obtained from, experiments on membrane transport in protoplasts and vacuoles isolated from protoplasts. We will begin by summarizing some of the methods used to obtain protoplasts and vacuoles and to measure solute fluxes therein. Then, the suitability of isolated protoplasts and vacuoles for such studies will be assessed in terms of the results obtained to date. Next, the transport of some specific solutes will be considered in greater detail. Finally, the use of protoplasts to purify the plasma membrane and the tonoplast membrane will be considered. We are not aware of any review which specifically deals with these subjects.

II. PROCEDURES FOR MEASURING MEMBRANE TRANSPORT

A. Isolated Protoplasts

1. Protoplast Preparation

Studies of membrane transport using protoplasts require the rapid isolation of large quantities of pure, viable protoplasts. It is important that the preparation be as free as possible from contamination by microorganisms, subcellular debris, vascular elements, undigested cells, and broken protoplasts. The yield of protoplasts should be in the range of about 10^6/g fresh tissue, and from 20 to 100 g of tissue should be readily available on a routine basis.

Viability of isolated protoplasts can be measured in several ways,[3-5] but the most common means is by light microscope examination. Intact, spherical cells with a surrounding "halo" (when viewed under phase contrast), showing even cytoplasmic distribution and cyclosis as examined under the light microscope, are assumed to be viable. Dye accumulation is another method often used to confirm protoplast viability. The vital stain neutral red indicates a lower pH in the vacuole relative to the cytoplasm and thus the maintenance of tonoplast and plasma membrane integrity. Fluorescein diacetate (FDA) is also used as a protoplast viability stain. Intact cells fluoresce as an enzymatic product of FDA, fluorescein, accumulates inside the plasma membrane. The exclusion of Evan's blue dye is utilized as an indication of intact membranes. Metabolic indicators of viability such as oxygen uptake, indicating respiratory metabolism, and oxygen evolution or CO_2 uptake, indicating photosynthetic activity, have also been used to determine protoplast viability, although much less frequently than the light microscope examination and dye accumulation methods.

2. Generalized Procedure for Protoplast Isolation

Virtually all procedures used for protoplast isolation employ enzymatic removal of the cell wall of plasmolyzed cells. To obtain protoplasts, the plant tissue is harvested, usually surface sterilized, sliced into 1- to 2-mm sections, held in a minimum saline solution (usually 0.1 to 1 m*M* $CaCl_2$), and rinsed several times. The washed tissue (or washed cells in the case of plant material obtained from cell culture) is added to a digestion medium containing: osmoticum; usually mannitol, sorbitol, or sucrose; cell wall digestive enzymes obtained from fungi; usually a cellulase and a pectinase; various buffers; Ca^{2+}; other salts; EDTA; BSA; and DTT. The exact composition and concentrations are experimentally determined for a given plant species and tissue.[6-11] The tissue is often, but not always, vacuum infiltrated with the digestion medium for several minutes and is allowed to digest with gentle agitation for several hours, after which the released protoplasts are separated from debris by gentle filtration through cheesecloth or nylon (algal) mesh. The debris is washed several times with osmoticum to maximize protoplast yield, and the combined eluants are centrifuged and washed several times at low *g* forces (100 to 200) for 5 to 10 min to pellet the protoplasts and to remove any remaining digestive enzymes. The final pellet is resuspended and layered on the bottom or top of a discontinuous Ficoll or Percoll gradient, and the gradients are centrifuged at low *g* forces (200 to 500) for 30 to 60 min. Protoplasts are then collected at the appropriate interface and are resuspended in buffered osmoticum. At this point, the protoplasts can be counted using a hemacytometer, chlorophyll and protein contents can be established, and transport studies can be initiated.

3. Specific Protocols for Measurement of Solute Transport

a. Inorganic Ion Transport

Several methods of measuring K^+ transport in protoplasts have been reported,[9-13] although all of them involve radiotracers and differ mainly in the method of separating protoplasts from the radioactive absorption medium. During the process of separating protoplasts from the absorption medium, it is crictical that excessive breakage of the fragile protoplast is avoided. Taylor and Hall[12] were the first to measure K^+ transport in protoplasts and used the first expanded leaves from maize and tobacco as their source of protoplasts. The isolated protoplasts were incubated in 0.6 *M* sorbitol containing 5 m*M* KCl with 5 to 7.5 μCi ^{86}RbCl ($^{86}Rb^+$ is commonly used as a tracer for K^+) adjusted to pH 6.0. Aliquots were removed at different times over a 5-hr period, transferred to a vacuum filter unit, and washed three times with 5, 10, and 30 mℓ of a cold (2°C) unlabeled absorption medium. Vacuum filter discs containing radioactive protoplasts were dried on planchettes and counted in a gas flow counter. A zero time sample was taken within 30 sec of isotope addition to determine the nonspecific adsorption of isotope by the protoplasts and the filter disk.

Leonard and colleagues[10,11] used protoplasts isolated from cultured tobacco cells and from cortical tissues of corn roots in their K^+ transport studies. Protoplasts were incubated in experimental media containing 0.1 μCi/flask ^{86}Rb (or ^{42}K) at 30°C with gentle agitation to keep the protoplasts suspended and to increase O_2 exchange. Samples of protoplasts were periodically removed from the absorption medium and layered on top of a discontinuous Ficoll gradient held on ice. The top layer of the gradient included 10 m*M* of the nonradioactive salt of the radioactive tracer. The protoplasts and nonabsorbed tracer mixed with the top layer. Cold temperature and the dilution of specific radioactivity effectively stopped any further tracer influx. The gradients were centrifuged to pellet the protoplasts, the supernatant was aspirated off, and the sides of the centrifuge tubes were gently rised to remove any residual radioactivity. The final pellet was resuspended in water, a sample was removed for protein determination, and the remaining solution was transferred to a scintillation vial for counting. Influx values were determined by a linear regression analysis of short-term absorption of tracer at 5, 15, and 30 min from the addition of the tracer.

Lin[9] incubated protoplasts in a radioactive tracer similar to that described above, but he used a different gradient for separating protoplasts from the radioactive absorption solution, from bottom to top: 50$\mu\ell$ silicone oil with a density of 1.045 g/mℓ; 50$\mu\ell$ 0.7 *M* mannitol, 0.2 m*M* $CaCl_2$, 25 m*M* MES, pH 6.0; 50 $\mu\ell$ silicone oil with a density of 1.044 g/mℓ; 200 $\mu\ell$ absorption solution containing 0.5 μCi ^{86}Rb and 5×10^6 protoplasts. Incubation was at 30°C, and uptake was terminated by pelleting the protoplasts through the gradient. The tips of the tubes containing the labeled protoplasts were cut off and counted in scintillation vials.

Studies of $H_2PO_4^-$ and Cl^- transport in isolated protoplasts have been conducted similarly to those of K^+ transport and by the same researchers.[9-12] A radioactive tracer, $KH_2{}^{32}PO_4$ or $K^{36}Cl$ was included in the incubation medium. After specific time periods, aliquots were removed, washed, pelleted, and counted.

b. Metabolite Transport

Using mesophyll protoplasts isolated from wheat and tobacco, Huber and Moreland[14,15] have studied sugar efflux, and $^{14}CO_2$ fixation and product distribution. For the efflux studies, protoplasts were incubated at 25°C with light in a reaction mixture containing osmoticum (not sucrose), buffers, and $NaNCO_3$. Aliquots were withdrawn, centrifuged for 15 sec at 200 g to pellet the protoplasts, and the supernatants were anlayzed enzymatically for sucrose and hexose (glucose and fructose). The experiments on the distribution of products of $^{24}CO_2$ fixation were conducted in a similar manner except that $NaH^{14}CO_3$ (2 to 5 μCi/μmol) was added to the incubation medium. Following the centrifugation step, the labeled products of fixation in the supernatant and pellet were separated into various fractions by addition of 1 *N* acetic acid followed by ionexchange and paper chromotography.

Guy et al.[16] incubated mesophyll protoplasts from *Pisum sativum* L. var. Dan with ^{14}C-labeled 3-*O*-methyl-D-glucose (OMG), a glucose analog. The labeled protoplasts were centrifuged briefly after incubation, and the supernatant was removed by suction. The pellet was resuspended in incubation medium containing ^3H-labeled OMG. After another centrifugation step, followed by resuspending the pellet in unlabeled OMG, samples were removed and asayed for ^{14}C and ^3H content. All experimentation was conducted at 27°C and under light.

Amino acid uptake by isolated protoplasts has been reported once.[17] Labeled amino acids and α-aminoisobutyric acid (AIB), an amino acid analog, were added in appropriate concentration to protoplasts in a suspension medium. After the uptake period, protoplasts were separated from the radioactive medium by rapid centrifugation through a 6-cm layer of a KCl-$CaCl_2$ mixture into a layer of dibutyl phthalate. Cell fragments and organelles did not enter this layer. The pellet was resuspended in water, and aliquots were then taken for scintillation counting.

Theoretically, protoplasts can be utilized to measure the uptake of any chemicals which might come in contact with a plant cell, i.e., herbicides, insecticides, antibiotics, fungicides, etc. Darmstadt and co-workers[13] have followed the uptake of 2,4-D and atrazine into protoplasts of corn roots. Radiolabeled compounds were added to the protoplast suspension, allowed to incubate for several minutes, and were pelleted through a layer of silicone oil which terminated absorption of the herbicides.

B. Isolated Vacuoles

1. Purity of the Vacuole Preparation

Most procedures,[18-27] with one notable exception,[19] for isolating vacuoles involve the controlled lysis of enzymatically generated protoplasts. Total vacuole yield from protoplasts is generally less than 30% (by number), and it is therefore somewhat difficult to obtain vacuoles in large quantities. The purity of the vacuole preparation can present significant problems for experiments on membrane transport. The two major contaminants in vacuole preparations are plastids and unbroken protoplasts. The numbers of protoplasts or plastids relative to vacuoles are usually 10% or less, but this is a deceptively high level of contamination on a total protein basis. For example, tobacco protoplasts have 10 to 15 times more protein than a vacuole even though the size is similar.[24]

Due to the presence of hydrolytic enzymes in vacuoles, and to the osmotic shock necessary to lyse protoplasts to release vacuoles, stability of the vacuoles from breakage after isolation can be a problem.[20] Vacuole stability can be improved by holding the isolated vacuoles at 4°C.[21,22] After 6 hr at this temperature, 85% of the vacuoles remained intact.[22] Addition of 1 to 10 m*M* mercaptobenzothiazole, low concentrations (1 m*M* of $CaCl_2$, and 0.1 to 1% BSA had little effect on vacuole stability. High concentrations (10 m*M* of $CaCl_2$, 25 m*M* K_2HOP_4, and all tried concentrations of $MgCl_2$ were found to rupture the vacuole. Relatively high concentrations of BSA (5% w/v) stabilized vacuoles slightly.[21]

2. Generalized Procedure for Vacuole Isolation

A variety of methods for isolating and purifying vacuoles from protoplasts have been described.[18-28] In most cases, vacuoles are released from protoplasts by osmotic shock, usually in a 0.1 to 0.3 *M* osmoticum such as mannitol or sorbitol. Chelating agents, reducing agents, and buffers (pH 6.5 to 7.5) are frequently added. The suspension is gently agitated at room temperature for 10 to 15 min and a major portion of the remaining protoplasm is removed by filtration. Vacuoles are then separated from protoplasts and other debris in several different ways, the simplest and probably least effective of which is by a series of centrifugation steps. More common is separation by a step gradient of Ficoll or Percoll which is centrifuged at low g forces (500 to 1000) for 20 to 45 min. Vacuoles are then collected at the appropriate interface.

Boudet et al.[22] spun isolated sweet clover protoplasts through a DEAE dextran and Ficoll gradient to release vacuoles. The released vacuoles were then passed through a layer containing dextran sulfate in order to neutralize any DEAE dextran which might bind to the tonoplast and be harmful to the vacuoles. The vacuoles banded at 5/20% Ficoll interface.

Vacuoles have also been isolated from protoplasts by layering the protoplasts under a silicone oil/sorbitol or a silicone oil/Percoll gradient by forcing them through the needle 5 cm × 0.1 mm) of a 250-μl syringe.[27] This caused lysis of about 40% of the protoplasts. The vacuoles floated to the surface. The total time required for this isolation and purification was 45 to 60 sec.

3. Specific Protocols for Measurement of Solute Transport

a. H^+ Transport

The electrical potential difference across the tonoplast has been described as being in the range 0 to +25 mV (vacuole positive relative to the cytoplasm), yet the pH of the vacuole is usually several units lower than that of the cytoplasm.[1] It appears that H^+ transport into the vacuole is an active, energy-requiring process. H^+ transport into intact vacuoles has been measured primarily by following the uptake of two different radiolabeled compounds. The first commonly used compound, methylamine, is a lipid-permeable weak base.[29-31] Intact vacuoles are incubated in 10 μM ^{14}C-methylamine, the vacuoles are pelleted out, and the radioactivity in them is determined. pH determinations are based on the assumption that the uncharged compounds can easily pass across the membrane whereas charged forms are impermeant. The ratio of charged to uncharged molecules can be calculated with the assumption that the uncharged form is equally distributed on both sides of the membrane. Since the ratio of charged to uncharged is governed by pH, this value can be deduced.

5′-5′-dimethyloxazolidene-2,4-dione (DMO), a weak acid, is another compound which can be used to determine H^+ transport across the tonoplast.[32] Vacuoles are incubated for 5 to 10 min in 2 μM DMO (^{14}C-labeled). The vacuoles are then separated from the medium by silicone oil centrifugation, and internal and external radioactivities are determined. Internal pH values can be calculated on the basis that the uncharged DMO passes through membranes and that the charged DMO is impermeable.

Fluorescent amines have recently been employed to measure H^+ gradients across vesicular membranes.[33,34] These types of probes have been successfully employed in various animal systems (liposomes, gastric microsomes, and sea urchin gametes), and they should be applicable to isolated vacuoles. The probes exhibit: (1) decrease in quantum yield in the presence of a pH gradient, (2) pH dependence, and (3) concentration-dependent shifts in their fluorescence spectra. Quinacrine, 9-aminoacridine, and acridine orange are some of the common probes used.

b. Malate Transport

Organic acids tend to accumulate in large quantities in many higher plant vacuoles. This phenomenon is observed in a variety of instances: when cation uptake exceeds anion uptake from an external medium, following organic acid synthesis by dark CO_2 fixation, when the anions of a salt taken up are metabolized, and during auxin-stimulated H^+/K^+ exchange (the acid growth hypothesis). The organic acids synthesized function as a pH-stat mechanism, as osmotic agents for turgor maintenance, and as a means of CO_2 storage in CAM or C_4 photosynthesis.

Organic acid accumulation, made up largely of malic acid (mal), in the leaf cells of plants having Crassulacean acid metabolism (CAM) can be envisioned as the most dramatic example of organic acid build-up in cells. Electrochemical investigation of intact tissue of CAM plants suggests that there is an active influx of H^+ and $Hmal^-$ into the vacuole, while mal^{2-} seems to be passively distributed across the tonoplast membrane.[35] Two attempts have been made to isolate and purify vacuoles from a CAM plant and to follow malic acid transport in the vacuoles.[36,37] In one report,[36] vacuoles were isolated following $^{14}CO_2$ incubation. A vacuole sample was immediately measured for radioactivity, and it was assumed that all radioactivity was due to malic acid. After several hours, vacuoles were isolated again from the same tissue preincubated with $^{14}CO_2$. The change in radioactivity of vacuoles was measured and assumed to be due to malic acid transport out of the vacuoles.

In the other investigation on this subject,[37] isolated vacuoles from a CAM plant were incubated in 0.67 μCi of ^{14}C-malate at 20°C with gentle agitation. The vacuoles were separated from the incubation medium by allowing them to settle across a discontinuous gradient. The gradients were frozen and then cut for liquid scintillation counting.

c. Sugar Transport

In sugar cane and in red beet, vacuoles accumulate the economically important product, sucrose, and the loading and storage of sugars in these tissues are of interest. The stalk parenchyma cells of sugarcane are particularly interesting as they can store sucrose concentrations of up to 20% fresh weight or 60% dry weight. Most studies of sucrose uptake into the vacuole have been conducted with suspension-cultured cells obtained from stalk parenchyma of sugarcane. In one study,[38] the cultured cells were depleted of sugars for 18 hr prior to the enzymatic release of protoplasts and vacuoles. This depletion reduced sucrose, glucose, and fructose concentrations by about 50%. Uptake into isolated vacuoles was measured using ^{14}C-sugars. Vacuoles were separated from the incubation medium by rapid centrifugation, and radioactivity was determined by scintillation spectrometry. Uptake rates were calculated from the slope of the linear portion of the time course curve. Other workers have used similar procedures to measure ^{14}C-sugar uptake in isolated vacuoles.[20]

III. CHARACTERISTICS OF INORGANIC AND ORGANIC ION TRANSPORT

A. Protoplasts

1. Membrane Potential Difference

An important assumption to be tested is that the removal of the cell wall by enzymic digestion does not significantly alter the transport properties of the protoplast. One parameter which is important in the study of transport phenomena is the membrane potential, which is both a result of and greatly influences the transport of charged and uncharged species.[1] Microelectrodes have frequently been utilized in the measurement of membrane potentials in giant algal and in higher plant cells and indicate potentials of −50 to −200 mV, inside negative. An inherent problem of using microelectrodes to measure membrane potentials in isolated protoplasts is that a method must be developed to hold the protoplast steady so impalement can be achieved. Racusen et al.[39] impaled protoplasts which had been embedded in agar to circumvent the problem, and other workers have used suction micropipettes to hold the protoplasts steady to allow impalement.[40] Initial reports indicated that isolated plant protoplasts had a positive membrane potential with respect to the outside.[39] Also, Heller et al.,[40] experimenting with *Acer pseudoplantanus* L. cells grown in liquid suspension, showed a membrane potential of −22 to −40 mV, inside negative, whereas protoplasts isolated from these same cells had a membrane potential of +10 mV, inside positive. These early results suggested that ion transport properties of plant cells were changed with protoplast isolation.

Briskin and Leonard[41] measured the membrane potential in isolated protoplasts from tobacco suspension cells and obtained a value of about −50 mV, inside negative, when measured for cells held steady by a suction micropipette using microelectrode impalement. This value was not significantly different from that of the cells from which the protoplasts were isolated, and the membrane potential of intact tobacco cells and isolated protoplasts was depolarized similarly by KCN. These results indicated that the enzymic removal of the cell wall caused no significant alteration in the transport properties of the tobacco cell protoplast.

Rubenstein[42] estimated the membrane potential of *Avena sativa* L. cv. "Garry" protoplasts using the lipophilic cation triphenylmethylphosphonium ($TPMP^+$). This procedure utilized the passive uptake of the lipophilic cation whose concentration ratio across the cell membrane at equilibrium is converted to membrane potential by the Nernst equation. Using this method, the membrane potential was estimated to be −62 mV, inside negative, in isolated protoplasts. Protoplasts isolated from *Nitella expansa* were also found to have negative membrane potentials.[43] Hence, it seems that earlier reports of positive membrane potentials resulting from protoplast isolation have not been confirmed by more recent studies. This important membrane transport characteristic of intact cells appears to be retained in isolated protoplasts.

2. Concentration Kinetics

The rate of inorganic ion transport into plant cells responds in a characteristic fashion to increasing concentrations of the ion. The interpretation of such kinetic data has been debated for more than 30 years.[2] Protoplasts isolated from cultured tobacco cells[44] and corn roots[2] retain their unique features with respect to concentration kinetics. This implies that such kinetic data are characteristic of transport through the plasma membrane of individual cells without cell walls and that the process of protoplast isolation has not produced fundamental changes in this aspect of membrane transport.

3. Selectivity

Plant cells also exhibit a preference for essential nutrient ions over similar but nutritionally unessential ones. As far as has been tested,[11,44] isolated protoplasts retain an ability to selectively absorb inorganic ions. For example, K^+ transport into tobacco protoplasts was not reduced by the presence of excess Na^+, while Cl^- transport was reduced by normally competitive Br^- or I^-.[44] As expected, $H_2PO_4^-$ did not inhibit Cl^- transport. In general, addition of Ca^{2+} to absorption solutions improved the selective properties of membrane transport in isolated protoplasts.[44]

4. Sensitivity to Temperature, pH, and Inhibitors

The transport of charged and uncharged ions and metabolites is directly or indirectly influenced by the membrane potential. This electrical potential difference across the plasma membrane is, for the most part, the result of the action of an energy-requiring, electrogenic H^+ pump.[1] Hence, conditions or substances which inhibit the action of the H^+ pump, or reduce the supply of metabolic energy to it, will also inhibit solute transport across the plasma membrane. Many researchers have studied the effects of temperature, pH, and various inhibitors on transport in isolated protoplasts.[9-12,15,16,44,45] In general, protoplasts have retained the expected characteristics with respect to the response of various membrane transport processes to changes in temperature or pH and to addition of the various inhibitors.

B. Vacuoles

The vacuole is commonly regarded as the major storage compartment in the cell for water, organic acids, some amino acids, sugars, phenolics (tannins, alkaloids, glycosides), some inorganics, and various hydrolytic enzymes.[46,47] Due to the apparent specificity of inorganic ions and compounds sequestered in the vacuole, selection at the tonoplast membrane must occur. Major transport functions have been suggested for the vacuole, but many remain only hypothetical as it is very difficult to study transport through the tonoplast membrane of intact cells. Presently, there is no well-established marker enzyme for the tonoplast membrane, so it is difficult to identify tonoplast vesicles in homogenates of plant cells.

It is well known that the pH in the vacuole is at or below 5.5 while that in the cytoplasm is 7.5 or above. Since the electrical potential difference across the tonoplast membrane is near zero or slightly positive, there is a sizable free energy gradient for H^+ between the cytoplasm and the vacuole. It appears that H^+ is accumulated in the vacuole by an ATP-driven H^+ pump in the tonoplast membrane.[1] Transport of other solutes through the tonoplast membrane may be facilitated by various carrier proteins which utilize energy conserved in the H^+ gradient.

The ability to purify intact vacuoles from protoplasts is very important for research on transport processes at the tonoplast membrane. However, initial results indicate that transport properties of isolated vacuoles are different from those expected from studies on transport in vacuoles of intact cells. In isolated vacuoles from sugarcane suspension cells, a membrane potential difference of −80 mV, inside negative, was observed.[32] This potential difference was independent of external pH and addition of a protonophore but was consistently reduced by increasing the external K^+ concentration. This contrasts with *in situ* results indicating a pH gradient of 1.3 units and a membrane potential difference of +50 mV, inside positive.[32] Presumably, removal of cytoplasm from around the tonoplast membrane has altered the transport properties of the vacuole. A similar conclusion was drawn when transport of sugars was studied by these same workers.[38] It may be necessary to suspend isolated vacuoles in a cytoplasm-like solution to study membrane transport properties of the tonoplast.

Other researchers have measured pH differences across the tonoplast membranes of lutoids (vacuo-lysosomal particles) isolated from *Hevea brasiliensis* latex.[29,30] They observed a pH difference of 0.9 to 1.0 unit, inside more acidic, and an electrical potential difference of −70 to −120 mV, inside negative, in isotonic media at physiological K^+ concentration (30 m*M*) and pH 7.0. The pH difference was fully accounted for by the Donnan potential. When 5 m*M* ATP was added to the external medium, an acidification of the internal space (0.7 to 1.0 pH unit) was seen along with a membrane depolarization by 60 mV, interior becoming less negative. This observation supports the existence of an ATP-driven, electrogenic H^+ pump located in the tonoplast membrane.

IV. SOME CASE STUDIES

A. K^+ Transport in Isolated Protoplasts

It has long been recognized that the rate of K^+ absorption into plant tissues as a function of increasing K^+ concentration responds as expected for a carrier-mediated process. However, the kinetic data for K^+ absorption are more complex than expected for a simple carrier-mediated process. The interpretation of these data has been a source of controversy and confusion.[48] The ability to mesure K^+ transport in isolated protoplasts has significantly influenced the understanding of the kinetic data for K^+ transport. For example, the complex kinetic data have been attributed to a diffusion limitation of ions which is a function of cells organized into a tissue.[48] However, recently the nature of the kinetic data for K^+ transport into protoplasts has been shown to be similar to that which has long been observed for various plant tissues. Hence, the complex kinetics of K^+ absorption is clearly due to uptake at the cellular level and is not a function of tissue organization.

B. Malate Transport in CAM Plants

Malate accumulation in the leaf cells of plants having CAM is a dramatic example of organic acid buildup in cells. Cellular compartmentation of synthesized malate into the vacuole must exist as the synthesizing cytoplasmic enzyme phosphoenolpyruvate carboxylase is subject to feedback inhibition by malate. Vacuole isolation from leaf mesophyll cells of CAM plants has strongly suggested that the bulk of malate is localized in the vacuoles.[49,50]

Lüttge and Ball[51] conducted an electrochemical investigation of malic acid transport at the tonoplast of a CAM plant using intact tissue of *Kalanchöe daigremontiana*. They observed that the membrane potential (vacuole to external medium) in the light and the dark is approximately −200 mV; it is reversibly depolarized by uncouplers and respiratory inhibitors; it shows light-dependent transient oscillations observed in all photosynthetic cells; it is independent of the amount of malic acid accumulated in the cells in the range 30 to 140 m*M*, and it is hyperpolarized by fusicoccin. Electrochemical gradients for H^+, $Hmal^-$, and mal^{2-} were calculated using the Nernst equation and assumptions concerning cytoplasmic pH and malate concentrations. These results suggested an active influx of H^+ and $Hmal^-$ into the vacuole and a passive transport of mal^{2-} across the tonoplast. The authors conclude that the most likely mechanism of malic acid accumulation in the vacuoles of CAM plants seems to be an active H^+ transport at the tonoplast coupled with passive movement of mal^{2-} possibly mediated by a translocator. $Hmal^-$ is subsequently formed within the vacuole.

Buser-Suter et al.[37] looked at malic acid transport into vacuoles isolated from the CAM plant *Bryophyllum daigremontiana*. Their vacuole preparation was contaminated less than 5% (by number) by protoplasts, and the calculated malic acid uptake by contaminating chloroplasts, three per vacuole in the worst case, was less than 5% of the rate of uptake. Rates of malic acid uptake in vitro were 25 times lower than that assumed to accumulate in vacuoles in vivo. However, conditions of transport in vitro are quite different from conditions within the living cell. Also, it took 24 hr of preparation to isolate pure vacuoles, and this may cause rates of uptake to decline.

The uptake measured with the isolated vacuoles was not dependent on ATP and was practically unaffected by uncouplers such as 2,4-dinitrophenol and carbonylcyanide *m*-chlorophenylhydrazone. The uncouplers did not affect vacuole stability. Transport was completely inhibited by 1 m*M* $HgCl_2$ which suggested the involvement of a carrier protein, and saturation kinetics of malic acid transport provided a further indication of the involvement of a permease. Malic acid transport in *Bryophyllum* vacuoles has features in common with arginine transport in yeast vacuoles. In yeast, arginine transport is due to the exchange of external arginine with vacuolar arginine. Similarly, the uptake of labeled malic acid by the CAM plant vacuoles seemed to reflect a catalyzed exchange diffusion across the tonoplast. The fact that ATP and uncouplers did not affect malic acid transport supports the hypothesis that an exchange reaction independent of an associated energy source is responsible for transport. The authors[37] postulated that the permease observed catalyzes a facilitated exchange diffusion across the tonoplast which does not result in a net uptake of malic acid into the vacuoles. They concluded that the mechanism generating the driving force for the accumulation and depletion of large amounts of malic acid in CAM plants still remains to be elucidated.

V. USE OF PROTOPLASTS FOR ISOLATION AND CHARACTERIZATION OF MEMBRANES

A. Plasma Membrane

The isolation and biochemical characterization of the plasma membrane is dependent on the separation of the membrane from other cellular components. Plasma membrane isolation is generally achieved by homogenization of plant tissue, or cultured cells, in a buffered osmoticum including various chemicals which serve to preserve the activities of marker enzymes. Upon cell disruption, the plasma membrane forms vesicles which are subsequently separated from other cellular components such as mitochondria, plastids and microbodies, and from other membrane vesicles, i.e., those formed from Golgi apparatus, endoplasmic reticulum, and tonoplast, by a series of density gradient centrifugation steps,[52] or, more recently, by a partitioning in a two-phase polymer system.[53,54] Positive indentification of vesicles in subcellular fractions as plasma membrane vesicles is difficult.[52] It would be desirable to specifically label the plasma membrane prior to the cell fractionation procedure. With isolated protoplasts, one should be able to radioactively label the plasma membrane so that it can be readily identified during cell fractionation.

Galbraith and Northcote[55] used diazotized ^{35}S-sulfanilic acid to label the plasma membrane of protoplasts isolated from soybean cells. This label was selected because the reaction occurs at neutral pH and shows broad substrate specificity without denaturation of the labeled proteins. They observed that most of the radioactivity incorporated into the protoplasts was recovered in particulate rather than soluble fractions of lysed protoplasts, suggesting that the label did not penetrate across the plasma membrane. Most of the radioactivity (83.1%) sedimented following a low-speed centrifugation, indicating that when protoplasts are lysed, most of the plasma membrane remains as large sheets or vesicles. The radioactivity remaining in the supernatant was apparently associated with smaller vesicles of plasma membrane which equilibrated to a buoyant density of about 1.14 g/cc in a sucrose gradient. The distribution of radioactively labeled plasma membrane vesicles correspond to the distribution of ATPase activity, indicating that this enzyme is localized on the plasma membrane. These basic findings were confirmed by Perlin and Spanswick.[56] Hence, it appears that with isolated protoplasts, it is possible to specifically label the plasma membrane so that it can readily be identified during subsequent cell fractionation.

There is a significant problem which must be overcome if isolated protoplasts are going to be used routinely to obtain plasma membrane vesicles in high yield. While gentle lysis of the protoplast preserves organelle integrity, it can also produce large vesicles of plasma membrane with significant amounts of cytoplasm and organelles trapped within. These large plasma membrane bags of cellular components confound attempts to purify plasma membrane vesicles. Apparently, this difficulty is not encountered with protoplasts from all plant sources because Ruesink and colleagues[57,58] obtained plasma membrane vesicles in good yield from cultured carrot cells. Nonetheless, this difficulty can limit the usefulness of protoplasts for isolating large quantities of plasma membrane vesicles.

B. Tonoplast Membrane

A specific biochemical marker for the tonoplast membrane has not, as yet, been discovered, so it is not possible to positively identify vesicles of tonoplast in membrane fractions obtained from conventional plant cell fractionation. However, tonoplast vesicles can be obtained from a suspension of intact, purified vacuoles, since it should be the only membrane present in such a vacuole preparation. This has proved to be more difficult than it would appear because vacuoles obtained from protoplasts are often contaminated by intact or broken protoplasts. Since there is much more membrane per protoplast than per vacuole, membranes pelleted from a vacuole preparation with an apparently low protoplast contamination by number can nonetheless be dominated by membranes which are not of tonoplast origin.[25,26] Hence, one cannot simply prepare vacuoles, break them, pellet the membranes, and automatically conclude that tonoplast vesicles have been purified. It is essential to first demonstrate that the vacuole preparation is not significantly contaminated by other cellular membranes. Unfortunately, this precaution has not been rigorously applied by many investigators. It is difficult to judge whether or not a parameter measured for membranes pelleted from a vacuole preparation is, in fact, characteristic of the tonoplast membrane. Relatively little information is available on the biochemical characteristics of membranes derived from a vacuole preparation.[59]

Briskin and Leonard[26] isolated vacuoles from protoplasts of cultured tobacco cells such that contamination by protoplasts was reduced to less than 1% (by number). Lipids of the vacuole membrane were labeled with ^{3}H-choline, and the sonicated vacuole preparation was centrifuged in a sucrose density gradient. Vesicles of tonoplast (identified by ^{3}H-choline labeling) had a peak density of about 1.12 g cm^{-3}. This indicates that tonoplast vesicles have a lower effective density in sucrose gradients than plasma membrane vesicles (1.14 to 1.17 g cm^{-3}) and are not likely to significantly contaminate plasma membrane preparations.

VI. SUMMARY AND FUTURE PROSPECTS

The ability to obtain large quantities of viable protoplasts from a variety of plant tissues and species has proven to be important for research on membrane transport in plants. Protoplasts have been used to study cellular mechanisms of ion and metabolite transport and as a starting material for the isolation of intact vacuoles. In the former instance, protoplasts offer a useful model system for research on the mechanism of solute transport through the plasma membrane. In the latter case, protoplasts represent the major way to study the storage and transport properties of the vacuole.

The results available, although not extensive in scope, indicate that transport properties of the plasma membrane are not changed in a fundamental way by protoplast isolation. Initial results indicated that the normally negative electrical membrane potential difference across the plasma membrane was completely depolarized to a small positive value during protoplast isolation. A change in membrane polarity would be expected to have a major influence on the transport of a charged solute. However, this result was not confirmed by later research and is not sustained by observations that the transport of ions into isolated protoplasts has characteristics (and rates) comparable to those for intact cells.

Surprisingly, it has proven to be difficult to isolate plasma membrane vesicles in large quantities from protoplasts. This problem has reduced the utility of protoplasts for studies on plasma membrane composition and structure. The development of procedures for purification of the plasma membrane from protoplasts will greatly increase the impact of research with isolated protoplasts on plasma membrane structure. It should also facilitate an assessment of alterations in the plasma membrane which might occur during preparation of protoplasts.

Virtually all procedures for purification of intact vacuoles from mature plant cells utilize isolated protoplasts as the starting material. Tonoplast vesicles can be prepared from isolated vacuoles, although contamination by other cellular membranes is a significant and sometimes unrecognized problem. The availability of protoplasts has had and will continue to have a significant impact on research on membrane transport processes at the tonoplast of the plant vacuole.

REFERENCES

1. Loenard, R. T., Membrane-associated ATPases and nutrient absorption by roots, in *Advances in Plant Nutrition*, Vol. 1, Tinker, P. B. and Läuchli, A., Eds., Praeger Scientific, New York, 1984, 209.
2. Kochian, L. V. and Lucas, W. J., Potassium transport in corn roots. II. The significance of the root periphery, *Plant Physiol.*, 73, 208, 1983.
3. Cocking, E. C., Plant cell protoplasts-isolation and development, *Annu. Rev. Plant Physiol.*, 23, 29, 1972.
4. Larkin, P. J., Purification and viability determinations of plant protoplasts, *Planta*, 128, 213, 1976.
5. Wagner, G. J., Butcher, H. C., and Siegelman, H. W., The plant protoplast: a useful tool for plant research and student instruction, *Bioscience*, 28, 95, 1978.

6. Evans, P. K., Keates, A. G., and Cocking, E. C., Isolation of protoplasts from cereal leaves, *Planta*, 104, 178, 1972.
7. Uchimiya, H. and Murashige, T., Evaluation of parameters in the isolation of viable protoplasts from cultured tobacco cells, *Plant Physiol.*, 54, 936, 1974.
8. Huber, S. C. and Edwards, G. E., An evaluation of some parameters required for the enzymatic isolation of cells and protoplasts with CO_2 fixation capacity from C_3 and C_4 grasses, *Physiol. Plant.*, 35, 203, 1975.
9. Lin, W., Corn root protoplasts: isolation and general characterization of ion transport, *Plant Physiol.*, 66, 550, 1980.
10. Gronwald, J. W. and Leonard, R. T., Isolation and transport properties of protoplasts from cortical cells of corn roots, *Plant Physiol.*, 70, 1391, 1982.
11. Mettler, I. J. and Leonard, R. T., Ion transport in isolated protoplasts from tobacco suspension cells. I. General characteristics, *Plant Physiol.*, 63, 183, 1979.
12. Taylor, A. R. D. and Hall, J. L., Some physiological properties of protoplasts isolated from maize and tobacco tissues, *J. Exp. Bot.*, 27, 383, 1976.
13. Darmstadt, G. L., Balke, N. E., and Schrader, L. E., Use of corn root protoplasts in herbicide absorption studies, *Pest. Biochem. Physiol.*, 19, 172, 1983.
14. Huber, S. C. and Moreland, D. E., Translocation: efflux of sugars across the plasmalemma of mesophyll protoplasts, *Plant Physiol.*, 65, 560, 1979.
15. Huber, S. C. and Moreland, D. E., Co-transport of potassium and sugars across the plasmalemma of mesophyll protoplasts, *Plant Physiol.*, 67, 163, 1981.
16. Guy, M., Reinhold, L., and Rahat, M., Energization of the sugar transport mechanism in the plasmalemma of isolated mesophyll protoplasts, *Plant Physiol.*, 65, 550, 1980.
17. Guy, M., Reinhold, L., and Laties, G. G., Membrane transport of sugars and amino acids in isolated protoplasts, *Plant Physiol.*, 61, 593, 1978.
18. Wagner, G. J. and Siegelman, H. W., Large scale isolation of intact vacuoles and isolation of chloroplasts from protoplasts of mature tissues, *Science*, 190, 1298, 1975.
19. Leigh, R. A. and Branton, D., Isolation of vacuoles from root storage tissue of *Beta vulgaris* L., *Plant Physiol.*, 58, 656, 1976.
20. Knuth, M. E., Keith, B., Clark, C., Garcia-Martinez J. L., and Rappaport, L., Stabilization and transport capacity of cowpea and barley vacuoles, *Plant Cell Physiol.*, 24, 423, 1983.
21. Kringstad, R., Kenyon, W. H., and Black, C. C., Jr., The rapid isolation of vacuoles from leaves of Crassulacean acid metabolism plants, *Plant Physiol.*, 66, 379, 1980.
22. Boudet, A. M., Canut, H., and Alibert, G., Isolation and characterization of vacuoles from *Melilotus alba* mesophyll, *Plant Physiol.*, 68, 1354, 1981.
23. Walker-Simmons, M. and Ryan, C. A., Immunological identification of proteinase inhibitors I and II in isolated tomato leaf vacuoles, *Plant Physiol.*, 60, 61, 1977.
24. Mettler, I. J. and Leonard, R. T., Isolation and partial characterization of vacuoles from tobacco protoplasts, *Plant Physiol.*, 64, 1114, 1979.
25. Boller, T. and Kende, H., Hydrolytic enzymes in the central vacuole of plant cells, *Plant Physiol.*, 63, 1123, 1979.
26. Briskin, D. P. and Leonard, R. T., Isolation of tonoplast vesicles from tobacco protoplasts, *Plant Physiol.*, 66, 684, 1980.
27. Kaiser, G., Martinoia, E., and Weimken, A., Rapid appearance of photosynthetic products in the vacuoles isolated from barley mesophyll protoplasts by a new fast method, *Z. Pflanzenphysiol.*, 107, 103, 1982.
28. Thom, M., Maretzki, A., and Komor, E., Vacuoles from sugarcane suspension cultures. I. Isolation and partial characterization, *Plant Physiol.*, 69, 1315, 1982.
29. Cretin, H., The proton gradient across the vacuo-lysosomal membrane of lutoids from the latex of *Hevea brasiliensis*. I. Further evidence for a proton-translocating ATPase on the vacuo-lysosomal membrane of intact lutoids, *J. Membr. Biol.*, 65, 175, 1982.
30. Marin, B., Marin-Lanza, M., and Komor, E., The protonmotive potential difference across the vacuo-lysosomal membrane of *Hevea brasiliensis* (rubber tree) and its modification by a membrane-bound adenosine triphosphatase, *Biochem. J.*, 198, 365, 1981.
31. Nishimira, M., pH in vacuoles isolated from castor bean endosperm, *Plant Physiol.*, 70, 742, 1982.
32. Komor, E., Thom, M., and Maretzki, A., Vacuoles from sugarcane suspension culture. III. Protonmotive potential difference, *Plant Physiol.*, 69, 1326, 1982.
33. Bennett, A. B. and Spanswick, R. M., Optical measurements of δ pH and δ π in corn root membrane vesicles: kinetic analysis of Cl^- effects on a proton-translocating ATPase, *J. Membr. Biol.*, 71, 95, 1983.
34. Sze, H., Proton-pumping adenosine triphosphatase in membrane vesicles of tobacco callus, *Biochem. Biophys. Acta*, 732, 586, 1983.

35. Lüttge, U., Smith, J. A. C., Marigo, G., and Osmond, C. B., Energetics of malate accumulation in the vacuoles of *Kalanchöe tubiflora* cells, *FEBS Lett.*, 126, 81, 1981.
36. Kringstad, R., Kenyon, W. H., and Black, C. C., Jr., The rapid isolation of vacuoles from leaves of Crassulacean acid metabolism plants, *Plant Physiol.*, 66, 379, 1980.
37. Buser-Suter, C., Weimken, A., and Matile, P., A malic acid permease in isolated vacuoles of a Crassulacean acid metabolism plant, *Plant Physiol.*, 69, 456, 1982.
38. Thom, M., Komor, E., and Maretzki, A., Vacuoles from sugarcane suspension cultures. II. Characterization of sugar uptake, *Plant Physiol.*, 69, 1320, 1982.
39. Racusen, R. H., Kinnersley, A. M., and Galston, A. W., Osmotically induced changes in electrical properties of plant protoplast membranes, *Science*, 198, 405, 1977.
40. Heller, R., Grignon, C., and Rona, J. P., Importance of the cell wall in the thermodynamic equilibrium of ions in free cells of *Acer psuedoplatanus*, in *Membrane Transport in Plants*, Zimmermann, J. and Dainty, J., Eds., Springer-Verlag, Berlin, 1974, 239.
41. Briskin, D. P. and Leonard, R. T., Ion transport in isolated protoplasts from tobacco suspension cells. III. Membrane potential, *Plant Physiol.*, 64, 959, 1979.
42. Rubenstein, B., Use of lipophilic cations to measure the membrane potential of oat leaf protoplasts, *Plant Physiol.*, 62, 927, 1978.
43. Abe, S., Takeda, J., and Senda, M., Resting membrane potential and action potential of *Nitella expansa* protoplasts, *Plant Cell Physiol.*, 21, 537, 1980.
44. Mettler, I. J. and Leonard, R. T., Ion transport in isolated protoplasts from tobacco suspension cells. II. Selectivity and kinetics, *Plant Physiol.*, 63, 191, 1979.
45. Lin, W., Inhibition of anion transport in corn root protoplasts, *Plant Physiol.*, 68, 435, 1981.
46. Matile, P., Biochemistry and function of vacuoles, *Annu. Rev. Plant Physiol.*, 29, 193, 1978.
47. Leigh, R. A., Methods, progress and potential for the use of isolated vacuoles in studies of solute transport in higher plant cells, *Physiol. Plant.*, 57, 390, 1983.
48. Kochian, L. V., and Lucas, W. J., Potassium transport in corn roots. I. Resolution of kinetics into a saturable and linear component, *Plant Physiol.*, 70, 1723, 1982.
49. Buser, C. and Matile, P., Malic acid in vacuoles isolated from *Bryophyllum* leaf cells, *Z. Pflanzenphysiol.*, 82, 462, 1977.
50. Kenyon, W. H., Kringstad, R., and Black, C. C., Jr., Diurnal changes in the malic acid content of vacuoles isolated from leaves of the Crassulacean acid metabolism plant, *Sedum telephium*, *FEBS Lett.*, 94, 281, 1978.
51. Lüttge, U. and Ball, E., Electrochemical investigation of active malic acid transport at the tonoplast into the vacuoles of the CAM plant *Kalanchöe daigremontiana*, *J. Membr. Biol.*, 47, 401, 1979.
52. Hall, J. L., Plasma membranes, in *Isolation of Membranes and Organelles from Plant Cells*, Hall, J. L. and Moore, A. L., Eds., Academic Press, London, 1983, chap. 3.
53. Widell, S., Lundborg, T., and Larsson, C., Plasma membranes from oats prepared by partition in an aqueous polymer two-phase system, *Plant Physiol.*, 70, 1429, 1982.
54. Uemura, M. and Yoshida, A., Isolation and identification of plasma membrane from light-grown winter rye seedlings (*Secale cereale* L. cv Puma), *Plant Physiol.*, 73, 586, 1983.
55. Galbraith, D. W. and Northcote, D. H., The isolation of plasma membrane from protoplasts of soybean suspension cultures, *J. Cell Sci.*, 24, 295, 1977.
56. Perlin, D. S. and Spanswick, R. M., Labeling and isolation of plasma membrane from corn leaf protoplasts, *Plant Physiol.*, 65, 1053, 1980.
57. Boss, W. F. and Ruesink, A. W., Isolation and characterization of concanavalin A-labeled plasma membranes of carrot protoplasts, *Plant Physiol.*, 64, 1005, 1979.
58. Randall, S. K. and Ruesink, A. W., Orientation and integrity of plasma membrane vesicles obtained from carrot protoplasts, *Plant Physiol.*, 73, 385, 1983.
59. Marty, F. and Branton, D., Analytical characterization of beetroot vacuole membrane, *J. Cell Biol.*, 87, 72, 1980.

Chapter 9

PROTOPLASTS IN VIROLOGY

Fusao Motoyoshi

TABLE OF CONTENTS

I. INTRODUCTION

In the field of plant virus studies, the introduction of plant protoplast systems, which are comparable to investigating bacteriophages with bacterial cells and animal viruses with tissue cultures, has made it possible to do more critical studies on the functions and interactions of plant viruses in living cells. Previously, only whole plants or excised tissues were available.

The potential usefulness of plant protoplasts for studies of virus infection was first demonstrated by Cocking[1,2] in his electron microscopic observations. These suggested that isolated tomato fruit locule protoplasts had the ability to take up an inoculum of tobacco mosaic virus (TMV). A few years later, evidence that TMV could multiply in plant protoplasts was presented by a number of authors.[3-5] These pioneering works were soon followed by the establishment of an excellent method for infection of tobacco mesophyll protoplasts with TMV RNA and TMV particles as described in the two papers by Aoki and Takebe,[6] and Takebe and Otsuki.[7] This method, which uses leaf mesophyll protoplasts, was so effective that only minor modifications were needed to apply it to other types of viruses or protoplasts of other species.

Following the successful infection of tobacco leaf mesophyll protoplasts with TMV,[6-8] a rod-shaped virus, some spherical viruses, such as cucumber mosaic virus (CMV) and cowpea chlorotic mottle virus (CCMV), and filamentous viruses such as potato virus X (PVX) were also shown to infect tobacco leaf mesophyll protoplasts.[9-12]

From the early 1970s, some viruses, such as brome mosaic virus (BMV), CCMV, alfalfa mosaic virus (AlMV), cowpea mosaic virus (CPMV), tobacco rattle virus (TRV), etc. attracted special attention, because these viruses are composed of more than one nucleoprotein component, into which the virus genome is divided. These viruses are also infectious in protoplast systems.[10,13-20]

Several viruses with specific characteristics were also successfully induced to infect protoplasts. Turnip yellow mosaic virus (TYMV) induced the formation of polyplasts (aggregates of chloroplasts) in inoculated *Brassica* protoplasts,[21] and the virus progeny produced appeared to associate with the chloroplast membrane.[22] Tobacco necrotic dwarf virus (TNDV), which is localized in the phloem of infected plants, could also infect tobacco leaf mesophyll protoplasts.[23] Cauliflower mosaic virus (CaMV) is a virus with potential usefulness as a vector in genetic engineering of higher plants, since it has a double-stranded DNA nucleoprotein. Its nucleoprotein and its naked DNA were shown to be infectious in turnip protoplasts.[24-26] Viroids are pathogens existing as free RNA molecules of 120,000 to 127,000 daltons and are single-stranded covalently closed circles. Some viroids have also successfully infected tomato protoplasts.[27]

Protoplast systems have been established for many virus species, and studies are now progressing to elucidate how viruses function in protoplasts and how protoplasts respond to viral infection. Actively replicating viral RNA species and translation products from viral genomes in infected protoplasts have been critically analyzed in several viruses. In addition, translation products in protoplasts have been compared to those produced in vitro, either in wheat germ extracts or rabbit reticulocyte lysates.

Molecular biological analysis of viral RNAs and proteins may help determine the mechanisms of virus behavior in living cells. In addition these findings in protoplast systems should lead us to understand what happens in tissues or whole plants after virus infection.

In order to see what roles these protoplast systems have played in the progress of plant virology, this chapter will concentrate on four subjects: (1) protoplast systems for plant virus studies; (2) functions of viruses in protoplasts; (3) virus-to-virus interaction; and (4) functions of virus resistance genes of hosts. Some of the subjects concerning virus infection of plant protoplast have been reviewed also by different authors.[28-30]

II. PROTOPLAST SYSTEMS FOR PLANT VIRUS STUDIES

A. Protoplast Sources

Leaf mesophyll cells are the most common source of protoplasts for virus studies. Freshly isolated protoplasts from expanded tobacco leaves using a two-step digestion method (see Chapter 1) provide a fine system for virus infection, since the protoplasts thus isolated are mostly derived from palisade cells which are uniform in size and physiological state.[3] In many other plants, however, mixtures of protoplasts from palisade and spongy mesophyll cells are usually used because of the difficulty in collecting palisade cells in the first steps of digestion. In addition to mesophyll cells, protoplasts isolated from cells in suspension cultures have recently been utilized.[31-35] Protoplasts from other sources have rarely been used. The only examples known so far are tomato fruit locule protoplasts[5] and tobacco leaf epidermal protoplasts.[36]

B. Methods of Inoculation

Inoculation is usually carried out by exposing protoplasts to a virus or viral RNA inoculum in a buffered mannitol solution in the presence of poly-L-ornithine.[7] Poly-L-ornithine that has a strong positive charge is usually essential because it binds negatively charged viruses or viral RNAs to the protoplast membrane, the surface of which is also negatively charged.[30] Some viruses such as pea enation mosaic virus (PEMV) and BMV, which are positively charged at the pH of inoculum, do not require poly-L-ornithine to infect protoplasts.[13,14,37]

An alternative method uses liposomes which encapsulate the viruses or viral RNAs and carry them into the protoplasts.[33-35,38,39] The liposomes become attached to the protoplast membrane by using procedures developed for cell fusion. This method is especially useful for inoculating protoplasts with viral RNAs, since the liposomes protect the RNA from enzymatic degradation.

C. Methods for Assaying Virus Amounts

The traditional local lesion assay is still important for estimating infectivities in extracts of infected protoplasts after a desired incubation period. Virus yields can also be assayed by sucrose gradient analysis. Crude extracts in an appropriate buffer are layered on top of sucrose density gradient columns, centrifuged, fractionated, and the virus quantities are calculated by measuring extinction at 260 nm.[40]

Microplate methods of enzyme-linked immunosorbent assay (ELISA), which were applied by Clark and Adams[41] for detecting some plant viruses, can be used to assay virus yield in infected protoplasts.[42] This "double antibody sandwich" form of ELISA is capable of detecting a low concentration of virus (1 to 10 ng/mℓ) in extracts. Another advantage of this method is that purification steps are not needed for its application, since almost all cell components in the extracts are washed away from the virus antigen, which is bound to the antibody that is attached to polystyrene microplates. Therefore, few contaminants are present when the alkaline phosphatase-conjugated antibody is bound to the antigen, and the phosphate substrate is added.

The method of fluorescent antibody staining, applied by Otsuki and Takebe[43] to detect virus in protoplasts, has become a standard method for assessing the proportions of virus-infected protoplasts in an inoculated protoplast population.

Electron microscopy is also useful in several respects. There have been a number of electron microscopic studies concerned with the location of viruses at early and late infection stages and with the ultrastructural changes induced by virus infection in protoplasts.[2,5,44-51] A special use of electron microscopy was developed by Otsuki and Takebe,[52] who observed mixed coated virus progeny in protoplasts infected with two different strains of TMV. They stained these virus particles with strain-specific immunoglobulin preparations for their electron microscopic observations.

D. Protoplasts as Systems for Studying Virus Function

In protoplast systems, virus particles infect a large proportion of cells simultaneously and proceed synchronously in one-step growth. Protoplast membranes are directly exposed to liquid media, so that they are able to take up small molecules in the culture medium easily. Hence it is easy to label newly synthesized RNAs and proteins with radioisotopes, such as inorganic ^{32}P, or ^{35}S, and ^{3}H or ^{14}C-labeled nucleotides or amino acids. "Pulse-chase" experiments examining the incorporation of radioisotopes into RNAs or proteins may be much easier to do with protoplasts than with whole plant tissues. In addition, metabolic inhibitors may act more efectively on protoplasts than on plant tissues.

Kinetic analysis of nucleic acid, protein, and virion synthesis can be made more precisely using protoplasts than with leaf tissues. However, viral suppression of host RNA and protein synthesis is usually not as complete in protoplasts as in plant tissues, confounding clear-cut differentiation of virus-specific RNAs and proteins from host RNAs and proteins.

Actinomycin D can be used effectively to determine virus-specific RNA synthesis. This antibiotic inhibits the DNA-dependent RNA synthesis in host cells, but does not substantially affect viral RNA synthesis of some viruses, such as CCMV and TMV, in protoplasts.[53,54] Hence it is very useful for identifying not only viral RNA but also its replicative form (RF) and its replicative intermediate (RI). These are detectable by polyacrylamide-gel electrophoresis of phenol extracts of protoplasts that are cultured in media containing ^{32}P-labeled phosphate or ^{3}H- or ^{14}C-labeled uridine.[53,54] This antibiotic is however rather ineffective in suppressing host protein synthesis.[55]

Several attempts have been made to identify virus-coded proteins in protoplasts. Sakai and Takebe[56] demonstrated that irradiation of TMV-inoculated and noninoculated protoplasts for a few minutes with UV light from a germicidal lamp reduced host protein synthesis and allowed them to detect TMV-coded proteins. This UV irradiation technique was successfully applied to several other virus-protoplast combinations,[57-59] but Paterson and Knight[60] reported that TMV-coded protein synthesis in tobacco protoplasts was decreased by UV irradiation almost to the same extent as host protein synthesis. They used only chloramphenicol, which did not inhibit virus multiplication, to suppress host protein synthesis, but the reduction in host protein synthesis was only 25 to 30%.

Subcellular fractionation of homogenates of virus-infected protoplasts facilitates the identification of virus-coded proteins when combined with the UV irradiation method.[61]

Another method to identify virus-coded proteins is to use protoplasts of two plant species, distant in relation to each other, but capable of being hosts of the virus in question.[62] This method is based on the assumption that these two types of protoplasts will produce disimilar host-coded proteins, which will exhibit distinct electropherogram patterns. Since the virus in both protoplasts is identical, its protein bands should be in identical positions. Tobacco and cowpea protoplasts are a suitable pair for analyzing viral protein synthesis of TMV.[62]

E. Special Uses of Protoplasts in Plant Virology

One interesting question is whether the host range of a virus species is wider at the level of protoplasts than at the level of intact plants. There is only one example where nonhost plant-derived protoplasts could be infected with a particular virus. This example was described by Furusawa and Okuno,[63] who found that mesophyll protoplasts isolated from Japanese radish, a nonhost BMV, could be infected with a standard strain of BMV. In general, it appears that if a plant is a host of a virus, then the protoplasts isolated from that plant can also host the virus. Conversely, if the plant is a nonhost, the protoplasts are usually nonhosts, although there may be many apparent exceptions. For example, tobacco plants are not common hosts of PEMV, but both protoplasts and plants can be infected with this virus, although infected plants do not express any symptoms.[13] Tobacco is usually a nonhost of BMV. Neither protoplasts nor leaves of tobacco could be infected with the Norwich culture of wild type BMV. With BMV V5, a variant of BMV, however, both protoplasts and leaves of tobacco were infected.[14] Both leaf mesophyll protoplasts and intact plants of tobacco could be infected with cucumber green mottle mosaic virus (CGMMV), although tobacco is not an usual host of this virus.[64] TYMV could infect Chinese cabbage protoplasts, but failed to infect tobacco protoplasts.[30]

Tobacco leaf mesophyll protoplasts are most commonly used in virus studies. Various viruses have been known to infect them, irrespective of whether tobacco is their main host or not. Using tobacco leaf mesophyll protoplasts, one can do comparative studies of the functions and behavior of quite a number of viruses in common host cells.

Since the infection frequency of individual cells is extremely high in protoplast systems compared to inoculation of leaf tissues, one can obtain a high proportion of protoplasts infected with two different viruses, either when the protoplasts are inoculated sequentially or with a virus mixtures.[65-67] Furthermore, triple infection of protoplasts with three different viruses is also possible.[65] Infecting protoplasts with two viruses which are genetically related, but have distinctly different host ranges, is one of the special uses of protoplasts.[67] This type of work would be difficult if leaf tissues were used, because of the low frequency of infection and the complexity of the system. Problems of virus-to-virus interaction in protoplasts will be discussed later.

III. FUNCTIONS OF VIRUSES IN PROTOPLASTS

A. Tobacco Mosaic Virus (TMV)

1. RNA replication

Before the establishment of protoplast systems, excised leaf tissues were the most useful materials for investigating virus functions in living cells. Nilsson-Tillgren et al.[68] demonstrated that tobacco leaves showing veinclearing symptoms contained a majority of cells in the same stage of infection. Using threse leaves, Nilsson-Tillgren[69,70] analyzed the RNAs produced in the cells, and detected two forms of presumable TMV RNA precursors, in addition to the single stranded TMV RNA, a replicative form (RF) and a replicative intermediate (RI). The former was a double-stranded RNA consisting of a full length plus-strand RNA and a full-length minus-strand RNA, and the latter was also a double-stranded RNA but had single-stranded tails of different length.

A critical analysis of the synthesis of these TMV-specific RNAs was subsequently carried out by Aoki and Takebe[54] using tobacco mesophyll protoplasts. This assured TMV infection in frequencies higher than 80%, and resulted in a single cycle of synchronous replication of the virus. They also found three types of TMV-specific RNA using polyacrylamide-gel electrophoresis, and identified them as single-stranded TMV RNA and double-stranded RF and RI, based on their behavior in cellulose chromatography, their solubility in NaCl solution, and the molecular sizes and structure of products resulting from denaturation with dimethylsulfoxide and digestion with RNase. Furthermore they followed the time course of the synthesis of these RNAs and virus particles. They found three successive phases with respect to viral RNA synthesis and particle formation. In the initial phase (0- to 10-hr postinoculation), both RNA synthesis and particle formation start and progress exponentially. However, the start of particle formation and its exponential increase are delayed about 4 to 5 hr behind the start of RNA synthesis and its exponential increase. In the intermediate phase (10- to 20-hr postinoculation), the rate of RNA synthesis becomes slower while virus particle synthesis continues to increase exponentially, and in the final phase (20- to 35-hr postinoculation), the rate of production of particles slows down and becomes parallel to those of the RNA. The time course of RF and RI synthesis is similar to that of viral RNA. This observation and the results from some pulse-chase experiments suggest that RF and RI are precursors of viral RNA in infected protoplasts.

Two other types of TMV-specific RNAs present in TMV-infected leaves have been reported. One of them is a small RNA of 350,000 daltons termed LMC (low-molecular-weight-component), which was first found by Jackson et al.[71] Beachy and Zaitlin[72] detected TMV-specific small RNAs of several sizes in fractions containing either membrane bound or cytoplasmic (free) polysomes, a considerable proportion of which were LMC, and noticed that they could serve as mRNA for the coat protein. Subsequent studies concerning in vitro translation of TMV-specific RNAs extracted from infected tobacco leaves revealed that LMC was the sole mRNA for the coat protein and was synthesized as 7mGppp-capped RNA fragments containing about 700 nucleotide residues having a sequence homology at the 3′ end of TMV RNA.[73,75]

Another type of RNA is I_2-RNA (intermediate-length RNA), which has been detected in purified preparations of TMV particles.[76,77] I_2-RNA contains about 15,000 nucleotide residues, including the 3′-terminus of TMV RNA,[75] and a reading frame for a protein of 29,987 daltons (30 K protein)[74,75] that may play important roles in the activity of TMV in living cells or tissues.

Confirmation of the existence of these TMV-specific small RNAs in protoplast systems has been considerably delayed. Aoki and Takebe[54] did not find any sign of production of these small RNAs in tobacco leaf mesophyll protoplasts infected with TMV. Recently, however, Ogawa et al[78] succeeded in detecting LMC and other minor components in extracts of TMV-infected tobacco mesophyll protoplasts by modifying the techniques of Aoki and Takebe.[54] The main modifications were as follows: (1) concentration of actinomycin D in protoplast culture medium was raised from 10 $\mu g/m\ell$ to 20 $\mu g/m\ell$; (2) tris-HCl buffer (pH 7.5) was used for RNA extraction instead of glycine buffer (pH 9), the latter being very ineffective in extracting LMC; (3) denaturation of the RNA samples with 4 *M* urea prior to electrophoresis greatly increased the yield of LMC; and (4) more sensitive devices were used to scan for the radioactivity of small RNA in gels. With these modifications, they could detect LMC in the total RNA extracted from TMV-infected protoplasts. LMC was also found in both free and membrane-bound polyribosomes, but it was present in free polyribosomes in much larger amounts, especially associated with small-sized polyribosomes such as monosomes to tetrasomes.

In further study, Ogawa and Sakai[78] examined the synthesis of LMC in relation to the production of other TMV-specific RNAs, proteins, and progeny particles, and found that the synthesis of RI, RF, and a high-molecular-weight protein (140 K) preceded the synthesis of LMC and TMV RNA by 4 to 6 hr. Synthesis of LMC was just a little earlier than TMV RNA synthesis in timing and was followed by the synthesis of coat protein and the formation of progeny particles with a time-lag of 6 to 8 hr. They also inferred from time course analysis of the occurrence of LMC in polyribosomes that LMC was actively functioning as the messenger for coat protein synthesis in protoplasts.

The occurrence of an RNA species characterized by the same electrophoretic mobility as I_2-RNA in TMV-inoculated protoplasts derived from tobacco cell cultures was recently demonstrated by Watanabe et al.[80] They identified conclusively that this RNA is the counterpart in protoplasts of I_2-RNA from purified TMV preparations by using Southern hybridization to TMV cDNA fragments including I_2 sequences. Furthermore, they observed that this RNA and its translation product, the 30 K protein, were transiently synthesized from 2- to 7- to 9-hr postinoculation, in contrast to other TMV-specific RNAs and proteins, which are more continuously synthesized.

2. Protein Synthesis

In protoplasts infected with TMV, three TMV-coded protein species, 183 K, 126 K, and 17.5 K (coat protein), have been repeatedly observed,[56,57,60,62,79-81] and the occurrence of another protein species, the 30 K protein, has been occasionally reported.[62,80,82] The molecular weight of the two larger species had been variously estimated by different authors to be 160,000 to 180,000 daltons and 130,000 to 160,000 daltons, respectively. However, their precise molecular weights were recently deduced from the nucleotide sequences of TMV RNA as being 183,253 daltons (183 K) and 125,941 daltons (126 K), respectively. In the same way, the molecular weight of the intermediate-sized species was deduced to be 29,892 daltons (30 K).[75]

In vitro translation studies revealed that the 183 K and 126 K proteins were translation products from genomic RNA or TMV; the larger having a common N terminus with the smaller, and is produced by reading through the termination codon (UAG) for the latter protein.[83] The coat protein and 30 K protein were translation products from LMC and I_2-RNA, respectively.[73,84] The time course synthesis of these proteins and its relation to RNA synthesis in protoplasts have been examined as mentioned earlier.[79,80]

It was hypothesized that 30 K protein has a function related to the ability of the virus to move from cell to cell.[84] The functions of the 126 K and its read-through 183 K product remain obscure.

3. Assembly of RNA and Coat Protein into Virus Particles

Although the assembly process has been studied in detail using in vitro systems, it is still essential to determine how the virus particles are formed in living cells.

Regarding in vivo assembly of TMV, however, there is only one brief abstract described by Otsuki.[85] In this study, he prepared samples containing putative assembly intermediates by sucrose-gradient fractionation of extracts from tobacco mesophyll protoplasts infected with the strain TMV T and incubated for 24 hr. He then incubated these samples in the presence of excess TMV OM coat protein under in vitro reconstitution conditions. After the incubation, the samples were treated with either TMV OM-specific antibody or TMV T-specific antibody and observed under the electron microscope to examine the location of these antibodies on the virus particles. A large number of the TMV particles thus obtained were composed of both of these two types of coat proteins. The OM protein was located at the 3′ end within 30 nm of the end, and the T protein occupied the remaining part. From this result he assumed that particles whose RNA remained uncoated at 3′ ends were assembly intermediates in protoplasts 24-hr postinoculation.

4. Cell-to-Cell Movement

Control of virus movement from cell to cell may be associated with host cell controlled mechanisms, while the virus may also have some control over its cell-to-cell movement in interaction with normal cell functions. However, little is known about the mechanisms of cell-to-cell movement of virus.

Nishiguchi et al.[86] used protoplasts to examine the ability of a virus to move from cell to cell. In this study, they used TMV Ls1, a *ts* mutant of TMV, and TMV L, the parent strain of Ls1. In tomato leaf discs, Ls1 can grow normally at low temperatures (22° C), but show only limited growth at higher temperatures (32° C). In leaf mesophyll protoplasts, however, this mutant showed no temperature sensitivity and multiplied at 32° C as rapidly as the L strain did. In their experiment, tomato leaf discs were prepared in which the mutant had just started multiplication at the permissive temperature. These discs were then incubated in medium for 12 hr either at 22 or at 32° C. As expected, the *ts* mutant showed limited growth at 32° C, but grew normally at 22° C. They isolated protoplasts from some discs before starting incubation, and also from other discs after the 12-hr incubation period. These protoplasts were incubated for 2 days to allow the virus to multiply and to accumulate sufficient antigen to be detected by fluorescent antibody. The protoplasts were then fixed and stained with fluorescent antibody to determine the percentage of protoplasts infected. These percentages were used to estimate the percentages of cells infected in the leaf discs at the beginning and end of incubation. The estimated percentage of infected cells for the *ts* mutant was substantially the same at the beginning and end of incubation at the nonpermissive temperature, while it was much higher at the end of incubation at the permissive temperature. The percent of infected cells was also higher at the end of incubation for TMV L at both temperatures. From these results, they concluded that the *ts* mutant could multiply in already infected cells, but could not spread from cell to cell at the nonpermissive temperature.

The malfunction of cell-to-cell movement of Ls1 at the nonpermissive temperature was also confirmed in epidermal tissues which were infected with this mutant and stained with fluorescent antibody.[87] Shalla et al.[88] showed that the number of plasmodesmata decreased when Ls1 infected leaves were exposed to the nonpermissive temperature. Thaliansky et al.[89] observed that Ls1 could move from cell to cell even at nonpermissive temperatures, if it was inoculated with a mixture of either normal TMV or PVX. That is, the defect of Ls1 could be complemented by the normal functions of TMV or PVX. Leonard and Zaitlin[84] analyzed products translated from Ls1 RNA and L RNA in vitro and found that the 30 K protein translated from Ls1 I_2-RNA had amino acid substitution(s). Therefore, the mechanism of TMV cell-to-cell movement may be related to the function of the 30 K protein.

B. Turnip Yellow Mosaic Virus (TYMV)

The genome of TYMV is a single-stranded RNA which is encapsidated to form one nucleoprotein species (B_1), and in addition, TYMV has empty protein shells (T), and minor nucleoproteins containing an efficient coat protein mRNA.[90] Sugimura and Matthews[22] investigated the time course of production of B_1, T, and the minor nucleoprotein components in Chinese cabbage protoplasts and obtained results supporting an hypothesis of Hatta and Matthews[91] which proposed that B_1 and T become assembled with protein subunits in different locations; the former are produced by encapsidation of progeny genomic RNA at vesicles in the outer chloroplast membrane formed by infection, and the latter are produced by spontaneous assembly at any site where there is a sufficient concentration of coat protein. They also reported that the minor nucleoprotein fraction known to contain a substantial proportion of coat mRNA appears to be assembled earlier than T, B_1, and the other minor components.

C. Viruses with Multipartite Genomes

Protoplast systems have been established for investigating a number of viruses with multipartite genomes, i.e., viruses composed of more than one species of nucleoprotein into which one virus genome is partitioned. RNA replication and protein synthesis in protoplasts have been studied in some of the typical viruses having multipartite genomes. One advantage in using protoplasts for the study of viruses with multipartite genomes is that protoplasts can be inoculated with individual nucleoproteins or RNA components, which have been separated by gradient centrifugation or electrophoresis. Using such techniques virus functions have been studied for several viruses including AlMV,[20,92,93] BMV,[94] CPMV,[95,98] cowpea severe mosaic virus (CPsMV),[99] tobacco black ring virus (TBRV),[100] and TRV.[59] Among these viruses, AlMV, CpMV, and TRV have been investigated in some detail.

1. Alfalfa Mosaic Virus (AlMV)

AlMV has a genome partitioned into three RNA components which are encapsidated into separate nucleoproteins with different sedimentability. RNA 1 is encapsidated into the bottom component (B), RNA 2 into the middle component (M), and RNA 3 into the top component b (Tb).[101] Besides these nucleoproteins, another component, top component a (Ta), is also present in AlMV preparations.[101] Ta contains RNA 4, which has a sequence identical to the part of RNA 3 at the 3′ end. RNA 4 functions as the messenger for coat protein translation.[102]

AlMV virions were produced in cowpea leaf mesophyll protoplasts either when protoplasts were inoculated with unfractionated virus or with a mixture of the three nucleoprotein components, B, M, and Tb.[20,92] No viral RNA synthesis, however, occurred when protoplasts were inoculated with AlMV RNAs 1, 2, and 3 without coat protein.[94] Thus the presense of coat protein is essential for initiating infection.

As in other single-stranded RNA viruses, AlMV RNAs appear to be synthesized via partially double-stranded forms composed of plus-strand and complementary minus-strand RNA. For detecting RNA synthesis in protoplasts, Nassuth and Bol[93] separated extracted RNA by agarose-gel electrophoresis and carried out Northern hybridization with ^{32}P-labeled cDNA of AlMV RNA, and ^{32}P-labeled viral genomic RNA as probes. The former probe was used for detecting AlMV RNA (plus-strand RNA) and the latter for detecting its minus-strand RNA.

Using this technique, Nassuth and Bol[93] found that protoplasts that were inoculated with unfractionated virus or a mixture of B, M, and Tb contained plus-strand RNAs primarily in single-stranded forms, and the amount of plus-strand RNAs in replicative structures and the amount of minus-strand RNAs were below the detection limits. In protoplasts inoculated with a mixture of only B and M, however, the synthesis of plus-strand RNAs was greatly reduced, and the production of minus-strand RNAs was much larger, almost equal to that of the plus-strand RNAs. No detectable viral RNA synthesis was observed when protoplasts were inoculated with mixtures of only B and Tb, or M and Tb. They concluded that the expression of genomic RNAs 1 and 2 resulted in the formation of replicase activity that produced roughly equal amounts of virus plus- and minus-strand RNAs, and that an RNA 3-encoded product, possibly the coat protein, was responsible for the switch to asymmetric production of viral plus-strand RNAs. They could detect no minus-strand RNA 4, and suggested that RNA 4 did not replicate but was transcribed from the RNA 3 minus-strand.

2. Cowpea Mosaic Virus (CPMV)

CPMV is an isometric virus that has single-stranded RNAs and belongs to the comovirus group. Its genome is divided into two RNA species which are encapsidated separately into two nucleoprotein components, middle (M) and bottom (B), that are separable by density gradient centrifugation.[103] The protein capsids of both the M and B components consist of two protein species.[104,105] Both of the RNA species, called M RNA and B RNA, have poly A at their 3′-termini and a virus-coded protein, VPg, covalently bound to their 5′-termini.[106]

In vivo virus-coded protein synthesis has been studied in detail by van Kammen's group using cowpea protoplasts. These results have been compared with those obtained from in vitro translation using either wheat germ extracts or rabbit reticulocyte lysates to deduce the mechanisms of protein synthesis. In protoplasts inoculated with CPMV, at least eight virus-specific proteins with molecular weights of 170,000, 110,000, 87,000, 84,000, 60,000, 37,000, 32,000, and 23,000 daltons have been detected.[95-98] In addition two proteins of 130,000 and 112,000 daltons were occasionally detected in very small amounts.[95] No virus specific proteins were detected in protoplasts inoculated with M alone, which should contain the codes for the capsid proteins of 37,000 and 23,000 daltons. However, all virus-specific proteins, except for the two capsid proteins, were synthesized in protoplasts inoculated with B alone.[96-98] From these results, the authors suggested that M RNA codes for both capsid proteins, while B RNA directs the synthesis of the proteins involved in the early events of virus replication, such as replicase or its subunit production.

Franssen et al.[107] showed that M RNA was translated into two overlapping polypeptides of 95,000 and 105,000 daltons in in vitro systems and that these polypeptides were further processed into four smaller polypeptides by treatment with a supernatant fraction (30,000 × g) from cowpea mesophyll protoplasts inoculated either with CPMV or its B component. Furthermore they found that the polypeptide of 60,000 daltons among these four polypeptides may be the precursor of both capsid proteins, since it reacted with antisera for both capsid proteins. From these results, two remarkable characteristics of CPMV protein synthesis were inferred: (1) both capsid proteins are produced by successive cleavages, probably from one large precursor polypeptide; and (2) proteolytic activity is associated with the B component-specific polypeptide(s).

Pelham[108] showed that in the rabbit reticulocyte system the primary translation product of B RNA is a polypeptide of approximately 205,000 daltons, which is cleaved into two polypeptides of 170,000 and 32,000 daltons. Rezelman et al.[98] confirmed this in vitro by observing the production of a large polypeptide of 200,000 daltons, which was cleaved into two polypeptides of 170,000 and 32,000 daltons. Based on the comparison of these products with B component-specific polypeptides produced in protoplasts, they concluded that the large protein of 200,000 daltons is probably a polyprotein from which all of the B component-specific proteins arise by successive cleavages. Goldbach et al.[109,110] indicated that VPg is encoded in the polypeptide of 170,000 daltons as well as in its cleavage product of 60,000 daltons.

3. Tobacco Rattle Virus (TRV)

The genome of the Campinas (CAM) strain of TRV is distributed into two RNA species, RNA 1 of 2.4×10^6 daltons and RNA 2 of 0.7×10^6 daltons.[111] These RNAs are encapsidated into nucleoprotein rods of length 197 and 51 nm, respectively.[112]

Synthesis of polypeptides induced by TRV in tobacco protoplasts was investigated by Mayo.[59] He found two polypeptides of 187,000 and 142,000 daltons in protoplasts that were inoculated either with a mixture of long and short rods or with long rods only. These polypeptides migrated to positions close to the in vitro translation products of RNA 1 in polyacrylamide-gel electrophoresis. Thus they are probably coded by RNA 1. The codes for these polypeptides must overlap in some way, because their combined molecular weight exceeds the coding capacity of RNA 1. In addition to these polypeptides, a polypeptide of 31,000 daltons was detected in protoplasts inoculated with long and short rods, but not in those inoculated with long rods only. This polypeptide comigrated with the TRV coat protein. Other RNA 2-specific polypeptides were not detected. He also demonstrated that translation of RNA 1 in protoplasts occurred much sooner after inoculation than did that of RNA 2.

IV. VIRUS-TO-VIRUS INTERACTION

A form of virus-to-virus interaction known as "interference" or "cross protection" occurs in which a virus strain prevents the multiplication or symptom expression of another virus strain in a plant. This has attracted great attention in plant virus studies, not only from a theoretical perspective but also with respect to its potential usefulness in crop protection. Crops might be inoculated with an avirulent virus strain which protects them against infection by virulent strains.

Interference usually has been found in the interaction between two strains of virus or between two related viruses. A number of hypotheses have been proposed regarding the mechanism of interference; for example, the two strains mutually exclude each other in the infection of individual cells resulting in the multiplication of one virus in a cell,[113] specific recognition of the RNA of the second virus and prevention of its replication by the first virus-coded replicase,[114] or inhibition of uncoating of the second virus by the free coat protein of the first virus.[115]

Otsuki and Takebe[66] used the tobacco protoplast system to study interaction of two TMV strains, one of which was OM, a tobacco strain, and the other T, a tomato strain. Protoplasts were inoculated with mixtures or sequentially with OM and T, incubated for 24 hr, and stained with OM-specific fluorescent antibody, T-specific antibody, or a mixture of both. With this procedure the percentage of doubly infected protoplasts could be estimated. It was found that T and OM could multiply simultaneously in one cell without apparent competition, as long as the protoplasts were inoculated with both strains in a 1:1 ratio. This finding contradicts the exclusion hypothesis proposed by Siegel.[113] An antagonistic interaction, however, occurred when protoplasts were inoculated with mixtures in which one strain was more common than the other. Furthermore they showed that protoplasts previously inoculated with one strain became resistant to infection by the other strain within a short period. Although these results suggest that interference between the two strains of TMV was occurring at the level of the protoplast, its mechanism still remains to be elucidated.

Watts and Dawson[67] demonstrated interaction in an interesting set of experiments using tobacco protoplasts doubly infected with two related bromoviruses, BMV V5 strain and CCMV. Both viruses synthesized an infectious set of RNAs, RNA 1, 2, and 3 as well as RNA 4, a subgenomic RNA generated from RNA 3, when the protoplasts were singly inoculated with either virus. In doubly inoculated protoplasts, BMV synthesized infectious progeny, but CCMV failed to synthesize detectable quantities of infectious particles, although CCMV capsids were abundantly produced. All attempts to rescue a temperature sensitive *ts* strain of CCMV when inoculated with BMV were unsuccessful.

Sakai et al.[116] investigated these phenomena more recently and reported that the CCMV capsids that were produced in the doubly infected protoplasts contained only RNA 3, but not RNA 1 or 2. In addition the RNA 3-coded CCMV coat protein and a protein of 35,000 daltons were found, the amounts of the latter protein sometimes exceeding the amounts of the corresponding protein coded for by BMV RNA. From these results they concluded that BMV may exploit the gene products of CCMV while suppressing the synthesis of infectious CCMV particles. Although this type of interference may be specific to BMV and CCMV, these experiments demonstrate the usefulness of protoplast systems for investigating the mechanism of interference between viruses on a moelcular basis.

V. FUNCTIONS OF VIRUS RESISTANCE GENES

A. Tobacco

Otsuki et al.[117] inoculated protoplasts from two tobacco varieties, Xanthi nc, which possesses the N gene responsible for TMV-specific necrotization, and Samsun, a susceptible variety in which TMV can spread systemically and causes mosaic symptoms. They observed that no necrotization of the Xanthi nc protoplasts occurred and the yield of virus was similar to that in Samsun. Consequently, they assumed that the expression of necrotization is related to the interaction of infected cells with neighboring cells in intact tissues.

Loebenstein et al.,[118] however, showed that the 2,4-D that was included in the media used by Otsuki et al.[117] enhanced TMV replication in protoplasts isolated from two necrotic-responding tobacco varieties, Samsun NN and Xanthi nc, but inhibited multiplication in protoplasts from a susceptible variety, Samsun. But they could not observe necrotization in protoplasts possessing the N gene and thus suggested that necrotization and the suppression of virus multiplication are separate processes.

Continuing investigation into the mechanism of the suppression of virus multiplication, Loebenstein and Gera[119] found that some substance(s) inhibiting TMV multiplication (IVR) was released into the media from TMV-infected protoplasts of Samsun NN, which is homozygous for the N gene. They partially purified IVR using the $ZnAc_2$ precipitation method for interferon purification, and found two peaks having inhibitory activity in Sephadex® G-75 filtrates of the partially purified material. These peaks corresponded to molecular weights of approximately 26,000 to 27,000 and 56,000 to 57,000 daltons. IVR was characterized to be active at extremely low concentrations, stable at pH 2.5, and not affected at relatively high concentrations of $CaCl_2$. Actinomycin D and chloramphenicol appeared to inhibit the production of IVR in TMV-infected protoplasts that were isolated form necrotic-responding tobacco varieties.[120] Though IVR may be one of the important substances that is involved in the mechanisms of TMV inhibition in the tobacco varieties possessing the *N* gene, however, it still remains to be determined what is the host-specific reaction triggered by primary contact of TMV with the cells and what mechanisms induce the necrotization.

Imaizumi et al.[121] examined the expression of other TMV-resistance factors in protoplasts isolated from tobacco leaves of two varieties, Ambalema and T. I. 245. The onset of virus increase in these protoplasts was somewhat delayed, compared to that in protoplasts from a susceptible variety, Bright Yellow. Both the proportion of infected protoplasts and virus yields in these protoplasts were lower than in protoplasts of the susceptible variety.

B. Tomato

Motoyoshi and Oshima[122,123] reported that the genes *Tm-2* and *Tm-2²* in tomato control strong resistance to TMV infection in intact plants, but fail to suppress virus replication in protoplasts inoculated with TMV L, a common tomato strain of TMV, and incubated in a medium. Similar results were reported by Stobbs and MacNeil,[124] who used protoplasts isolated from plants homozygous for *Tm-2.*

In intact plants, *Tm-2* and *Tm-2²* may confer hypersensitivity to TMV, because these genes sometimes cause necrosis on inoculated leaves or systemic necrosis in response to some strains of TMV under certain genotypic and environmental conditions.[125,126] Such a necrotic response is, however, not always visible, even when the plants manifest strong resistance to TMV. Taliansky et al.[127] observed that *Tm-2* tomato lines became susceptible and the virus spread from inoculated leaves to upper leaves after preinfection with a helper virus (PVX). Based on this result, they concluded that the function of *Tm-2* is possibly to block the cell-to-cell movement of the virus. However it is also possible that the blockage of virus movement occurs from a hypersensitive response of *Tm-2* induced at the time of TMV infection. The essential function of *Tm-2* therefore still remains unknown.

The gene *Tm-1* induces tolerance in intact tomato plants to infection by TMV.[125,126]

Low levels of TMV multiplication can occur, but symptom expression was suppressed. Motoyoshi and Oshima[123] however observed that protoplasts and leaf discs prepared from homozygous *Tm-1* plants showed very strong resistance to TMV L and no progeny TMV L was detected after 3 days of incubation.

These authors further investigated the expression of *Tm-1* in protoplasts using mainly two TMV strains, L and CH2. CH2 is a strain that is not affected by *Tm-1* and multiplies to similar levels in both types of protoplasts which are isolated from the susceptible plants and from the resistant plants possessing *Tm-1*.[123]

The results that Motoyoshi and Oshima[123,128-130] obtained account for several characteristics of the *Tm-1* gene: (1) *Tm-1* is dominant in expression of TMV L resistance. In heterozygous protoplasts, however, the resistance was somewhat weakened and allowed TMV L to multiply to a very low level. (2) The expression of resistance is not inhibited by actinomycin D added into the medium at the onset of incubation. It is possible that *Tm-1* is constitutively expressing its effects in protoplasts as well as in intact plants. (3) In *Tm-1* protoplasts, virus progeny are not detectable when inoculated with L RNA and L particles, whereas a large amount of virus progeny is produced when inoculated with CH2 RNA and CH2 particles. Therefore *Tm-1* does not appear to inhibit virus uncoating or the processes prior to virus uncoating. (4) Neither infective RNA nor coat protein accumulates in detectable amounts in *Tm-1* protoplasts. The effects of *Tm-1* therefore may act on the inoculum RNA itself or on some stage of RNA replication. (5) CH2 could not help L (or Ls1, a *ts* mutant isolated from L) to multiply in the presence of *Tm-1* when protoplasts were doubly inoculated with both strains. It is thus possible that *Tm-1* acts directly on L RNA (or Ls1 RNA) rather than on products translated from L RNA (or Ls1 RNA).

C. Cowpea

Beier et al.[131] observed that replication curves of CPMV strain SB were similar in two types of cowpea protoplasts, when the protoplasts were inoculated just after isolation. One host was a systemic host, Blackeye 5, and the other a local lesion host, Chinese red x Iron. However, when protoplasts were isolated from leaves of the local lesion host that were already inoculatd with CPMV at least 45 hr earlier and incubated for an additional 40 hr, the virus infectivity associated with the protoplasts at the time of isolation was less. From these results they concluded that some kind of hypersensitivity was induced in the local lesion host that influenced virus replication in protoplasts after isolation.

In addition to the hypersensitive variety, they also examined CPMV infection of protoplasts isolated from cowpea varieties immune to the virus. First they tested for CPMV resistance in protoplasts isolated from 55 lines of cowpea that showed immunity to mechanical inoculation with CPMV SB. They found that protoplasts from 54 of these lines permitted the virus to multiply to high yields, but one cowpea line, Arlington, was highly resistant to the virus strain even at the level of protoplasts.[132] In further detailed study, Beier et al.[131] demonstrated that the growth curves of CPMV in protoplasts isolated from the three immune cowpea lines, Guarentana, Brabham(2), and Black, were similar to that in protoplasts isolated from a susceptible cowpea line, Blackeye 5, except that the onset of virus increase was delayed by 10 to 12 hr. However, in Arlington, only a low level of replication occurred in protoplasts, and infectivity associated with the protoplasts at 24-hr postinoculation was only about 1% that in a similarly inoculated susceptible line, Blackeye 5.

D. Cucumber

Coutts and Wood[133] used leaf mesophyll protoplasts isolated from a CMV-resistant variety, China K, and a susceptible variety, Ashley, to examine virus-resistance expression at the protoplast level. In protoplasts from the resistant variety, they could not recover either infectious RNA or nucleoprotein at 48-hr postinoculation, while infectivity was recovered from protoplasts prepared from the susceptible variety. This result showed that the resistance of China K to CMV was expressible in protoplasts as well as in intact plants.

VI. CONCLUSION

The protoplast systems discussed in this chapter are especially useful now in studying plant pathology of virus disease and the molecular biology of viruses and plants.

In plant pathological studies, the isolated protoplasts from leaf tissues are excellent materials for studying virus infection, since they probably maintain some of the properties of the tissues from which they were isolated. In addition, protoplasts can be directly inoculated with virus and these can be regarded as a model of single-cell infection. Protoplasts can also be isolated from leaves that are already infected with virus, and these may be useful for studying the response of the host to the virus, since these cells have probably undergone some metabolic changes in response to virus infection in the leaves. However disease expression after virus infection, such as the manifestation of mosaic symptoms, deforming, stunting, etc. are caused by systemic processes in whole plants, so protoplasts have limits as experimental systems in completely understanding the mechanisms of disease expression. On the positive side, it is noteworthy that almost all products in the virus-specific genetic codes and at least a part of the host products specifically induced by virus infection are possibly produced in infected protoplasts. Some of these host products may play important roles in the expression of or, conversely, in the defense against diseases. More comprehensive studies at the levels of protoplasts, tissues, organs, and whole plants will be needed to understand disease and defense mechanisms.

From the viewpoint of molecular biology, some of the unknown functions of the molecular components in plant cells may be elucidated in the progress of virus studies. For example, the molecular analysis of viral genomic and subgenomic RNAs will contribute to the understanding of the structure and function of the messenger RNAs of higher plants, since virus plus-strand RNAs and the messenger RNAs of higher plants have similar structures, but the former can be more easily manipulated than the latter. Viroid RNAs are also interesting materials in their molecular biological functions,[134] and are probably related to some of the host RNA components.[135]

Finally the usefulness of virus mutants should be noted, although there have been only a few studies in which virus mutants were used in protoplast systems.[86,130,136,137] The use of mutants, especially temperature-sensitive mutants and mutants overcoming host resistance may facilitate our understanding of the mechanisms of virus functions and host-cell interaction.

REFERENCES

1. Cocking, E. C., Ferritin and tobacco mosaic virus uptake; and nuclear cytoplasmic relationships in isolated tomato fruit protoplasts, *Biochem. J.*, 95, 28, 1965.
2. Cocking, E. C., An electron microscopic study of the initial stages of infection of isolated tomato fruit protoplasts by tobacco mosaic virus, *Planta*, 68, 206, 1966.
3. Takebe, I., Otsuki, Y., and Aoki, S., Isolation of tobacco mesophyll cells in intact and active state, *Plant Cell Physiol.*, 9, 115, 1968.

4. Hibi, T., Yora, K., and Asuyama, H., TMV infection of *Petunia* protoplasts, *Ann. Phytopathol. Soc. Jpn.*, 34, 1968.
5. Cocking, E. C. and Pojnar, E., An electron microscopic study of the infection of isolated tomato fruit protoplasts by tobacco mosaic virus, *J. Gen. Virol.*, 4, 305, 1969.
6. Aoki, S. and Takebe, I., Infection of tobacco mesophyll protoplasts by tobacco mosaic virus ribonucleic acid, *Virology*, 39, 439, 1969.
7. Takebe, I. and Otsuki, Y., Infection of tobacco mesophyll protoplasts by tobacco mosaic virus, *Proc. Natl. Acad. Sci. U.S.A.*, 64, 843, 1969.
8. Coutts, R. H. A., Cocking, E. C., and Kassanis, B., Infection of tobacco mesophyll protoplasts with tobacco mosaic virus, *J. Gen. Virol.*, 17, 289, 1972.
9. Otsuki, Y. and Takebe, I., Infection of tobacco mesophyll protoplasts by cucumber mosaic virus, *Virology*, 52, 433, 1973.
10. Motoyoshi, F., Bancroft, J. B., Watts, J. W., and Burgess, J., The infection of tobacco protoplasts with cowpea chlorotic mottle virus and its RNA, *J. Gen. Virol.*, 20, 177, 1973.
11. Shalla, T. A. and Petersen, L. J., Infection of isolated plant protoplasts with potato virus X, *Phytopathology*, 63, 1125, 1973.
12. Otsuki, Y., Takebe, I., Honda, Y., Kajita, S., and Matsui, C., Infection of tobacco mesophyll protoplasts by potato virus X., *J. Gen. Virol.*, 22, 375, 1974.
13. Motoyoshi, F. and Hull, R., The infection of tobacco protoplasts with pea enation mosaic virus, *J. Gen. Virol.*, 24, 89, 1974.
14. Motoyoshi, F., Bancroft, J. B., and Watts, J. W., The infection of tobacco protoplasts with a variant of brome mosaic virus, *J. Gen Virol.*, 25, 31, 1974.
15. Motoyoshi, F., Hull, R., and Flack, I. H., Infection of tobacco mesophyll protoplasts by alfalfa mosaic virus, *J. Gen. Virol.*, 27, 263, 1975.
16. Beier, H. and Bruening, G., The use of an abrasive in the isolation of cowpea leaf protoplasts which support the multiplication of cowpea mosaic virus, *Virology*, 64, 272, 1975.
17. Hibi, T., Rezelman, G., and van Kammen, A., Infection of cowpea mesophyll protoplasts with cowpea mosaic virus, *Virology*, 64, 308, 1975.
18. Okuno, T., Furusawa, I., and Hiruki, C., Infection of barley protoplasts with brome mosaic virus, *Phytopathology*, 67, 610, 1977.
19. Kubo, S., Harrison, B. D., Robinson, D. J., and Mayo, M. A., Tobacco rattle virus in tobacco mesophyll protoplasts: infection and virus multiplication, *J. Gen. Virol.*, 27, 293, 1975.
20. Alblas, F. and Bol, J. F., Factors influencing the infection of cowpea mesophyll protoplasts by alfalfa mosaic virus, *J. Gen. Virol.*, 36, 175, 1977.
21. Renaudin, J., Bové, J. M., Otsuki, Y., and Takebe, T., Infection of *Brassica* leaf protoplasts by turnip yellow mosaic virus, *Mol. Gen. Genet.*, 141, 59, 1975.
22. Sugimura, Y. and Matthews, R. E. F., Timing of the synthesis of empty shells and minor nucleoproteins in relation to turnip yellow mosaic virus synthesis in *Brassica* protoplasts, *Virology*, 112, 70, 1981.
23. Kubo, S. and Takanami, Y., Infection of tobacco mesophyll protoplasts with tobacco necrotic dwarf virus, a phloem-limited virus, *J. Gen. Virol.*, 42, 387, 1979.
24. Furusawa, I., Yamaoka, N., Okuno, T., Yamamoto, M., Kohno, M., Kunoh, H., Infection of turnip protoplasts with cauliflower mosaic virus, *J. Gen. Virol.*, 48, 431, 1980.
25. Yamaoka, N., Furusawa, I., and Yamamoto, M., Infection of turnip protoplasts with cauliflower mosaic virus DNA, *Virology*, 122, 503, 1982.
26. Sakai, F. and Shohara, Y., Transfection of turnip protoplasts by cauliflower mosaic virus DNA, in *Proc. 5th Int. Congr. Plant Tissue Cell Culture, Plant Tissue Culture 1982*, Fujiwara, A., Ed., Maruzen Co., Tokyo, 1982, 669.
27. Mühlbach, H.-P., Camacho-Henriquez, A., and Sänger, H. L., Infection of tomato protoplasts by ribonucleic acid of tobacco mosaic virus and by viroids, *Phytopathol. Z.*, 90, 289, 1977.
28. Zaitlin, M. and Beachy, R. N., The use of protoplasts and separated cells in plant virus research, *Adv. Virus Res.*, 19, 1, 1974.
29. Sarker, S., Use of protoplasts for plant virus studies, in *Methods in Virology*, Vol. 6, Maramorosch, K. and Koprowski, H., Eds., Academic Press, New York, 1977, 435.
30. Takebe, I., Protoplasts in the study of plant virus replication, in *Comprehensive Virology*, Vol. 11, Fraenkel-Conrat, H. and Wagner, R. R., Eds., Plenum Publishing, New York, 1977, 237.
31. Jarvis, N. P. and Murakishi, H. H., Infection of protoplasts from soybean cell culture with southern bean mosaic and cowpea mosaic viruses, *J. Gen. Virol.*, 48, 365, 1980.
32. Lesney, M. S. and Murakishi, H. H., Infection of soybean protoplasts from cell suspension culture with bean pod mottle virus, *J. Gen. Virol.*, 57, 387, 1981.
33. Fukunaga, F., Nagata, T., and Takebe, I., Liposome-mediated infection of plant protoplasts with tobacco mosaic virus RNA, *Virology*, 113, 752, 1981.
34. Kikkawa, H., Nagata, T., Matsui, C., and Takebe, I., Infection of protoplasts from tobacco suspension cultures by tobacco mosaic virus, *J. Gen. Virol.*, 63, 451, 1982.

35. Watanabe, Y., Ohno, T., and Okada, Y., Virus multiplication in tobacco protoplasts inoculated with tobacco mosaic virus RNA encapsulated in large unilamellar vesicle liposomes, *Virology*, 120, 478, 1982.
36. Fannin, F. F. and Shaw, J. G., Infection of tobacco leaf epidermal protoplasts with tobacco mosaic virus, *Virology*, 123, 323, 1982.
37. Okuno, T. and Furusawa, I., Modes of infection of barley protoplasts with brome mosaic virus, *J. Gen. Virol.*, 38, 409, 1978.
38. Nagata, T., Okada, K., Takebe, I., and Matsui, C., Delivery of tobacco mosaic virus RNA into plant protoplasts mediated by reverse-phase evaporation vesicles (liposomes), *Mol. Gen. Genet.*, 184, 161, 1981.
39. Rollo, F. and Hull, R., Liposome-mediated infection of turnip rosette virus and RNA, *J. Gen. Virol.*, 60, 359, 1982.
40. Motoyoshi, F., Bancroft, J. B., and Watts, J. W., A direct estimate of the number of cowpea chlorotic mottle virus particles observed by tobacco protoplasts that become infected, *J. Gen. Virol.*, 21, 159, 1973.
41. Clark, M. F. and Adams, A. N., Characteristics of the microplate method of enzyme-linked immunosorbent assay for the detection of plant viruses, *J. Gen. Virol.*, 34, 475, 1977.
42. Mayo, M. A. and Barker, H., Defective multiplication of a pseudorecombinant of tomato black ring virus in tobacco protoplasts, *J. Gen. Virol.*, 64, 649, 1983.
43. Otsuki, Y. and Takebe, I., Fluorescent antibody staining of tobacco mosaic antigen in tobacco mesophyll protoplasts, *Virology*, 38, 497, 1969.
44. Otsuki, Y., Takebe, I., Honda, Y., and Matsui, C., Ultrastructure of infection of tobacco mesophyll protoplasts by tobacco mosaic virus, *Virology*, 49, 188, 1972.
45. Hibi, T. and Yora, K., Electron microscopy of tobacco mosaic virus infection in tobacco mesophyll protoplasts, *Ann. Phytopathol. Soc. Jpn.*, 38, 350, 1972.
46. Burgess, J., Motoyoshi, F., and Flemming, E. N., Effect of poly-L-ornithine on isolated tobacco mesophyll protoplasts: evidence against stimulated pinocytosis, *Planta*, 111, 199, 1973.
47. Burgess, J., Motoyoshi, F., and Flemming, E. N., The mechanism of infection of plant protoplasts by viruses, *Planta*, 112, 323, 1973.
48. Burgess, J., Motoyoshi, F., and Flemming, E. N., Structural changes accompanying infection of tobacco protoplasts with two spherical viruses, *Planta*, 117, 133, 1974.
49. Honda, Y., Matsui, Y., Otsuki, Y., and Takebe, I., Ultrastructure of tobacco mesophyll protoplasts inoculated with cucumber mosaic virus, *Phytopathology*, 64, 30, 1974.
50. Honda, Y., Kajita, S., Matsui, C., Otsuki, Y., and Takebe, I., An ultrastructural study of the infection of tobacco mesophyll protoplasts by potato virus X, *Phytopathol. Z.*, 84, 66, 1975.
51. Harrison, B. D., Kubo, S., Robinson, D. J., and Hutcheson, A. L., The multiplication cycle of tobacco rattle virus in tobacco mesophyll protoplasts, *J. Gen. Virol.*, 33, 237, 1976.
52. Otsuki, Y. and Takebe, I., Production of mixedly coated particles in tobacco mesophyll protoplasts doubly infected by strains of tobacco mosaic virus, *Virology*, 84, 162, 1978.
53. Bancroft, J. B., Motoyoshi, F., Watts, J. W., and Dawson, J. R. O., Cowpea chlorotic mottle and brome mosaic viruses in tobacco protoplasts, in *2nd John Innes Symp. Modification of the Information Content of Plant Cells*, Markham, R., Davis, D. R., Hopwood, D. A., and Horne, R. W., Eds., North-Holland, Amsterdam, 1975, 133.
54. Aoki, S. and Takebe, I., 1975, Replication of tobacco mosaic virus RNA in tobacco mesophyll protoplasts inoculated *in vitro*, *Virology*, 65, 343, 1975.
55. Sakai, F. and Takebe, I., RNA and protein synthesis in protoplasts isolated from tobacco leaves, *Biochim. Biophys, Acta*, 224, 531, 1970.
56. Sakai, F. and Takebe, I., A non-coat protein synthesized in tobacco mesophyll protoplasts infected by tobacco mosaic virus, *Mol Gen. Genet.*, 118, 93, 1972.
57. Sakai, F., Dawson, J. R. O., Watts, J. W., and Bancroft, J. B., Synthesis of proteins in tobacco protoplasts infected with cowpea chlorotic mottle virus, *J. Gen. Virol.*, 34, 285, 1977.
58. Sakai, F., Dawson, J. R. O., and Watts, J. W., Synthesis of proteins in tobacco protoplasts infected with brome mosaic virus, *J. Gen. Virol.*, 42, 323, 1979.
59. Mayo, M. A., Polypeptides induced by tobacco rattle virus during multiplication in tobacco protoplasts, *Intervirology*, 17, 240, 1982.
60. Paterson, R. and Knight, C. A., Protein synthesis in tobacco protoplasts infected with tobacco mosaic virus, *Virology*, 64, 10, 1975.
61. Sakai, F. and Takebe, I., Protein synthesis in tobacco mesophyll protoplasts induced by tobacco mosaic virus infection, *Virology*, 62, 426, 1974.
62. Huber, R., Proteins synthesized in tobacco mosaic virus infected protoplasts, in *Mededelingen Landbouwhogeschool*, Wageningen, No. 79-15, 1, 1979.
63. Furusawa, I. and Okuno, T., Infection with BMV of mesophyll protoplasts isolated from five plant species, *J. Gen. Virol.*, 40, 489, 1978.

64. Sugimura, Y. and Ushiyama, R., Cucumber green mottle mosaic virus and its bearing on cytological alterations in tobacco mesophyll protoplasts, *J. Gen. Virol.*, 29, 93, 1975.
65. Otsuki, Y. and Takebe, I., Double infection of isolated tobacco mesophyll protoplasts by unrelated plant viruses, *J. Gen. Virol.*, 30, 309, 1976.
66. Otsuki, Y. and Takebe, I., Double infection of isolated leaf protoplasts by two strains of tobacco mosaic virus, in *Biochemistry and Cytology of Plant-Parasite Interaction*, Tomiyama, K., Daly, J. B., Uritani, I., Oku, H., and Ouchi, S., Eds., Kodansha, Tokyo, 1976, 213.
67. Watts, J. W. and Dawson, J. R. O., Double infection of tobacco protoplasts with brome mosaic virus and cowpea chlorotic mottle virus, *Virology*, 105, 501, 1980.
68. Nilsson-Tillgren, T., Kolehmainen-Seveus, L., and von Wettstein, D., Studies on the biosynthesis of TMV. I. A system approaching a synchronized virus synthesis in a tobacco leaf, *Mol Gen. Genet.*, 104, 124, 1969.
69. Nilsson-Tillgren, T., Studies on the biosynthesis of TMV. II. On the RNA synthesis of infected cells, *Mol. Gen. Genet.*, 105, 191, 1969.
70. Nilsson-Tillgren, T., Studies on the biosynthesis of TMV. III. Isolation and characterization of the replicative form and the replicative intermediate RNA, *Mol. Gen. Genet.*, 109, 246, 1970.
71. Jackson, A. O., Zaitlin, M., Siegel, A., and Francki, R. I. B., Replication of tobacco mosaic virus. III. Viral RNA metabolism in separated leaf cells, *Virology*, 48, 655, 1972.
72. Beachy, R. N. and Zaitlin, M., Replication of tobacco mosaic virus. VI. Replicative intermediate and TMV-RNA-related RNAs associated with polyribosomes, *Virology*, 63, 84, 1975.
73. Hunter, T. R., Hunt, T., Knowland, J., and Zimmern, D., Messenger RMA for the coat protein of tobacco mosaic virus, *Nature (London)*, 260, 759, 1976.
74. Hunter, T., Jackson, R., and Zimmern, D., Multiple proteins and subgenomic m RNAs may be derived from a single open reading frame on tobacco mosaic virus RNA, *Nucl. Acids Res.*, 11, 801, 1983.
75. Goelet, P., Lomonossoff, G. P., Butler, P. J. G., Akam, M. E., Gait, M. J., and Karn, J., Nucleotide sequence of tobacco mosaic virus RNA, *Proc. Natl. Acad. Sci. U.S.A.*, 79, 5818, 1982.
76. Bruening, G., Beachy, R. N., Scalla, R., and Zaitlin, M., *In vitro* and *in vivo* translation of the ribonucleic acids of a cowpea strain of tobacco mosaic virus, *Virology*, 71, 498, 1976.
77. Beachy, R. N. and Zaitlin, M., Characterization and *in vitro* translation of the RNAs from less-than-full-length, virus-related, nucleoprotein rods present in tobacco mosaic virus preparations, *Virology*, 81, 160, 1977.
78. Ogawa, M., Sakai, F., and Takebe, I., A messenger RNA for tobacco mosaic virus coat protein in infected tobacco mesophyll protoplasts, *Phytopathol. Z.*, 107, 146, 1983.
79. Ogawa, M. and Sakai, F., A messenger RNA for tobacco mosaic virus coat protein in infected tobacco mesophyl protoplasts. II. Time course of its synthesis, *Phytopathol. Z.*, 109, 193, 1984.
80. Watanabe, Y., Emori, Y., Ooshika, I., Meshi, T., Ohno, T., and Okada, Y., Synthesis of TMV-specific RNAs and proteins at the early stage of infection in tobacco protoplasts: transient expression of the 30K protein and its m RNA, *Virology*, 133, 18, 1984.
81. Siegel, A., Hari, V., and Kolacz, K., The effect of tobacco mosaic virus infection on host and virus-specific protein synthesis in protoplasts, *Virology*, 85, 494, 1978.
82. Beier, H., Mundry, K. W., and Issinger, O. G., *In vivo* and *in vitro* translation of the RNAs of four tobamoviruses, *Intervervirology*, 14, 299, 1980.
83. Pelham, H. R. B., Leaky UAG termination codon in tobacco mosaic virus RNA, *Nature (London)*, 272, 469, 1978.
84. Leonard, D. A. and Zaitlin, M., A temperature-sensitive strain of tobacco mosaic virus defective in cell-to-cell movement generates an altered viral-coded protein, *Virology*, 117, 416, 1982.
85. Otsuki, Y., *In vivo* particle formation of tobacco mosaic virus, *Ann. Phytopathol. Soc. Jpn.*, 47, 420, 1981.
86. Nishiguchi, M., Motoyoshi, F., and Oshima, N., Behaviour of a temperature sensitive strain of tobacco mosaic virus in tomato leaves and protoplasts, *J. Gen. Virol.*, 39, 53, 1978.
87. Nishiguchi, M., Motoyoshi, F., and Oshima, N., Further investigation of a temperature-sensitive strain of tobacco mosaic virus: its behaviour in tomato leaf epidermis, *J. Gen. Virol.*, 46, 497, 1980.
88. Shalla, T. A., Petersen, L. J., and Zaitlin, M., Restricted movement of a temperature-sensitive virus in tobacco leaves is associated with a reduction in numbers of plasmodesmata, *J. Gen. Virol.*, 60, 353, 1982.
89. Taliansky, M. E., Malyshenko, S. I., Pshennikova, E. S., Kaplan, I. B., Ulanova, E. F., and Atabekov, J. G., Plant virus-specific transport function. I. Virus genetic control required for systemic spread, *Virology*, 122, 318, 1982.
90. Matthews, R. E. F., Portraits of viruses: turnip yellow mosaic virus, *Intervirology*, 15, 12, 1981.
91. Hatta, T. and Matthews, R. E. F., Sites of coat protein accumulation in turnip yellow mosaic virus-infected cells, *Virology*, 73, 1, 1976.
92. Nassuth, A., Alblas, F., and Bol, J. F., Localization of genetic information involved in the replication of alfalfa mosaic virus, *J. Gen. Virol.*, 53, 207, 1981.

93. Nassuth, A. and Bol, J. F., Altered balance of the synthesis of plus-strand RNAs induced by RNAs 1 and 2 of alfalfa mosaic virus in the absence of RNA 3, *Virology*, 124, 75, 1983.
94. Kiberstis, P. A., Loesch-Fries, L. S., and Hall, T. C., Viral protein synthesis in barley protoplasts inoculated with native and fractionated brome mosaic virus RNA, *Virology*, 112, 804, 1981.
95. Rottier, P. J. M., Rezelman, G., and van Kammen, A., Protein synthesis in cowpea mosaic virus-infected cowpea protoplasts: detection of virus-related proteins, *J. Gen. Virol.*, 51, 359, 1980.
96. Rottier, P. J. M., Rezelman, G., and van Kammen, A., Protein synthesis in cowpea mosaic virus-infected cowpea protoplasts: further characterization of virus-related protein synthesis, *J. Gen. Virol.*, 51, 373, 1980.
97. Goldbach, R., Rezelman, G., and van Kammen, A., Independent replication and expression of B-component RNA of cowpea mosaic virus, *Nature (London)*, 286, 297, 1980.
98. Rezelman, G., Goldbach, R., and van Lammen, A., Expression of bottom component RNA of cowpea mosaic virus in cowpea protoplasts, *J. Virol.*, 36, 366, 1980.
99. Beier, H., Issinger, O. G., Deuschle, M., and Mundry, K. W., Translation of the RNA of cowpea severe mosaic virus in vitro and in cowpea protoplasts, *J. Gen. Virol.*, 54, 379, 1981.
100. Robinson, D. J., Barker, H., Harrison, B. D., and Mayo, M. A., Replication of RNA-1 of tobacco black ring virus independently of RNA-2, *J. Gen. Virol.*, 51, 317, 1980.
101. van Vloten-Doting, L. and Jaspers, E. M. J., Plant covirus systems: three component systems, in *Comprehensive Virology*, Vol. 11, Fraenkel-Conrat, H. and Wagner, R. R., Eds., Plenum Publishing, 1977, 237.
102. Gould, A. R. and Symons, R. H., Alfalfa mosaic virus RNA. Determination of the sequence homology between the four RNA species and a comparison with the four RNA species of cucumber mosaic virus, *Eur. J. Biochem.*, 91, 269, 1978.
103. van Kammen, A., Plant viruses with a divided genome, *Ann. Rev. Phytopathol.*, 10, 125, 1972.
104. Wu, G.-J. and Bruening, G., Two proteins from cowpea mosaic virus, *Virology*, 46, 596, 1971.
105. Geelen, J. L. M. C., van Kammen, A., and Verduin, B. J. B., Structure of the capsid of cowpea mosaic virus. The chemical subunit: molecular weight and number of subunits per particles, *Virology*, 49, 205, 1971.
106. Stanley, J., Goldbach, R., and van Kammen, A., The genome-linked protein of cowpea mosaic virus is coded by RNA from the bottom component, *Virology*, 106, 180, 1980.
107. Franssen, H., Goldbach, R., Broekhuijsen, M., Moerman, M., and van Kammen, A., Expression of middle-component RNA of cowpea mosaic virus: *in vitro* generation of a precursor to both capsid proteins by a bottom-component RNA-encoded protease from infected cells, *J. Virol.*, 41, 8, 1982.
108. Pelham, H. R. B., Synthesis and proteolytic processing of cowpea mosaic virus proteins in reticulocyate lysates, *Virology*, 96, 463, 1979.
109. Goldbach, R., Rezelman, G., Zabel, P., and van Kammen, A., Expression of the bottom-component RNA of cowpea mosaic virus: evidence that 60-kilodalton VPg precursor is cleaved into single VPg and a 58-kilodalton polypeptide, *J. Virol.*, 630, 1982.
110. Goldbach, R. and Rezelman, G., Orientation of the cleavage map of the 200-kilodalton polypeptide encoded by the bottom-component RNA of cowpea mosaic virus, *J. Virol.*, 614, 1983.
111. Cooper, J. I. and Mayo, M. A., Some properties of the particles of three tobravirus isolates, *J. Gen. Virol.*, 16, 285, 1972.
112. Harrison, B. D. and Woods, R. D., Serotypes and particle dimensions of tobacco rattle viruses from Europe and America, *Virology*, 28, 610, 1966.
113. Siegel, A., Mutual exclusion of strains of tobacco mosaic virus, *Virology*, 8, 170, 1959.
114. Gibbs, A. J., Plant virus classification, *Adv. Virus Res.*, 14, 263, 1969.
115. Sherwood, J. L. and Fulton, R. W., The specific involvement of coat protein in tobacco mosaic virus cross protection, *Virology*, 119, 150, 1982.
116. Sakai, F., Dawson, J. R. O., and Watts, J. W., Interference in infection of tobacco protoplasts with two bromoviruses, *J. Gen. Virol.*, 64, 1347, 1983.
117. Otsuki, Y., Shimomura, T., and Takebe, I., Tobacco mosaic virus multiplication and expression of the N gene in necrotic responding tobacco varieties, *Virology*, 50, 45, 1972.
118. Loebenstein, G., Gera, A., Barnett, A., Shabtai, S., and Cohen, J., Effect of 2,4-dichlorophenoxyacetic acid on multiplication of tobacco mosaic virus in protoplasts from local-lesion and systemic-responding tobaccos, *Virology*, 100, 110, 1980.
119. Loebenstein, G. and Gera, A., Inhibitor of virus replication released from tobacco mosaic virus-infected protoplasts a local lesion-responding tobacco cultivar, *Virology*, 114, 132, 1981.
120. Gera, A., Loebenstein, G., and Shabtai, S., Enhanced tobacco mosaic virus production and suppressed synthesis of a virus inhibitor in protoplasts exposed to antibiotics, *Virology*, 127, 475, 1983.
121. Imaizumi, S., Takanami, Y., and Kubo, S., TMV infection and multiplication in protoplasts isolated from TMV resistant tobacco varieties, *Ann. Phytopathol. Soc. Jpn.*, 45, 129, 1979.
122. Motoyoshi, F. and Oshima, N., Infection with tobacco mosaic virus of leaf mesophyll protoplasts from susceptible and resistant lines of tomato, *J. Gen. Virol.*, 29, 81, 1975.

123. Motoyoshi, F. and Oshima, N., Expression of genetically controlled resistance to tobacco mosaic virus infection in isolated tomato leaf mesophyll protoplasts, *J. Gen. Virol.*, 34, 499, 1977.
124. Stobbs, L. W. and MacNeil, B. H., Response to tobacco mosaic virus of a tomato cultivar homozygous for gene Tm-2, *Can. J. Plant Pathol.*, 2, 5, 1980.
125. Pelham, J., Resistance in tomato to tobacco mosaic virus, *Euphytica*, 15, 258, 1966.
126. Pelham, J., Strain-genotype interaction of tobacco mosaic virus in tomato, *Ann. Appl. Biol.*, 71, 219, 1972.
127. Taliansky, M. E., Malyshenko, D. I., Pshennikova, E. S., and Atabekov, J. G., Plant virus-specific transport function. II. A factor controlling virus host range, *Virology*, 122, 327, 1982.
128. Motoyoshi, F. and Oshima, N., Resistance and susceptibility of tomato protoplasts to tobacco mosaic virus, in *International Virology*, Vol. 4, Abstracts of the 4th Int. Congr. Virol. Centre for Agricultural Publishing and Documentation, Wageningen, 1978, 263.
129. Motoyoshi, F. and Oshima, N., Standardization in inoculation procedure and effect of a resistance gene on infection of tomato protoplasts with tobacco mosaic virus RNA, *J. Gen. Virol.*, 44, 801, 1979.
130. Motoyoshi, F., Characterization of Tm-1 gene which confers resistance to tobacco mosaic virus in tomato plants and protoplasts, in *Proc. 5th Int. Congr. Plant Tissue Cell Culture, Plant Tissue Culture 1982*, Fujiwara, A., Ed., Maruzen Co., Tokyo, 1982, 659.
131. Beier, H., Bruening, G., Russell, M. L., and Tucker, C. L., Replication of cowpea mosaic virus in protoplasts isolated from immune lines of cowpea, *Virology*, 95, 165, 1979.
132. Beier, H., Siler, D. J., Russell, M. L., and Bruening, G., Survey of susceptibility to cowpea mosaic virus among protoplasts and intact plants from *Vigna sinensis* lines, *Phytopathology*, 67, 917, 1977.
133. Coutts, R. H. A. and Wood, K. R., Inoculation of leaf mesophyll protoplasts from a resistant and a susceptible cultivar with cucumber mosaic virus, *FEMS Microbiol. Lett.*, 121, 1977.
134. Mühlbach, H. P. and Sänger, H. L., Viroid replication is inhibited by α-amanitin, *Nature (London)*, 278, 185, 1979.
135. Diener, T. O., Are viroids escaped introns?, *Proc. Natl. Acad. Sci. U.S.A.*, 78, 5014, 1981.
136. Dawson, J. R. O., Motoyoshi, F., Watts, J. W., and Bancroft, J. B., Production of RNA and coat protein of a wild type isolated and a temperature-sensitive mutant of cowpea chlorotic mottle virus in cowpea leaves and tobacco protoplasts, *J. Gen. Virol.*, 29, 99, 1975.
137. Sarachu, A. N., Nassuth, A., Roosien, J., van Vloten-Doting, L., and Bol, J. F., Replication of temperature-sensitive mutants of alfalfa mosaic virus in protoplasts, *Virology*, 125, 64, 1983.

Chapter 10

PLANT PROTOPLASTS AND GENETIC VARIATION

Richard I. S. Brettell and Philip J. Larkin

TABLE OF CONTENTS

I. GENERAL INTRODUCTION

Evidence has accumulated over recent years that enhanced genetic variation may be observed in plants derived from cells that have been exposed to a tissue or cell culture environment without a deliberate mutagenic treatment.[1-3] The term "somaclonal variation" has been proposed to describe these genetic events, and possible mechanisms responsible for the variation are currently the subject of much discussion.

The phenomenon of somaclonal variation is seen in a wide range of plant species and does not appear to be confined to a single type of explant. However, the supposition that a single mechanism is involved would be dangerous at this stage and is certainly not supported by the available data. In this chapter we will consider the genetic variation observed in plants derived from protoplasts. Protoplasts offer the advantage of being discrete cellular plating units which may be manipulated in large and uniform populations.

II. SOME SPECIFIC EXAMPLES

In a previous paper, Larkin et al.[4] concluded that protoplasts are not fundamentally different from other types of cell or tissue explant with respect to somaclonal variation. This conclusion is derived from largely circumstantial evidence owing to the difficulties in making a direct comparison between different experimental techniques, particularly when applied to a range of genotypes in different laboratories. Nevertheless, it is still worthwhile to compare the published data within a given species. The majority of results have been obtained with potato (*Solanum* spp.) and tobacco (*Nicotiana* spp.) and will be reviewed in this section before wider implications are considered.

A. Potato

Potato represents one of the few examples of plant regeneration from protoplasts of a major crop plant. Shepard and co-workers first described somaclonal variation in plants derived from protoplasts of the cultivar Russet Burbank.[5-7] Variation was observed in a number of characters including growth habit, maturity date, tuber morphology, and reaction to early and late blight diseases, caused, respectively, by *Alternaria solani* and *Phytophthora infestans.* Many of the variants were reported to be stable through subsequent vegetative propagation. Russet Burbank is a long established line which arose some 80 years ago as a sport from the cultivar Burbank. Unfortunately, its sterility has precluded a formal genetic analysis of the variation.

The problem of sterility extends to the cultivar Bintje which, like Russet Burbank, is also a long established genotype of potato and can be readily manipulated in culture.[8] Again many alterations were observed in the morphology of the plants regenerated from protoplasts,[9] but much of this variation was correlated with changes in ploidy. Ploidy levels were tentatively determined in 489 regenerant plants according to a score of the number of chloroplasts in stomatal guard cells. Chromosome counts were then made in root tip cells of 43 selected plants. For the 31 plants classified as showing variation according to morphological characters, most showed a predominance of aneuploid chromosome counts.[9] Shepard et al.[5,6] made no detailed cytological analysis of their protoplast regenerants, but, following rigorous selection for agronomically superior types (60 of 1700 clones), assumed that most plants with gross chromosomal abnormalities had been eliminated. Examination of five clones that were selected for enhanced tuber yields gave a normal tetraploid count of 48 chromosomes.[5]

The question of whether polyploidy is a prerequisite for somaclonal variation is raised by the work of Wenzel et al.[10] on plants derived from two dihaploid lines ($2n = 2x = 24$) of *Solanum tuberosum.* No variation was seen in plants rapidly regenerated from mesophyll protoplasts as judged by leaf, stem, and flower color as well as tuber morphology.

Plants obtained from protoplasts of the cultivar Maris Bard were examined by Thomas et al.,[11] who found considerable variation in 22 out of 23 clones potted in soil and assessed for 3 morphological characters. Maris Bard is a much more recent cultivar then either Russet Burbank or Bintje, indicating that an accumulation of mutations following prolonged vegetative propagation is not necessarily the cause of the observed variation. That the variation did occur during the culture phase is supported by the fact that variation was seen between plants obtained from one callus clone derived from a single protoplast (see also Reference 12).

Further support for the contention that alterations in chromosomal status contribute to somaclonal variation in potato is provided by the study of Karp et al.[13] on protoplast-derived plants of Maris Bard and Fortyfold, the former being the same plants described by Thomas et al.[11] High chromosome numbers, up to 93, and a wide range of aneuploidy were seen in the Maris Bard regenerants. One third of the Fortyfold calluses gave plants with a normal count of 48 chromosomes while the rest, with one exception, gave aneuploids in the range 46 to 49. A direct comparison between the two cultivars is aggravated by the use of different protocols for protoplast isolation and culture. Interestingly the Fortyfold regenerants showed less morphological variation than the plants derived from Maris Bard.[13] The results of a further experiment by Wenzel et al.[10] should be considered in this context. Plants were generated from protoplasts obtained from one of the *S. tuberosum* dihaploid lines, but on this occasion from cells cultured in suspension over an extended period. In contrast to the plants regenerated from leaf protoplasts, phenotypic variation was observed in these and was associated with a higher frequency of aneuploidy.

From these studies we conclude that it is yet to be resolved whether alterations in chromosome complement are a cause or consequence of somaclonal variation in potato. The answer will perhaps await the chromosomal mapping of defined somaclonal mutations, and their genetic analysis in the progeny of the primary regenerants.

B. Tobacco

A number of species of *Nicotiana* have proved to be very amenable to protoplast culture and the subsequent regeneration of intact plants.[14] Indeed *Nicotiana tabacum* was the first species from which plants were successfully obtained from leaf protoplasts.[15] In the following year Dulieu[16] made the prescient remark that finer changes at the intragenic level might occur in tissue culture in addition to gross chromosomal changes. Since certain tobacco species are self-fertile, they allow a much easier genetic analysis of somaclonal variation than is possible with sterile cultivars of *Solanum tuberosum*. Research has thus tended toward an examination of the behavior of a small number of defined loci.

Barbier and Dulieu[17] made a detailed study in which they compared plants derived from cotyledon protoplasts with plants obtained from cotyledon tissue cultures. They used an intervarietal hybrid of *N. tabacum* which was heterozygous at two loci involved in chloroplast differentiation (a_1^+/a_1, yg^+/yg). In contrast to the genetic stability seen in seed-derived plants, many somaclonal variants appeared in the plants regenerated from both cotyledon protoplasts and cotyledon tissue cultures and included morphological variants as well as phenotypes variant for leaf color. Of 1666 regenerants from culture, 53 showed a mutation at the a_1 locus only, 55 at the yg locus only, and 5 at both loci. Green leaf color variants were crossed to the parental lines, a_1/a_1 and yg/yg, as well as being self-pollinated. The results indicated that the mutations could be ascribed to deletions and reversions at the loci under study. Cytological analysis of the variant plants showed that all were euploid except for a single plant with 46 chromosomes. Of 96 variant plants, 53 had 48 chromosomes and 43 had 96 chromosomes. The authors concluded that the culture conditions could induce some doubling of the chromosome number from the normal allotetraploid complement of 48, but that this doubling occurred independently from the genetic variation that was scored.

In an analogous series of experiments Lörz and Scowcroft[18] used the leaf color mutant, sulfur, to assess variation in protoplast derived plants of *N. tabacum*. Heterozygous *Su/su* plants were used as donor material and 2222 colonies were grown from isolated protoplasts. Several shoots were regenerated from each of the morphogenic colonies and scored for leaf color. On the basis of phenotype, the homozygous (*su/su* green; *Su/Su* albino) and heterozygous (*Su/su* yellow-green) classes could be distinguished. Treatments included the addition of chemical mutagen to the culture medium and prolonged culture, both of which enhanced the frequency of variants obtained. Of the 2156 morphogenic colonies, 79 gave rise to one or more variant plants. A quarter of the 79 colonies gave plants which were all scored as nonparental. These were interpreted as representing preexisting variation in the cells of the leaves taken for protoplast isolation. The other 59 colonies gave both variant and parental regenerants, reflecting the variation occurring during the culture phase.

Barbier and Dulieu[19] have recently followed up their earlier experiments in order to define more precisely the timing of the events responsible for the observed genetic variation. As in the experiments described above, they used plants heterozygous at the a_1 and *yg* loci for protoplast isolation. In one experiment they obtained 1048 colonies from leaf protoplasts. The colonies were divided into four pieces from each of which a plant was regenerated. Sectoring, when some but not all of the shoots from a colony were variant, was observed for variant green shoots in 13.1% of the colonies and for variant yellow shoots in 3.8%. A further experiment was made with cotyledon protoplasts from which 110 colonies were obtained. From 85 of these, plants were regenerated after a single subcloning and frequencies similar to those measured for leaf protoplasts were observed in the appearance of green and yellow sectors. To test the effect of prolonged culture, the other 25 colonies were subcloned 3 times and plants regenerated from each of the 1600 resulting callus pieces; 21 of the original colonies revealed no variation, but the remaining 4 yielded 32 green and 32 yellow variants, together representing one quarter of the subclones originating from the 4 colonies. The changes had already appeared following the first period of culture, and no new variation had been induced by further culture in contrast to the results of Lörz and Scowcroft.[18] The results were interpreted as indicating that the genetic changes may have been due to preexisting lesions in the DNA of the protoplasts. Following culture, somatic segregation in the first or second divisions after isolation of the protoplasts could lead to the expression of variant phenotypes.

There are possible objections to all the experiments so far described for tobacco in that donor cells were heterozygous at defined loci and that any changes seen at these loci might represent gene conversion events not representative of somaclonal variation in homozygous donor cultures. In addition, *Nicotiana tabacum* is an allotetraploid where genetic analysis may be confounded because many of the structural genes are likely to be duplicated. The experiments of Prat[20] were designed to examined the effect of protoplast culture on true-breeding lines using the diploid tobacco *Nicotiana sylvestris*. Two sources were used for protoplast isolation: a line of *N. sylvestris* maintained for seven generations by self-pollination and an androgenetic line obtained from the same original line after five consecutive rounds of pollen culture; 172 plants were regenerated from 82 calluses; 72 plants were diploid while the rest were found to be polyploid rather than aneuploid. Selfed progenies were obtained from the diploid regenerants and screened for the presence of mutations. Segregating mutant phenotypes were seen in the progenies of 9 of 21 plants for a number of characters including height, leaf color, and male-sterility. These 21 plants were derived from 12 different calluses, but different segregating mutations were seen in the progenies of plants regenerated from the same callus line, which would strongly suggest that the mutations had arisen during the culture phase. In addition to the segregating mutations, the progenies also showed a general depression in a number of quantitative characters when compared to the source lines.

C. Other Species

In this review we have concentrated on the genetic variation seen in species of tobacco and potato. Evidence for similar phenomena is accumulating for other species such as lettuce where methods for protoplast isolation and culture have recently been described.[21] Variation was observed among the progeny of 119 lettuce plants derived from mesophyll protoplasts of the cultivar 'climax'.[22] Many of the variants showed segregation that conformed to Mendelian expectations for single gene changes. This follows earlier work by Sibi,[23] who observed apparent nuclear and cytoplasmic mutations following cotyledon culture in lettuce. Somaclonal variation has been reported in a wide range of species from different plant genera,[2] and we would predict that as methods for protoplast culture and the regeneration of plants are perfected for these species, then analogous and heritable genetic variation will be seen in the progeny of plants derived from protoplasts.

III. DISCUSSION OF POSSIBLE MECHANISMS RESPONSIBLE FOR THE OBSERVED VARIATION

The genetic variation in plants regenerated from protoplasts, and their progeny, is reflected in phenotypic alterations which have now been described in a number of species. For potato, none of these phenotypic alterations has, as far as we can ascertain, been correlated with changes at a defined locus or even to a given pair of chromosomes. Tobacco has provided better opportunities for genetic analysis where alterations have been examined for various leaf color mutations.[17,19] Yet since somatic recombination can occur normally in tobacco tissues for such leaf color mutations,[24,25] albeit at a lower frequncy than would account for the variation observed following protoplast culture, the changes at these particular loci in tobacco may not be entirely representative of somaclonal variation. Given these constraints it would be foolhardy to make too many generalizations from these examples regarding the mechanisms responsible for variation in protoplast-derived plants, however we will briefly consider particular mechanisms which may generate genetic variation in the culture environment.

A. Aneuploidy

The occurrence of karyotypic alterations in cultured plant cells has been well documented.[26,27] From the recent results of work on plants obtained from potato protoplasts it has become increasingly clear that chromosomal variation is a critical factor in the analysis of phenotypic variation following culture. The data of Karp et al.[13] with the cultivars Maris Bard and Fortyfold, and Sree Ramulu et al.[9] with the cultivar Bintje have shown that tetraploid potato can certainly tolerate deviations from the normal chromosome complement of 48 without necessarily producing gross morphological abnormalities. It is therefore unfortunate that earlier work by Shepard and co-workers[5,6] on the cultivar Russet Burbank did not include a more rigorous cytological analysis of the regenerant clones. Although much of the observed variation might be ascribed to changes in chromosome number, we would contend that a part of the variation is also due to more subtle changes than simple loss or addition of intact chromosomes. In a study using suspensions of *Brachycome dichromosomatica,* a species of Compositae with a diploid complement of four chromosomes, Gould[28] concluded that chromosome number was a poor estimator of chromosome stability in culture. Structural changes can include translocations (reciprocal and nonreciprocal), inversions, and chromosome segment deletion, all of which have been documented in somaclones of oat,[29] garlic,[30] *Haworthia,*[31] and wheat.[32] In addition to these, there could occur fine structural changes which would influence the regulation and expression of one or a small number of genes on a given chromosome. An assessment of the relative contribution of gross chromosomal or fine structural events will depend on a more intensive examination of the mutants obtained from culture.

B. The Need for Heterozygosity

The results of experiments on tobacco where plants heterozygous at defined loci were used as donor material for protoplast isolation have revealed that changes at these loci occur as a consequence of culture.[17-19] As mentioned above, these may not necessarily be representative of somaclonal variation as a whole; however, they do provide an insight into genetic events which may occur following protoplast isolation. The work of Prat[20] with rigorously established true-breeding lines of *Nicotiana sylvestris* amply demonstrates that heterozygosity is not an absolute prerequisite for somaclonal variation, yet it is interesting to note that many of the plants for which somaclonal variation has been described have been polyploid and/or vegetatively propagated, which would allow for interaction between distinct but related genes. Thus we could postulate that variation is enhanced by the operation of mechanisms which effect gene conversion or mis-match correction as has been described in yeast.[33] That interaction between related pieces of DNA may have a role in generating novel variants is further suggested by the fact that high levels of variation are often seen following the formation of somatic hybrids.[34-35] In somatic hybrids of *Nicotiana tabacum* and *N. knightiana* phenotypic instability has been attributed to chromosome segregation,[36] although rearrangements of mitochondrial DNA which may also contribute to the variation were found to be a feature of this combination.[37]

A comparison has been made of hybrids of *Nicotiana tabacum* and *N. nesophila* produced by ovule culture and by protoplast fusion.[38] The sexual hybrids produced by ovule culture could be obtained only when *N. nesophila* was used as the female parent. No significant variation was seen between the sexual hybrids for a number of characters including pollen viability, flower morphology, leaf morphology, and trichome density. However, statistically significant variation in these characters was found in 5 somatic hybrid clones which had the amphiploid chromosome complement of 96. Although this does not exclude the possibility of chromosome substitution among the four genomes of the octoploid somatic hybrid, the variation could not be attributed to alterations in chromosome number, but might rather be explained by small chromosome rearrangements or by events such as recombination of organelle DNA that could occur as the result of the mixing of disparate cytoplasms. Further examples of enhanced variation in somatic hybrids have been provided in combinations of *Nicotiana tabacum* with *N. otophora* and *N. sylvestris*[39] and *Solanum nigrum* with *S. tuberosum.*[40]

C. Amplification of Specific DNA Sequences

The induction of genetic changes in certain flax varieties under particular environmental conditions was first described by Durrant.[41] The changes were shown to be heritable, and stable variant lines were termed genotrophs. An examination of the DNA of flax genotrophs revealed that differences in phenotype were associated with alterations in the amounts of ribosomal DNA[42] as well as total nuclear DNA.[43] If similar changes in amount of specific DNA sequences occurred during differentiation or in cells subjected to an in vitro environment then they might consequently be maintained in plants regenerated from culture and their progeny.

The progenies of plants obtained by Prat from protoplasts of *Nicotiana sylvestris* showed, in addition to segregating mutations which might be ascribed to single genes, a general depression in a number of quantitative characters such as leaf dimension, height, and greenhouse input weight.[20] A similar depression was observed in the progenies of *N. sylvestris* obtained following pollen culture and chromosome doubling of the resulting haploid plants.[44] These phenotypic alterations were correlated with increases in the total amount of DNA and further analysis suggested that this was attributable to an increase in the proportion of highly repeated sequences.[45] It has been established in *Drosophila* that gene expression may be influenced by rearrangements of heterochromatic DNA,[46] suggesting that selective amplification or depletion of repeated sequences should be considered as a mechanism contributing to the heritable genetic changes observed as a consequence of protoplast culture.

D. The Induction of Latent Instabilities

In a number of instances, an increase in the instability of expression of a gene has been observed following protoplast culture. Barbier and Dulieu[17] found a number of plants which showed increased leaf-spotting and variegation among those derived from cotyledon protoplasts and cotyledon tissue cultures of *Nicotiana tabacum* heterozygous at two loci affecting leaf color. An increase in leaf-spotting has also been observed in plants derived from protoplasts of tobacco heterozygous for the sulfur mutation. Evans et al.[39,47] described one clone, obtained following fusion of *Nicotiana tabacum* (*Su/Su*) with *N. sylvestris,* which was designated 'Superspot'. Lörz and Scowcroft[18,48] obtained two regenerants from protoplasts of *N. tabacum* heterozygous at the sulfur locus which showed a dramatic increase in the number of dark green, albino, and twin spots (Figure 1). Genetic analysis has shown that the tendency towards a high level of spotting is heritable and is transmitted through both male and female parents, although it does not conform to simple Mendelian expectations. Further work is in progress to assess the extent to which changes at other loci are associated with the increased mutability of the sulfur locus.

Genetic instability is known in a number of plant species. The best characterized system is in maize and has been explained by the operation of controlling elements.[49] These operate in an analogous manner to bacterial transposons and have recently been shown to have a similar molecular basis.[50] Transposition of genetic elements would provide an attractive explanation for some of the variation seen as a consequence of protoplast culture. Excision or insertion of pieces of DNA will readily affect expression of structural genes but can also lead to modifications of chromosome architecture.

It is possible that the rearrangements of genetic material that are responsible for somaclonal variation do not occur entirely in random fashion. Further molecular analysis of somaclonal variants will no doubt provide a clearer insight into those directed changes in the genome that occur in the culture environment.

IV. CONCLUDING REMARKS

The data we have been able to assemble on genetic variation as a consequence of protoplast culture is confined to a relatively small number of experiments on a handful of plant species. It is therefore difficult to deduce a common pattern of genetic events. Nevertheless, the information suffices to demonstrate that somaclonal variation is an

FIGURE 1. Enhanced leaf-spotting in a plant of *Nicotiana tabacum*. This individual was regenerated from protoplasts isolated from a plant which was heterozygous at the sulfur locus (*Su*/*su*). The donor for protoplast isolation was itself derived from a protoplast regenerant showing instability for the leaf color phenotype.

important feature of protoplast culture. Although genetic variability may limit the application of protoplast culture as a means of faithful clonal propagation, it does allow the selection of useful mutants. Indeed it is probable that somaclonal variation has generated efficient genetic heterogeneity for the selection of heritable mutants in a number of successful experiments. In only a few cases have mutagens been used systematically with zero mutagen controls. One example is Gebhardt et al.,[51] who selected auxotrophic and temperature-sensitive clones using haploid protoplasts of *Hyoscy-*

amus muticus (see Chapter 11). In other experiments, stable mutants have been obtained without a specific mutagenic treatment, such as those described by Thanutong et al.,[52] where tobacco plants resistant to *Alternaria* and *Pseudomonas* toxins were selected from protoplast-derived calluses.

Finally, the enhanced variation seen among somatic hybrid plants may offer valuable prospects for enhanced alien gene introgression, particularly for combinations of genomes from species which are sexually incompatible. For instance, Dudits et al.[53] fused protoplasts from a suspension culture of albino *Daucus carota* with leaf protoplasts of *Aegopodium podagraria*. The resulting green plants appeared to have only *Daucus* chromosomes but expressed a number of *Aegopodium* characters in addition to the correction of the albinism.

ACKNOWLEDGMENTS

The authors wish to thank Bill Scowcroft, Sarah Ryan, and colleagues at CSIRO and the Australian National University for ideas, discussion, and helpful suggestions during the preparation of the manuscript.

REFERENCES

1. Skirvin, R. M., Natural and induced variation in tissue culture, *Euphytica*, 27, 241, 1978.
2. Larkin, P. J. and Scowcroft, W. R., Somaclonal variation — a novel source of variability from cell cultures for plant improvement, *Theor. Appl. Genet.*, 60, 197, 1981.
3. Larkin, P. J. and Scowcroft, W. R., Somaclonal variation and crop improvement, in *Genetic Engineering of Plants*, Kosuge, T., Meredith, C. P., and Hollaender, A., Eds., Plenum Plublishing, New York, 1983, 289.
4. Larkin, P. J., Brettell, R., Ryan, S., and Scowcroft, W. R., Protoplasts and variation from culture, in *Protoplasts 1983*, Proc. 6th Int. Protoplast Symp., Potrykus, I., Harms, C. T., Hinnen, A., Hütter, R., King, P. J. and Shillito, R. D., Eds., Birkhauser Verlag, Basel, 1983, 51.
5. Shepard, J. F., Bidney, D., and Shahin, E., Potato protoplasts in crop improvement, *Science*, 208, 17, 1980.
6. Secor, G. A. and Shepard, J. F., Variability of protoplast-derived potato clones, *Crop Sci.*, 21, 102, 1981.
7. Shepard, J. F., Protoplasts as sources of disease resistance in plants, *Ann. Rev. Phytopathol.*, 19, 145, 1981.
8. Bokelmann, G. S. and Roest, S., Plant regeneration from protoplasts of potato (*Solanum tuberosum* cv. Bintje), *Z. Pflanzenphysiol.*, 109, 259, 1983.
9. Sree Ramulu, K., Dijkhuis, P., and Roest, S., Phenotypic variation and ploidy level of plants regenerated from protoplasts of tetraploid potato *Solanum tuberosum* L. cv. 'Bintje'), *Theor. Appl. Genet.*, 65, 329, 1983.
10. Wenzel, G., Schieder, O., Przewozny, T., Sopory, S. K., and Melchers, G., Comparison of single cell culture derived *Solanum tuberosum* L. plants and a model for their application in breeding programs, *Theor. Appl. Genet.*, 55, 49, 1979.
11. Thomas, E., Bright, S. W. J., Franklin, J., Lancaster, V. A., Miflin, B. J., and Gibson, R., Variation amongst protoplast-derived potato plants (*Solanum tuberosum* cv. 'Maris Bard') *Theor. Appl. Genet.*, 62, 65, 1982.
12. Austin, S. and Cassells, A. C., Variation between plants regenerated from individual calli produced from separated potato stem callus cells, *Plant Sci. Lett.*, 31, 107, 1983.
13. Karp, A., Nelson, R. S., Thomas, E., and Bright, S. W. J., Chromosome variation in protoplast-derived potato plants, *Theor. Appl. Genet.*, 63, 265, 1982.
14. Chupeau, Y., Bourgin, J.-P., Missonier, C., Dorion, N., and Morel, G., Préparation et culture de protoplastes de divers *Nicotiana*, *C. R. Acad. Sci. Ser. D.*, 278, 1565, 1974.
15. Nagata, T. and Takebe, I., Plating of isolated tobacco mesophyll protoplasts on agar medium, *Planta*, 99, 12, 1971.

16. Dulieu, H., The combination of cell and tissue culture with mutagenesis for the induction and isolation of morphological or developmental mutants, *Phytomorphology*, 22, 283, 1972.
17. Barbier, M. and Dulieu, H. L., Effets génétiques observés sur des plantes de Tabac régénérées à partir de cotyledons par culture *in vitro*, *Ann. Amelior. Plantes*, 30, 327, 1980.
18. Lörz, H. and Scowcroft, W. R., Variability among plants and their progeny regenerated from protoplasts of *Su/su* heterozygotes of *Nicotiana tabacum*, *Theor. Appl. Genet.*, 66, 67, 1983.
19. Barbier, M. and Dulieu, H., Early occurrence of genetic variants in protoplast cultures, *Plant Sci. Lett.*, 29, 201, 1983.
20. Prat, D., Genetic variability induced in *Nicotiana sylvestris* by protoplast culture, *Theor. Appl. Genet.*, 64, 223, 1983.
21. Engler, D. E. and Grogan, R. G., Isolation, culture and regeneration of lettuce leaf mesophyll protoplasts, *Plant Sci. Lett.*, 28, 223, 1983.
22. Engler, D. E. and Grogan, R. G., Variation in lettuce plants regenerated from protoplasts, *J. Hered.*, 75, 426, 1984.
23. Sibi, M., La notion de programme génétique chez les végétaux supérieures, *Ann. Amelior. Plantes*, 26, 523, 1976.
24. Dulieu, H., de Boelpaepe, R., and Deshayes, A., Sur l'existance spontanee de recombinaisons somatiques chez un mutant de *Nicotiana xanthi* et leur induction par le rayonnement γ; premières études génétiques, *C. R. Acad. Sci. Ser. D*, 272, 3287, 1971.
25. Evans, D. A. and Paddock, E. F., Comparisons of somatic crossing-over frequency in *Nicotiana tabacum* and three other crop species, *Can. J. Genet. Cytol.*, 18, 57, 1976.
26. Bayliss, M. W., Origin of chromosome number variation in cultured plant cells, *Nature (London)*, 246, 529, 1973.
27. D'Amato, F., Chromosome number variation in cultured cells and regenerated plants, in *Frontiers of Plant Tissue Culture 1978*, Proc. 4th. Int. Cong. Plant Tissue and Cell Culture, Thorpe, T. A., Ed., University of Calgary, Calgary, 1978, 287.
28. Gould, A. R., Chromosome instability in plant tissue cultures studied with banding techniques, in *Plant Tissue Culture 1982*, Proc. 5th Int. Congr. Plant Tissue Culture, Fujiwara, A., Ed., Japanese Association for Plant Tissue Culture, Tokyo, 1982, 431.
29. McCoy, T. J., Phillips, R. L., and Rines, H. W., Cytogenetic analysis of plants regenerated from oat (*Avena sativa*) tissue cultures; high frequency of partial chromosome loss, *Can. J. Genet. Cytol.*, 24, 37, 1982.
30. Novák, F. J., Phenotype and cytological status of plants regenerated from callus cultures of *Allium sativum* L., *Z. Pflanzenzuecht.*, 84, 250, 1980.
31. Ogihara, Y., Tissue culture in *Haworthia*. IV. Genetic characterization of plants regenerated from callus, *Theor. Appl. Genet.*, 60, 353, 1981.
32. Karp, A. and Maddock, S. E., Chromosome variation in wheat plants regenerated from cultured immature embryos, *Theor. Appl. Genet.*, 67, 249, 1984.
33. Mikus, M. D. and Petes, T. D., Recombination between genes located on non-homologous chromosomes in *Saccharomyces cerevisiae*, *Genetics*, 101, 369, 1982.
34. Dudits, D., Hadlaczky, G., Lázár, G. and Haydu, Z., Increase in genetic variability through somatic cell hybridization of distantly related plant species, in *Plant Cell Cultures: Results and Perspectives*, Sala, F., Parisi, B., Cella, R., and Ciferri, O., Eds., Elsevier/North-Holland Biomedical Press, Amsterdam, 1980, 207.
35. Harms, C. T., Somatic hybridisation by plant protoplast fusion, in *Protoplasts 1983*, Proc. 6th Int. Protoplast Symp., Potrykus, I., Harms, C. T., Hinnen, A., Hütter, R., King, P. J., and Shillito, R. D., Eds., Birkhäuser Verlag, Basel, 1983, 69.
36. Maliga, P., Kiss, Z. R., Nagy, A. H., and Lazar, G., Genetic instability in somatic hybrids of *Nicotiana tabacum* and *Nicotiana knightiana*, *Mol. Gen. Genet.*, 163, 145, 1978.
37. Nagy, F., Török, I., and Maliga, P., Extensive rearrangements in the mitochondrial DNA in somatic hybrids of *Nicotiana tabacum* and *Nicotiana knightiana*, *Mol. Gen. Genet.*, 183, 437, 1981.
38. Evans, D. A., Flick, C. E., Kut, S. A., and Reed, S. M., Comparison of *Nicotiana tabacum* and *Nicotiana nesophila* hybrids produced by ovule culture and protoplast fusion, *Theor. Appl. Genet.*, 62, 193, 1982.
39. Evans, D. A., Bravo, J. E., Kut, S. A., and Flick, C. E., Genetic behaviour of somatic hybrids in the genus *Nicotiana: N. otophora + N. tabacum* and *N. sylvestris + N. tabacum*, *Theor. Appl. Genet.*, 65, 93, 1983.
40. Binding, H., Jain, S. M., Finger, J., Mordhorst, G., Nehls, R., and Gressel, J., Somatic hybridisation of an atrazine resistant biotype of *Solanum nigrum* with *Solanum tuberosum*. I. Clonal variation in morphology and in atrazine sensitivity, *Theor. Appl. Genet.*, 63, 273, 1982.
41. Durrant, A., The environmental induction of heritable changes in *Linum*, *Heredity*, 17, 27, 1962.
42. Cullis, C. A., Environmentally induced changes in ribosomal RNA cistron number in flax, *Heredity*, 36, 73, 1976.

43. Evans, G. M., Durrant, A., and Rees, H., Associated nuclear changes in the induction of flax genotrophs, *Nature (London)*, 212, 697, 1966.
44. De Paepe, R., Bleton, D., and Gnangbe, F., Basis and extent of genetic variability among doubled haploid plants obtained by pollen culture in *Nicotiana sylvestris, Theor. Appl. Genet.*, 59, 177, 1981.
45. De Paepe, R., Prat, D., and Huguet, T., Heritable nuclear DNA changes in doubled haploid plants obtained by pollen culture of *Nicotiana sylvestris, Plant. Sci. Lett.*, 28, 11, 1983.
46. Hilliker, A. J. and Appels, R., Pleiotropic effects associated with the deletion of heterochromatin surrounding rDNA on the X chromosome of *Drosophila, Chromosoma*, 86, 469, 1982.
47. Evans, D. A., Bravo, J. E., Kut, S. A., and Flick, C. E., Genetic behaviour of somatic hybrids in the genus *Nicotiana: N. otophora + N. tabacum* and *N. sylvestris + N. tabacum*, in *Protoplasts 1983, Poster Proc.*, Proc. 6th Int. Protoplast Symp., Potrykus I., Harms, C. T., Hinnen, A., Hütter, R., King, P. J., and Shillito, R. D., Eds., Birkhäuser Verlag, Basel, 1983, 98.
48. Lörz, H. and Scowcroft, W. R., personal communication, 1983.
49. McClintock, B., Controlling elements and the gene, *Cold Spring Harbor Symp. Quant. Biol.*, 21, 197, 1956.
50. Sutton, W. D., Gerlach, W. L., Schwartz, D., and Peacock, W. J., Molecular analysis of *Ds* controlling element mutations at the *Adh1* locus of maize, *Science*, 223, 1265, 1984.
51. Gebhardt, C., Schnebli, V., and King, P. J., Isolation of biochemical mutants using haploid mesophyll protoplasts of *Hyoscyamus muticus*. II. Auxotrophic and temperature-sensitive clones, *Planta*, 153, 81, 1981.
52. Thanutong, P., Furusawa, I., and Yamamoto, M., Resistant tobacco plants from protoplast-derived calluses selected for their resistance to *Pseudomonas* and *Alternaria* toxins, *Theor. Appl. Genet.*, 66, 209, 1983.
53. Dudits, D., Hadlaczky, G., Bajszár, G., Koncz, C., Lázár, G., and Horváth, G., Plant regeneration from intergeneric cell hybrids, *Plant Sci. Lett.*, 15, 101, 1979.

Chapter 11

PLANT MUTANT ISOLATION VIA PROTOPLASTS

Anne D. Blonstein and Patrick J. King

TABLE OF CONTENTS

I. GENERAL INTRODUCTION

Beginning with the isolation of streptomycin-resistant and methionine sulfoxine-resistant mutants of *Nicotiana tabacum* by Maliga et. al.[1] and Carlson,[2] respectively, the application of tissue culture techniques has led to the isolation of a series of biochemical mutants of a type which would be found only with great difficulty, if at all, using whole plants. The major advantages of tissue culture for mutant isolation are clearly the high number of cells available, making a search for dominant mutants feasible, and the ease of application of biochemical selection agents. Even when the best available tissue cultures are slow growing, complex multicellular callus cultures with limited potential for plant regeneration, dominant resistant mutants have been isolated, for example, lysine/threonine resistance in *Zea mays.*[3]

Haploid protoplast cultures are a further refinement of the in vitro selection technique: unlike established tissue cultures, which often lose the capacity for plant regeneration and can rarely be maintained in the haploid state, populations of freshly isolated haploid leaf protoplasts can be immediately exposed to mutagens and induced to express their morphological potential. Furthermore, the division of single, isolated protoplasts leads to clone formation, which is an essential step in the isolation of auxotrophs and is also advantageous in resistance mutant isolation, to avoid negative interactions between cells in mixed aggregates. The formation of cell clones also simplifies quantification of mutagen effects on viability and mutation rate.

There are, unfortunately, few species for which haploids are available whose leaves* can be used reproducibly to produce large populations of protoplasts. The following species, all from the Solanaceae, are the most widely used: *N. tabacum,*[4,5] perhaps of restricted value because of its amphidiploidy; *Petunia hybrida,*[6] a species with the additional advantage of a well-characterized genetic system — the "Mitchell" haploid line is being increasingly used;[7] *N. sylvestris,*[8-10] a protoplast system that has proved unreproducible in some laboratories; *Daturia innoxia,*[11] used only infrequently; *N. plumbaginifolia,*[12,13] a very efficient protoplast system in use in several laboratories; *Hyoscyamus muticus,*[14,15] a system in which plant regeneration needs to be improved.

The most conspicuous problem in plant protoplast technology now is the recalcitrance of cereal protoplasts. Sustained division of cereal mesophyll protoplasts has not yet been observed. Plant regeneration from protoplasts prepared from morphogenic cereal cell cultures has been claimed but not unequivocally demonstrated.[16] Crop species in general are proving difficult to approach at the protoplast level.[17]

The mutants isolated using plant cell or protoplast culture have been reviewed many times recently.[18-22] It is not our intention in this review to present the same information yet again. Rather, we will concern ourselves with the methodology of mutant isolation using protoplasts. Mutagens will be discussed (Section II) from the point of view of modes of action, application, and effectiveness in protoplast systems. There is an urgent need for basic information on optimum conditions for mutation induction in cultured plant cells and for standardization of procedures and their quantification.

Selection techniques for a limited range of mutants have been investigated in a preliminary way (e.g., for resistance to chlorate, valine, threonine/lysine, and varous antibiotics, and for auxotrophs, temperature-sensitive and pigment mutants). The techniques and problems associated with their use will be discussed in Section III. Unlike mutagen application during the initial stages of protoplast culture, selection is usually applied to aggregated, protoplast-derived cells and will not differ in principle from selection applied to normal cell or tissue cultures. However, there is a need for some standardization of the methodology and also of ways of expressing results.

* The highest yield of viable protoplasts is usually to be obtained from leaves.

II. MUTAGENS

A. Introduction

Since 1973, when the first genetically confirmed mutants were obtained from plant cells in vitro, all of the major chemical and physical agents known as mutagens in other systems have been applied to protoplasts, although their use has not been very systematic: the mutagen, the dose, the cell type, cell growth conditions, and time of application frequently seem to have been chosen quite arbitrarily. In many reports the mutagenesis procedure is not adequately described and it is often not clear whether or not mutagen application increased mutant yield.

Where it is possible to obtain high protoplast yields and plating efficiency which thus permit the screening of several million viable colonies, careful consideration should be given to applying selection without prior mutagenesis of the protoplast population. For example, the frequency of spontaneous nitrate reductase minus (NR^-) or pigment-deficient mutants in *N. plumbaginifolia* arising during the first 10 days to 2 weeks of culture exceeds 10^{-5} per colony tested. Such mutants can easily be recovered without mutagen treatment, thus avoiding secondary metabolic complications, such as the decline in morphogenic capacity associated with high-kill mutagen doses. Similarly, it is to be expected that a broad screen for amino-acid auxotrophs, in which mutations in several hundred genes are sought, would yield mutants without applying mutagens. Gebhardt et al.[15] found such auxotrophs in the absence of mutagen application and at low mutagen (MNNG) doses at a frequency of about 10^{-4} per viable colony tested.

Furthermore, there is an increasing body of evidence which suggests that in vitro conditions themselves are mutagenic (see Chapter 10) and that mutations may accumulate in the leaf cells from which the protoplasts are derived.[23,24]

Although there has been extensive characterization of the basic mechanicsms by which chemical and physical agents can bring about alterations in DNA, some fundamental questions remain unanswered. The importance of error-prone repair of inactivating DNA alterations for the fixation of mutations has only recently become clear and it is not yet entirely understood how these repair mechanisms operate.[25-28] Different mutagens appear to "induce" different repair processes and, for any given mutagen, the exact process may differ between organisms. Therefore, extrapolation of mutagen data from one system to another is often not possible.

The schemes that have been presented for both the direct (mispairing) and indirect (misrepairing) mechanisms of mutation imply that the effect initially should be on one DNA strand with the other strand remaining in the nonmutant state. Sectoring of colonies for color mutations (so called "mosaic" colonies), predicted on the basis of single-strand mutations, can be seen in fungi: half sectors being presumably the result of mutations arising prior to DNA replication and quarter sectors the result of mutations in replicated cells. However, pure mutant colonies are also found, the proportion of pure to mosaic colones varying with the mutagen and/or mutagen dose.[29,30] The existence of the pure colonies requires an explanation and it is also necessary to investigate under which conditions and with which mutagens the highest frequency of pure mutant clones can be obtained.

Total isolation techniques (with or without replica plating) will not rescue auxotrophic plant cells that are part of chimeric colonies. Furthermore, the idea to use single-cell mutagenesis in vitro to avoid the problem of chimerism in plant breeding is too simplistic. In both cases the system must make use of single haploid cells, in an unreplicated state when exposed to a mutagen which causes mutation by a replication-independent repair process, the mutation being communicated to both DNA strands prior to DNA replication. However, the latter condition need not be satisfied if, for example, death of the sister cell, which is not carrying the mutation of interest, occurs

Table 1
FREQUENCY OF INDUCED MUTATION AFTER VARIOUS MUTAGENIC TREATMENTS IN G1 CELLS OF *Saccharomyces cerevisiae*

Mutagenic agent	Treatment	Survival (%)	No. of colonies observed	No. of mutants/10⁴survivors[a] C	M	Total	% Complete mutants
MNNG	10 min	52.7	33,996	15.30	12.40	27.70	55.9
	40 min	4.1	17,300	16.20	11.00	27.20	60.0
NA[b]	10 min	67.5	40,540	1.50	2.50	4.00	37.5
	25 min	18.1	213,021	3.40	12.10	15.50	21.8
EMS	10 min	67.2	34,307	2.60	12.80	15.40	16.9
MMS[c]	10 min	61.0—71.1	106,395	0.85	0	0.85	100.0
	25 min	8.9	15,900	1.88	0.63	2.51	75.0
UV	250 ergs	59.0—67.4	24,622	4.87	0	4.87	100.0
	600 ergs	5.2—13.2	35,595	7.86	1.69	9.55	82.4

[a] C, complete or pure mutant clones; M, mosaic clones.
[b] NA: nitrous acid.
[c] MMS: methyl methane sulfonate.

From Nasim, A., Hannan, M. A., and Nestmann, E. R., *Can. J. Genet. Cytol.*, 23, 73, 1981. With permission.

after the first mitosis due to the presence of a lethal mutation. If such a lethal-hit mechanism is operating to produce pure clones it is to be expected that the proportion of pure clones will rise with increasing mutagen dose. With haploid *Saccharomyces cerevisiae* cells in G1, UV, and methyl methane sulfonate produced mainly pure clones, ethyl methane sulfonate and nitrous acid produced mainly mosaics and nitrosoguanidine produced about an equal number of each.[30] (See Table 1).

Following on from the above discussion, an important consideration in protoplast mutagenesis experiments is the proportion of G1 and G2 cells in populations of mesophyll protoplasts. In some cases the majority of mesophyll cells in mature leaves are in G1[31] (Figure 1) but the cell cycle distribution can vary with the developmental stage of the leaf and the growth conditions. Leaves from in vitro cultures of *N. plumbaginifolia* have a higher proportion of cells in G2 than leaves of plants grown in the soil.[31]

Obviously the timing of mutagen application is also critical during the early stages of protoplast culture as cells undergo DNA replication.

The occurrence of chimeric colonies is not so critical when selecting for dominant resistant mutants using a strong selective agent, although the survival of resistant cells existing as a minority in a dying colony may be reduced..

It is clear that successful mutant selection requires the standardization of the source and preparation of leaf protoplasts as well as of the timing of mutagenesis and application of selection pressures. The inherent variability of protoplast preparations requires that visible or biochemical markers be used to plot the progress of protoplasts in each experiment.

B. Chemical Mutagens

1. Ethyl Methane Sulfonate (EMS)

Alkylating agents such as EMS comprise the largest group of mutagens and induce

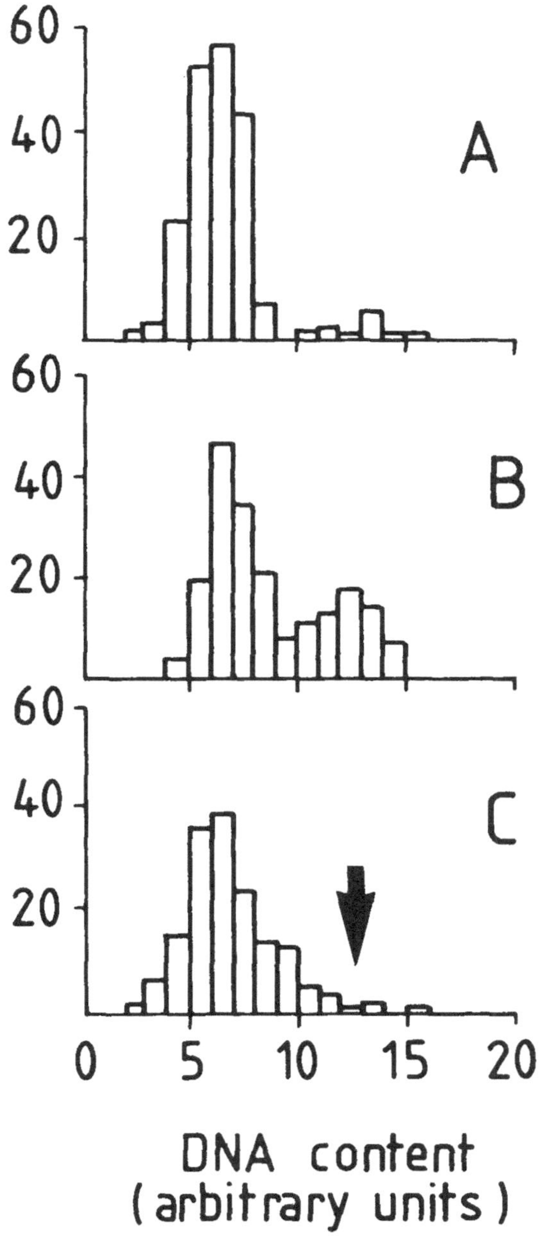

FIGURE 1. Frequency distributions of nuclear DNA contents of mesophyll protoplasts from leaves of in vitro cultured shoots. (A) *N. sylvestris* on agar medium. (B) *N. plumbaginifolia* on agar medium. (C) *N. plumbaginifolia* on sterilized turf. The arrow indicates the DNA value of 4C mitotic nuclei. (From Magnien, E., Dalschaert, X., and Faraoni-Sciamanna, P., *Plant Sci. Lett.*, 25, 291, 1982. With permission.)

point mutations, chromosome breaks, and chromosome mutations. The point mutations are mainly the result of GC/AT transitions following guanine alkylation.[25] EMS treatment produced mainly mosaic clones in *S. cerevisiae*[30] suggesting that mutations are single stranded and/or dependent upon replication for their fixation. EMS is relatively stable in water (half-life = 11.5 hr) but highly volatile. Special apparatus has been developed for prolonged application to plant cells in culture.[32] In the barley seedling test, EMS reaches very high mutation frequencies at the LD_{50} dose.[33] In other plants there are significant deleterious effects on growth. Myo-inositol has been reported to increase survival of plants if added during EMS seed treatment.[34] In experiments with animal cells, EMS, in contrast to several other mutagens, consistently induced mutations in the absence of cell killing.[35]

Although EMS has frequently been chosen as a mutagen for plant cells or protoplasts in culture,[2,36,37] it has not yet been demonstrated unequivocally to be mutagenic. In two instances EMS clearly failed to increase mutant recovery.[38,39]

2. N-Methyl-N′-Nitro-N-Nitrosoguanidine (MNNG)

This alkylating agent is one of the most potent in various cellular systems,[25] causing mutation by mispairing (giving transitions) or misrepair. The proportion of mutations produced by the two mechanisms varies between organisms.[26] It is more unstable than EMS, being rapidly denaturated at high or low pH (e.g., half-life at pH 7 = 7.5 hr). Its maximum stability is in the pH range 4.5 to 6, and pH 6 is optimal for induction of point mutations. It is also readily decomposed by light.[40] Evidence from microorganisms indicates enhanced mutagenicity of MNNG at replication forks where it may produce a series of closely linked mutations.[41] However, in *S. pombe* MNNG produces about equal numbers of pure and mosaic clones[29] and thus is able to induce some double-stranded mutations prior to DNA replication. Although there is controversy about the effectiveness of MNNG at different times in the yeast cell cycle,[42] it would be expected to be mutagenic applied to both resting and dividing cells.

MNNG has been found to be nonmutagenic in barley but is, however, a highly effecient mutagen in *Arabidopsis thaliana*. Though not a widely used mutagen in plants, MNNG is active when applied to protoplasts. The frequency of recovery of auxotrophs and temperature-sensitive lines was increased from about 2×10^{-4} to 1.1×10^{-3} at the highest tested dose of 20 $\mu g \ell^{-1}$ for 30 min (95% kill) applied to 2-day-old *Hyoscyamus muticus* haploid mesophyll protoplasts.[15,43,44] Treatment of freshly isolated (nondividing) protoplasts was ineffective. MNNG applied to protoplasts is also reported to have produced valine resistance in diploid *N. tabacum*[38] and lines with altered pigmentation in haploid *Datura innoxia*.[45]

MNNG is recognized as a very potent carcinogen and must be used with extreme caution.

3. N-Ethyl-N-Nitrosourea (NEU)

NEU is another alkylating agent far more efficient than MNNG as a mutagen in the barley test system and together with its methyl analog is sometimes referred to as a "supermutagen".[38] It is unstable in aqueous solution, decaying completely within 48 hr, and it is, therefore, not necessary to wash cells after treatment.

NEU apparently increased the frequency of recovery of mutants from haploid *N. plumbaginifolia* protoplasts resistant to lincomycin,[46] streptomycin,[47] and chlorate.[48] Maliga reports that NEU doses above 0.1 m*M* induced sterility in regenerated lincomycin-resistant plants.[49]

4. Other Chemicals

The three compounds described above are, so far as we know, the only chemical mutagens reported to have been used in protoplast experiments. Methyl methane sulfonate[50] and sodium azide[51] have been applied successfully to other types of plant cell culture systems. There are several other compounds known to be mutagenic in plants[33,52] and microorganisms that should be investigated as soon as a reproducible in vitro mutation test system is available.

C. Radiation

1. UV Irradiation

UV is one of the best-characterized and most easily applied mutagens. It causes discrete gene mutations predominantly as the result of error-prone excision and repair of the pyrimidine dimers it produces. UV irradiation is less likely to cause gross chromosomal mutations than the ionizing radiation of X- and γ-rays.[25] The precise biochemical mechanisms by which error-prone repair after UV damage causes mutation are not clear. However, Nasim et al.[30] have shown UV to result in mainly pure mutant colonies in *S. cerevisiae,* which is indicative of a pre-DNA replication double-stranded mutational event. This observation recommends UV as a mutagen for application to nondividing cells such as freshly isolated protoplasts. Rapid photo-reactivation of UV-induced DNA damage is found in most organisms and makes it essential to culture cells in the dark for some time after UV exposure.

The high UV absorbance, scattering and reflectance of tissue components (e.g., DNA, RNA, proteins, flavanoids, cuticle) make UV an unsuitable mutagenic agent for use with whole plants and there is little mention of UV in the plant breeding literature. Isolated protoplasts are ideal for UV treatment but even here the concentration and localization of organelles and other UV-absorbing materials can drastically influence the UV dose reaching cell nuclei.[53] However, the ease and comparative safety of UV application make it an obvious choice of mutagen for protoplasts. All the disturbing washing and medium changes associated with chemical mutagens can be avoided.

UV was probably mutagenic when applied in protoplast selection experiments for chlorate resistance in *N. plumbaginifolia*[13] and NAA resistance in *N. tabacum.*[54] UV was also applied to protoplasts in experiments in which the following variants were isolated: AEC resistance in *N. sylvestris*[55] and *N. plumbaginifolia,*[22] valine resistance in *N. tacacum,*[56] lysine plus threonine resistance in *N. tabacum*[57] and *N. sylvestris,*[22] chlorate resistance,[58] and amino acid auxotrophs[59] in *N. plumbaginifolia.*

The mutagenic efficiency of protoplast UV irradiation has been investigated by Grandbastien et al.[60] by selection of valine resistant mutants from haploid *N. tabacum* mesophyll-protoplast-derived cells. The plating efficiency of irradiated cells decreased with increasing UV dose applied 24 hr after protoplast isolation in a manner suggesting single-hit killing. The LD_{50} was approximately 1000 ergs mm^{-2} using a germicidal lamp (G30T8) with major emission at 254 nm and an incident dose rate of 33 ergs mm^{-2} sec^{-1}. (For an introduction to UV sources and measurement consult Jagger.[61]) The protoplasts were cultured at a density of 7×10^4 cells per milliliter in 10-mℓ aliquots in 90-mm diameter plastic petri dishes. After optimizing mutant recovery conditions in reconstruction experiments using mesophyll protoplasts from a previously isolated valine resistant mutant, these authors also investigated the induced mutant frequency at increasing UV doses. The spontaneous mutant frequency was 10^{-6} to 10^{-7} per cell submitted to selection and the frequency increased to a maximum of 1.6×10^{-4} per cell following a dose of 1500 ergs mm^{-2}. Variability in the results, however, increased with increasing doses and the authors recommend treatments reducing plating efficiency by less than 50% (Table 2).

2. X-Irradiation

Ionizing radiation causes inactivating DNA changes either by direct DNA hits in dehydrated tissues such as seeds and spores or by the indirect effects of reactive radicals produced in hydrated tissues which cause the alteration and liberation of DNA bases as well as backbone breakages. X-rays thus cause few point mutations but instead

Table 2
FREQUENCY OF VAL^r COLONIES AS RELATED TO MUTAGENIC TREATMENT

	UV dose (erg mm^{-2})				
	250	500	750	1000	1500
No. of independent experiments performed	2	5	2	8	4
Mean mutant frequency[a] ($\times 10^{-5}$)	1.25 ± 0.75	2.3 ± 2.0	6.9 ± 5.0	38.0 ± 3.0	53.0 ± 31.0

[a] Frequencies were calculated as the ratio of resistant colonies to plated dividing cells submitted to selection.

Adapted from Grandbastien, M. A., Bourgin, J. P., and Caboche, M., *Genetics*, 109, 409, 1985.

induce single- and double-strand breaks whose repair leads to chromosome mutation or loss by exchange reactions.[25] Following ionizing irradiation there appear to be differences in the capacity of different plant tissues for unscheduled DNA synthesis needed for repair.[62,63] The efficiency of mutagenesis by X-rays alters with O_2 partial pressure and temperature.[25,53]

Despite the induction of more chromosome aberrations than gene mutations, ionizing radiation has been preferred by plant breeders because of ease of application to multicellular organs, good penetration and reproducibility, and the high mutation frequencies achievable.[52] However, when working with single cells or protoplast cultures, where penetration of chemicals is no problem, ionizing radiation would seem less desirable than the use of point-mutation producing chemicals, or UV irradiation.

Preliminary experiments to investigate the sensitivity of haploid and diploid *N. tabacum* protoplasts to X-ray irradiation showed a peak of sensitivity 2 days after isolation, before the onset of mitoses (day 3) and cell division (day 4) and during the time in which DNA synthesis is in progress in a large proportion of the population.[64] Doses ranged from 750 to 1500 R at a rate of 750 R min^{-1}. Haploid cells were always more sensitive than diploid cells. Krumbiegel[45] demonstrated an exponential decrease in cell viability with increasing X-ray dose for mesophyll protoplasts of haploid and diploid *Datura innoxia* and *Petunia hybrida*. Doses between 250 and 1500 R affected haploid and diploid *Datura* cells equally; beyond 1500 R diploid cells were clearly more resistant to irradiation. For *Petunia*, doses up to 1000 R reduced survival of both haploid and diploid protoplasts to about 50%. Further increases in dose (up to 3000 R) had a greater effect on haploid than diploid cells. Unfortunately, neither of the two studies described above appear to have led to experiments in which mutants were isolated.

Schieder[11] irradiated haploid mesophyll protoplasts of *D. innoxia* with X-rays, immediately after isolation. About 10^5 protoplasts were given a dose of 1000 R (50% kill). Of the 2.5×10^4 calli produced, 10 showed altered pigment patterns that were also expressed in leaves or shoots regenerated from the cultures. No alterations were detected among 10^5 calli produced from nonirradiated cells. Steffen and Schieder[65] also report the isolation of four nitrate reductase deficient (NR^-) cell lines following X-irradiation (1000 R) of freshly isolated mesophyll protoplasts from *Petunia* "Mitchell" haploid plants. The frequency of mutants is given as 1×10^{-5} but it is not clear whether the X-rays induced the mutants. The background (spontaneous) frequency of NR^- lines in *N. plumbaginifolia* is 5.8×10^{-5}.[48]

Table 3
SOME DRUG-RESISTANT PLANTS ISOLATED USING PROTOPLASTS

Drug	Species	Mutagen	Mutant frequency[a]	Inheritance[b]	Ref.
Valine	*N. tabacum*	UV	1.6×10^{-4}	R(2)	60
Lysine/ threonine	*N. tabacum*	UV	ND	D	57
NAA[c]	*N. tabacum*	UV	1.0×10^{-4}	ND	54
Lincomycin	*N. plumbaginifolia*	NEU	6.0×10^{-4}	M	46
Chlorate	*N. plumbaginifolia*	UV	1.3×10^{-5}	R	13

[a] Mutant frequencies are given as resistant lines isolated per colonies tested. ND no data.
[b] D, dominant; R(2) recessive, two loci; R, recessive; M, maternal.
[c] NAA, 1-napthaleneacetic acid.

3. Gamma-Irradiation

Ionizing γ-rays have been used extensively for mutation induction in plant breeding for similar reasons as mentioned above for X-rays.[52] Studies of the γ-radiosensitivity of protoplasts have indicated a period shortly after protoplast isolation, just before the onset of DNA replication, when cells are at their maximum sensitivity.[66] The data are very similar to that given for X-irradiation by Galun and Raveh.[64]

Exposure of *N. plumbaginifolia* mesophyll protoplasts to ^{60}Co γ-rays at the Biological Research Centre in Szeged has resulted in the isolation of various mutants and putative mutants: an isoleucine dependent auxotroph,[67] nitrate reductase deficient (NR^-) cell lines,[48] and pigment-deficient mutants.[68] The frequency of chlorate resistant (NR^-) lines among the clones tested increased from 5.8×10^{-5} to 1.5×10^{-3} after irradiation reducing protoplast viability by 92%. Similar data were recorded for the pigment deficient lines.

III. SELECTION

Several biochemical mutants have been isolated using plant callus or cell suspension cultures (see the review by Maliga et al.[47]). The metabolism of protoplasts being essentially the same as that of cells in culture, the protoplast state itself does not lead to the application of any special selection pressures. In any case, in most mutant isolation experiments beginning with protoplasts, selection is applied to cells derived from protoplasts and not directly to the protoplasts themselves. Thus the reasons for using protoplasts in selection experiments are not so much related to the selection strategy but to stable ploidy levels, ability to regenerate plants, and ease of cloning, as described in Section I.

A. Positive Selection

Lines resistant to toxic amino acids, amino-acid analogs, antibiotics, hormones, and herbicides have been isolated from protoplast-derived cell suspensions of *Nicotiana* species (see Table 3 and the review by Bourgin[19]). In most cases plants were regenerated and the inheritance of the resistance trait studied. Where mutagens were applied, in most cases it was possible to demonstrate increased induction of mutations.

UV irradiation of haploid *N. tabacum* protoplasts induced mutations giving resistance to L-valine.[69] Grandbastien et al.[60] (see also Section II.C.1) using optimized selection conditions found Val^r mutants at a frequency of 1.6×10^{-4} per plated cell. The protoplasts were plated at 10^4 cells mℓ^{-1} and exposed to 2 m*M* valine after 1 to 2 days postirradiation incubation. Only 25% of the initial isolates were stably valine resistant and there was some evidence of chimerism among the developing colonies. Valine resistance was shown to be the result of reduced uptake and controlled by mutations at two independent loci. However, Bourgin et al.[70] has demonstrated the existence of other Val^r mutants which show no reduced uptake and single gene inheritance.

Using a similar procedure, resistance to a toxic combination of lysine and threonine (LT) was found in *N. tabacum* protoplast populations.[57] Protoplasts were incubated for 1 month after UV irradiation and then plated in amino-acid-containing medium. Two mutant plants were regenerated from resistant clones. In contrast to wild-type shoots, shoot explants from LT-resistant lines cultured on 1 m*M* lysine plus 1 m*M* threonine developed normally. Progeny testing by seed germination in the presence of 4 m*M* LT showed the trait to be dominant.

Cséplö and Maliga[46] treated freshly isolated protoplasts with NEU at 0.1 m*M* and then plated out surviving colonies after 3 weeks onto medium containing the antibiotic lincomycin hydrochloride (1 mg mℓ^{-1}). In the first experiment, 2 out of 1.5×10^4 colonies remained green and proved to be stably resistant. The trait was expressed by leaf tissue of regenerated plants and shown to be maternally inherited. In later experiments the spontaneous frequency of 10^{-4} per cell tested could be increased to 6×10^{-4} by NEU treatment.

With the availability of several practical protoplast systems and the repeated demonstrations of the isolation of resistance mutants, investigations of the factors limiting mutant recovery became possible. Grandbastien et al.[60] recognized the need for cell survival at low cell densities after mutagen treatments and used special media manipulations[71] that eliminate the dependence of plating efficiency on plating density between 1 cell mℓ^{-1} and 2×10^4 cells mℓ^{-1}. Using valine resistance as a marker, they measured and optimized the recovery of mutant cells in reconstruction experiments with *N. tabacum*. The cell density, the time of valine addition, and seasonal variations affecting the mother plants were the most critical factors. There appear to be some disadvantages to the use of valine resistance as a marker: a low (30% or less) proportion of the resistant lines produce stable resistant plants and several different mechanisms for resistance have been reported.[70] Resistance to lysine plus threonine has been suggested as a more suitable marker.[22]

B. Negative Selection

Conditional lethal mutants, for example auxotrophs and temperature-sensitive mutants that do not divide or that die under the conditions in which their phenotypes are expressed, usually cannot be selected directly. A good example of direct selection for a specific auxotroph is the use of chlorate to select NR$^-$ mutants (see Section III. A) but positive selection methods of this type are rare. Conditional lethals are usually isolated by clone testing (so called "total isolation" [72]). The labor of testing clones (essentially done by dividing each clone between permissive and restrictive conditions) can be reduced in two ways: (1) automated multiple clone testing by a replica plating device. The type originally designed for bacteria by Lederberg and Lederberg[73] is unsuitable for plant cells. A net plating procedure was described for suspension cultures by Schulte and Zenk[74] but so far has not found wide acceptance. (2) An enrichment process in which a nonspecific selective agent destroys most of the wild-type cells under restrictive conditions when wild-type cells divide and mutant cells do not. The surviving cells are plated in permissive conditions and the reduced number of clones tested individually. Several enrichment agents are being tested with plant cell cultures (see Section B.2).

Due largely to the development of suitable cell and protoplast culture systems (see Section I), recent attempts to isolate auxotrophs using direct selection, total isolation, and enrichment have been successful and reproducible, the latter in the sense that mutants were consistently isolated although the phenotypes were not always specified at the outset of the experiment.

The problems of regenerating plants from auxotrophic lines that are not temperature sensitive are fairly obvious. Unfortunately temperature-sensitive mutants are still infrequent and those so far reported are growth inhibited at the restrictive temperature for largely unknown reasons. However, a temperature-sensitive line of *H. muticus* isolated by Gebhardt et al.[15] has recently been shown to be auxotrophic for auxin at the high temperature.[75]

1. Total Isolation

Manual testing of clones derived from haploid cell cultures[37,76] or, more frequently, from haploid mesophyll protoplasts[15,44,67] to date has yielded 17 auxotrophic or temperature-sensitive lines with well-characterized phenotypes (Table 4). This appears to be the outcome of individually testing about 50,000 clones. The basic procedure involves culturing protoplasts in permissive conditions until the derived cell clusters reach a size which allows them to be picked with fine forceps and transferred as clones to a growth medium. As soon as the mass of individual clones is great enough, the calli are subdivided and cultured in permissive and restrictive conditions. Clones showing depressed growth in restrictive conditions are examined further. While the initial clone isolation can be achieved at the rate of 1000 clones per person per day, further clone testing requires much more time. For this reason Gebhardt et al.[15] picked clones directly into restrictive conditions and only retested the small percentage of clones unhappy in those conditions. This immediate testing can be carried out by placing clones in a regular array on medium in 9-cm petri dishes or, to avoid possible cross-feeding, in the wells of multi-well dishes. Such immediate testing should be carried out with clones as large as possible to avoid random failure of growth simply due to small clone size. There is the danger that clones with rapidly lethal phenotypes will not be recovered. The permissive conditions suitable for nutritional mutants are generally unknown before the mutants are isolated and some guesswork is involved in the construction of recovery media. It is interesting that the optimum amino-acid concentration for growth of histidine and tryptophan auxotrophs of *Hyoscyamus muticus* is close to that chosen originally for the rescue medium.[77] The effort needed for total isolation points to the need to use mutagens known to be effective for the protoplast systems in use. Because of the time required to establish the stable auxotrophy or temperature sensitivity of clones, it is advisable to use a plant regeneration medium from the very start as the basis of the recovery medium, to avoid loss of morphogenic potential.

2. Enrichment

5-Bromodeoxyuridine (BUdR) has been widely used in enrichment experiments with cultured plant cells (Table 5). BUdR is incorporated instead of thymine into DNA during replication. Because of the lability of this halogenated base, especially when irradiated with X-rays, UV, or fluorescent light, BUdR causes inactivating as well as mutagenic DNA alterations. Nonreplicating cells are relatively safe from BUdR exposure. The basic procedure has been to culture protoplasts in permissive conditions following mutagenesis to allow mutation expression and then to transfer the protoplast-derived cells to restrictive conditions. After an interval to allow the cessation of growth of mutant cells, BUdR is added for a period equal to at least 1.5 cell cycles to ensure that all dividing cells incorporate some BUdR. After washing, the population is plated in permissive conditions. In theory, only mutant cells will form colonies. Usually, in-

Table 4
TOTAL ISOLATION EXPERIMENTS PRODUCING AUXOTROPHIC AND TEMPERATURE-SENSITIVE CLONES

Species	Culture type	Mutagen	Dose	Survival (%)	No. of clones tested	No. of auxotrophs isolated	Phenotypes of lines recovered	Ref.
Datura innoxia	62% haploid cell culture	EMS	0.05% 1.0 hr	81	2,370	1	$Pant^-$	37
						1	$Aden^-$	76
Hyoscyamus muticus	Haploid mesophyll protoplasts	MNNG	20 $mg\ell^{-1}$ 0.5 hr	10	3,070	1	NR^-	44
Hyoscyamus muticus	Haploid mesophyll protoplasts	MNNG[a]	—	100	5,001	1	His^-	15
			5 $mg\ell^{-1}$	60	9,047	1	NR^-/Leu^-[c]	
			10 $mg\ell^{-1}$	23	10,140	5	NR^-, Trp^-, His^-, Nic^- (X2)	
			20 $mg\ell^{-1}$	8	4,684	5	NR^-, His^-, Nic^-, ts (X2)	
Nicotiana plumbaginifolia	Haploid mesophyll protoplasts	γ-rays[b]	13	57	3,727	0	—	67
			16	23	3,705	0	—	
			19	15	2,377	0	—	
			23	11	1,240	2	$Ileu^-$, $Uracil^-$	

[a] Treatment time = 0.5 hr.
[b] Dose expressed as Jkg^{-1} at a dose rate of 0.042 Jkg^{-1} sec^{-1}.
[c] A line showing both a requirement for leu and the absence of nitrate reductase activity.

Table 5
AUXOTROPHS ISOLATED IN EXPERIMENTS IN WHICH ENRICHMENT CONDITIONS WERE APPLIED

Species	Culture type	Mutagen	Dose	Survival (%)	Enrichment conditions[a]	Evidence for enrichment	No. of clones tested	Phenotypes of lines recovered	Ref.
Nicotiana tabacum	Haploid cell culture	EMS	0.25% 1 hr	46—67	BUdR S=96 hr C=10^{-5} *M* T=36 hr	None	119	6[b]	78
Nicotiana tabacum	Haploid cell culture	None	—	—	BUdR S=48 hr C=10^{-5} *M* T=72 hr	None	84	9[c]	79
Datura innoxia	Haploid cell culture	EMS	0.5% 1.5 hr	Not given	Arsenate S=48 hr C=2 m*M* T=24 hr	None[d]	13	Ileu/Val[e]	80
Datura innoxia	Haploid cell culture	EMS	0.5% 1.5 hr	Not given	Arsenate S=48 hr C=10 m*M* T=21 hr	None[d]	8370	NR^-,Thr^-	81
Nicotiana plumbaginifolia	Haploid mesophyll protoplasts	UV	Not given	88	BUdR S=96 hr C=1.25 × 10^{-5} *M* T=48 hr	None	1620	1[f]	59
		UV	Not given	77[g]	S=96 hr C=5 × 10^{-5} *M* T=48 hr	None	9282	5[h]	59
		UV	Not given	75	S=168 hr S=3.3 × 10^{-5} *M* T=72 hr	None	1796	0	59
Hyoscyamus muticus	Haploid mesophyll protoplasts	MNNG	20 mgℓ^{-1} 0.5 hr	15	BUdR S=96 hr C=2 × 10^{-5} *M* T=96 hr	None	375	His^-	82

Table 5 (continued)

AUXOTROPHS ISOLATED IN EXPERIMENTS IN WHICH ENRICHMENT CONDITIONS WERE APPLIED

[a] S = starvation time; C = concentration; T = treatment time.

[b] Lines required hypoxanthine, biotin, p-aminobenzoic acid, arginine, lysine or proline but all were leaky. Published data on nutrient requirements are not very convincing.

[c] Only two of the lines were characterized.

[d] Although in the isolation experiment there was no evidence for enrichment, the experiments were preceded by reconstruction experiments to establish the optimal conditions for rescuing auxotrophs from wild-type populations.

[e] A line with a double requirement for Ileu and Val.

[f] Possibly a requirement from among His, Ser, Pro, or Gly. No data published.

[g] Mean survival from 3 experiments.

[h] Possibly requirements for Ileu; two other clones grow on the group Ileu, Leu, Val, Cys; a Trp^- clone; and a further clone growing on the group = Lys, Thr, Meth, Arg.

termediate conditions are used that increase the proportion of mutant cells in the population (= enrichment). Although several experiments in which BUdR was applied have yielded auxotrophs or temperature sensitives there is no real evidence that BUdR enriched for mutant cells; the overall results so far are 8 to 10 lines recovered after testing about 22,000 clones (compare Table 4 with Table 5). In no case were control populations examined for frequency of mutants before BUdR treatment.

According to Polacco,[83] another potential enrichment agent, arsenate, most likely acts as a phosphate analog causing uncoupling of substrate level and oxidative phosphorylation, thus severely reducing the energy charge in growing cells below a "point of no return". Resting cells maintain the minimum ATP levels to reinitiate growth. Arsenate has been used in several reconstruction experiments using previously isolated auxotrophs[80,81] and conditions established in which auxotrophic cells are protected from an arsenate treatment killing the majority of growing cells. Horsch and King[81] further discovered that feeder layers enhanced the recovery of surviving cells at low density. This "low density" effect probably contributes significantly to the fluctuation in survival of nongrowing cells after arsenate[80] and BUdR[82] treatment.

Auxotrophs have been isolated in experiments using arsenate (Table 5) but as with BUdR there is no indication of enrichment. Further experiments are required to establish the effectiveness of enrichment procedures particularly when specific auxotrophic phenotypes are being sought.

REFERENCES

1. Maliga, P., Sz.-Breznovits, A., and Márton, L., Streptomycin-resistant plants from callus culture of haploid tobacco, *Nature New Biol.*, 244, 29, 1973.
2. Carlson, P. S., The use of protoplasts for genetic research, *Proc. Natl. Acad. Sci. U.S.A.*, 10, 598, 1973.
3. Hibberd, K. A. and Green, C. E., Inheritance and expression of lysine plus threonine resistance selected in maize tissue culture, *Proc. Natl. Acad. Sci. U.S.A.*, 79, 559, 1982.
4. Nagata, T. and Takebe, I., Plating of isolated tobacco mesophyll protoplasts on organic medium, *Planta*, 99, 12, 1971.
5. Caboche, M., Nutritional requirements of protoplast-derived, haploid tobacco cells grown at low cell densities in liquid medium, *Planta*, 149, 7, 1980.
6. Binding, H., Regeneration von haploiden Pflanzen aus Protoplasten von *Petunia hybrida*, *Z. Pflanzenphysiol.*, 74, 327, 1974.
7. Ausubel, F. M., Bahnsen, K., Hanson, M., Mitchell, A., and Smith, H. J., Cell and tissue culture of haploid and diploid *Petunia* "Mitchell", *Plant Mol. Biol. Newsl.*, 1, 26, 1980.
8. Bourgin, J.-P., Missonier, C., and Chupeau, T., Culture de protoplasts de mesophylle de *Nicotiana sylvestris* Spegazzini et Comes haploid et diploide, *C.R. Acad. Sci. (Paris) Ser. D.*, 282, 1853, 1976.
9. Nagy, J. I. and Maliga, P., Callus induction and plant regeneration from mesophyll protoplasts of *Nicotiana sylvestris*, *Z. Pflanzenphysiol.*, 78, 453, 1976.
10. Durand, J., High and reproducible plating efficiencies of protoplasts isolated from *in vitro* grown haploid *Nicotiana sylvestris* Spegaz. et Comes, *Z. Pflanzenphysiol.*, 93, 283, 1979.
11. Schieder, O., Isolation of mutants with altered pigments after irradiating haploid protoplasts from *Datura innoxia* Mill. with X-rays, *Mol. Gen. Genet.*, 149, 251, 1976.
12. Bourgin, J.-P., Chupeau, Y., and Missonier, C., Plant regeneration from mesophyll protoplasts of several *Nicotiana* species, *Physiol. Plant.*, 45, 288, 1979.
13. Negrutiu, I., Dirks, R., and Jacobs, M., Regeneration of fully nitrate reductase-deficient mutants from protoplast culture of *Nicotiana plumbaginifolia* (Viviani), *Theor. Appl. Genet.*, 66, 341, 1983.
14. Wernicke, W., Lörz, H., and Thomas, E., Plant regeneration from leaf protoplasts of haploid *Hyoscyamus muticus* produced via anther culture, *Plant Sci. Lett.*, 15, 239, 1979.

15. Gebhardt, C., Schnebli, V., and King, P. J., Isolation of biochemical mutants using haploid mesophyll protoplasts of *Hyoscyamus muticus* II Auxotrophic and temperature-sensitive clones, *Planta*, 153, 81, 1981.
16. Vasil, V. and Vasil, I. K., Isolation and culture of cereal protplasts. II. Embryogenesis and plantlet formation from protoplasts of *Pennisetum americanum, Theor. Appl. Genet.*, 56, 97, 1980.
17. Thomas, E., King, P. J., and Potrykus, I., Improvement of crop plants via single cells *in vitro* — an assessment, *Z. Pflanzenzücht.*, 82, 1, 1979.
18. King, P. J., From single cells to mutant plants, *Oxford Surv. Plant Mol. Cell Biol.*, 1, 1, 1984.
19. Bourgin, J.-P., Protoplasts and the isolation of plant mutants, in *Protoplasts 1983 Lecture Proc.*, Experientia Supplementum Vol. 46, Potrykus, I., Harms, C. T., Hinnen, A., Hütter, R., King, P. J., and Shillito, R. D., Eds. Birkhäuser Verlag, Basel, 1983, 43.
20. Maliga, P., Protoplasts in mutant selection and characterization, in *International Review of Cytology*, Suppl. 16, Giles, K. L., Ed., Academic Press, New York, 1983, 161.
21. Chaleff, R. S., *Genetics of Higher Plants. Applications of Cell Culture*, Cambridge University Press, Cambridge, 1981.
22. Negrutiu, I., Jacobs, M., and Caboche, M., Advances in somatic cell genetics of higher plants — the protoplast approach in basic studies on mutagenesis and isolation of biochemical mutants, *Theor. Appl. Genet.*, 67, 289, 1984.
23. Lörz, H. and Scrowcroft, W. R., Variability among plants and their progeny regenerated from protoplasts of *Su/su* heterozygotes of *Nicotiana tabacum, Theor. Appl. Genet.*, 66, 67, 1983.
24. Thomas, E., Bright, S. W. J., Franklin, J., Lancaster, V. A., Miflin, B. J., and Gibson, R., Variation amongst protoplast-derived potato plants (*Solanum tuberosum* cv Maris Bard), *Theor. Appl. Genet.*, 62, 65, 1982.
25. Freese, E., Molecular mechanisms of mutations, in *Chemical Mutagens. Principles and Methods for their Detection*, Vol. 1., Hollaender, A., Ed. Plenum Press, New York, 1971, chap. 1.
26. Drake, J. W. and Baltz, R. H., The biochemistry of mutagenesis, *Annu. Rev. Biochem.*, 45, 11, 1976.
27. Witkin, E. M., Ultraviolet mutagenesis and inducible DNA repair in *Escherichia coli, Bacteriol. Rev.*, 40, 869, 1976.
28. Bridges, B. A., Recent advances in basic mutation research, *Mutat. Res.*, 44, 149, 1977.
29. Nasim, A. and Auerbach, C., The origin of complete and mosaic mutants from mutagenic treatment of single cells, *Mutat. Res.*, 4, 1, 1967.
30. Nasim, A., Hannan, M. A., and Nestmann, E. R., Pure and mosaic clones — a reflection of differences in mechanisms of mutagenesis by different agents in *Saccharomyces cerevisiae, Can. J. Genet. Cytol.*, 23, 73, 1981.
31. Magnien, E., Dalschaert, X., and Faraoni-Sciamanna, P., Transmission of a cytological heterogeneity from the leaf to the protoplasts in culture, *Plant Sci. Lett.*, 25, 291, 1982.
32. Shillito, R., Robinson, N. E., and Street, H. E., Isolation and characterization of mutant cell lines via plant cell cultures, in *Experimental Mutagenesis in Plants*, Filev, K., Ed., Bulgarian Academy of Sciences, Sofia, 1978, 275.
33. Ehrenberg, L., Higher plants, in *Chemical Mutagens. Principles and Methods for their Detection*, Vol. 2, Hollaender, A., Ed., Plenum Press, New York, 1971, 365.
34. Rao, S. V., Myo-inositol induced growth in ethyl methanesulphonate treated tobacco, *Theor. Appl. Genet.*, 67, 203, 1984.
35. Fox, M., Factors affecting the quantitation of dose-response curves for mutation induction in V79 Chinese hamster cells after exposure to chemical and physical mutagens, *Mutat. Res.*, 29, 449, 1975.
36. Aviv, D. and Galun, E., An attempt at isolation of nutritional mutants from cultured tobacco protoplasts, *Plant Sci. Lett.*, 8, 299, 1977.
37. Savage, A. D., King, J., and Gamborg, O. L., Recovery of a pantothenate auxotroph from a cell suspension culture of *Datura innoxia* Mill., *Plant Sci. Lett.*, 16, 367, 1979.
38. Caboche, M. and Muller, J.-F., Use of a medium allowing low cell density growth for *in vitro* selection experiments: isolation of valine-resistant clones from nitrosoguanidine-mutagenized cells and gamma-irradiated tobacco plants, in *Plant Cell Cultures: Results and Perspectives*, Sala, F., Parisi, B., Cella, R., and Ciferri, O., Eds. Elsevier/North-Holland Biomedical Press, Amsterdam, 1980, 133.
39. Maliga, P., Lázár, G., Svab, Z., and Nagy, F., Transient cycloheximide resistance in a tobacco cell line, *Mol. Gen. Genet.*, 149, 267, 1976.
40. Ehrenberg, L. and Wachtmeister, C. A., Handling of mutagenic chemicals: experimental safety, in *Handbook of Mutagenicity Test Procedures*, Kilbey, B. J., Ed., Elsevier, Amsterdam, 1977, 411.
41. Cerdá-Olmedo, E., Hanawalt, P. C., and Guerola, N., Mutagenesis of the replication point by nitrosoguanidine: map and pattern of replication of the *Escherichia coli* chromosome, *J. Mol. Biol.*, 33, 705, 1968.

42. Carter, B. L. A. and Dawes, I. W., Nitrosoguanidine mutagenesis during the yeast cell cycle, *Mutat. Res.*, 51, 289, 1978.
43. Gebhardt, C., Strauss, A., and King, P. J., Isolation of auxotrophic and temperature-sensitive variants using haploid plant protoplasts, in *Induced Mutations — A Tool in Plant Research*, International Atomic Energy Agency, Vienna, 1981, 383.
44. Strauss, A., Bucher, F., and King, P. J., Isolation of biochemical mutants using haploid mesophyll protoplasts of *Hyoscyamus muticus*. I. A NO_3^- non-utilizing clone, *Planta*, 153, 75, 1981.
45. Krumbiegel, G., Response of haploid and diploid protoplasts from *Datura innoxia* Mill. and *Petunia hybrida* L. to treatment with X-rays and a chemical mutagen, *Environ. Exp. Bot.*, 19, 99, 1979.
46. Cséplö, Á., and Maliga, P., Lincomycin resistance, a new type of maternally inherited mutation in *Nicotiana plumbaginifolia*, *Curr. Genet.*, 6, 105, 1982.
47. Maliga, P., Menczel, L., Sidorov, V., Marton, L., Csepelo-Medyesy, P., Manh Dung, T., Lázár, G., and Nady, F., Cell culture mutants and their use, in *Plant Improvement and Somatic Cell Genetics*, Vasil, I. K., Frey, K. L., and Scrowcroft, W. G., Eds., Academic Press, New York, 1982, 231.
48. Marton, L., Manh Dung, T., Mendel, R. R., and Maliga, P., Nitrate reductase deficient cell lines from haploid protoplast cultures of *Nicotiana plumbaginifolia*, *Mol. Gen. Genet.*, 182, 301, 1982.
49. Maliga, P., Cell culture procedures for mutant selection and characterization in *Nicotiana plumbaginifolia*, in *Cell Culture and Somatic Cell Genetics*, Vasil, I. K., Ed., Academic Press, New York, in press.
50. Heinz, D. J., Krishnamurthi M., Nickell, L. G., and Maretzki, A., Cell tissue and organ culture in sugarcane improvement, in *Applied and Fundamental Aspects of Plant Cell, Tissue, and Organ Culture*, Reinert, J. and Bajaj, Y. P. S., Eds., Springer-Verlag, Berlin, 1977, 3.
51. Hibberd, K. A. and Green, C. E., Inheritance and expression of lysine plus threonine resistance selected in maize tissue culture, *Proc. Natl. Acad. Sci. U.S.A.*, 79, 559, 1982.
52. Broertjes, C. and van Harten, A. M., *Application of Mutation Breeding Methods in the Improvement of Vegetatively Propagated Crops*, Elsevier, Amsterdam, 1978.
53. Howland, G. P. and Hart, R. W., Radiation biology of cultured plant cells, in *Applied and Fundamental Aspects of Plant Cell, Tissue and Organ Culture*, Reinert, J. and Bajaj, Y. P. S., Eds., Springer-Verlag, Berlin, 1977, 731.
54. Muller, J. F. and Caboche, M., Isolation of tobacco clones resistant to naphthaleneacetic acid and affected in root morphogenesis, in *Protoplasts 1983 Poster Proc.*, Experientia Suppl. Vol. 45, Potrykus, I., Harms, C. T., Hinnen, A., Hütter, R., King, P. J., and Shillito, R. D., Eds. Birkhäuser Verlag, Basel, 1983, 156.
55. Negrutiu, I., Cattoir-Reynearts, A., Verbrugger, I., and Jacobs, M., Lysine-over-production in an S-aminoethylcysteine-resistant line, isolated in protoplast culture of *Nicotiana sylvestris*, *Arch. Int. Physiol. Biochim.*, 89, 188, 1981.
56. Vunsh, R., Aviv, D., and Galun, E., Valine resistant plants derived from mutated haploid and diploid protoplasts of *Nicotiana sylvestris* and *N. tabacum*, *Theor. Appl. Genet.*, 64, 51, 1982.
57. Bourgin, J.-P., Chupeau, M.-C., and Missonier, C., Amino acid-resistant plants from tobacco cells selected *in vitro*, in *Variability in Plants Regenerated from Tissue Culture*, Earle, E. D. and Demarly, Y., Eds., Praeger, New York, 1982, 163.
58. Caboche, M. and Marion-Poll, A., personal communication, 1984.
59. Negrutiu, I., Isolation of amino acid-requiring lines by negative selection in haploid protoplasts of *Nicotiana plumbaginifolia* (Viviani), in *Protoplasts 1983 Poster Proc.*, Experientia Suppl. Vol. 45, Potrykus, I., Harms, C. T., Hinnen, A., Hütter, R., King, P. J., and Shillito, R. D., Eds. Birkhäuser Verlag, Basel, 1983, 158.
60. Grandbastien, M. A., Bourgin, J. P., and Caboche, M., Valine-resistance a potential marker in plant cell genetics. II. Optimization of UV-mutagenesis and selection of valine-resistant colonies from tobacco mesophyll protoplasts, *Genetics*, 109, 409, 1985.
61. Jagger, J., *Introduction to Research in Ultraviolet Photobiology*, Prentice-Hall, Englewood Cliffs, N.J., 1967.
62. Painter, R. B. and Wolff, S., Apparent absence of repair replication in *Vicia faba* after X-irradiation, *Mutat. Res.*, 19, 133, 1973.
63. Howland, G. P., Hart, R. W., and Yette, M. L., Repair of DNA strand breaks after gamma-irradiation of protoplasts isolated from cultured wild carrot cells, *Mutat. Res.*, 27, 81, 1975.
64. Galun, E. and Raveh, D., *In vitro* culture of tobacco protoplasts: survival of haploid and diploid protoplasts exposed to X-ray radiation at different times after isolation, *Radiat. Bot.*, 15, 79, 1975.
65. Steffen, A. and Schieder, O., Selection and characterization of nitrate reductase deficient mutants of *Petunia*, in *Protoplasts 1983 Poster Proc.*, Experientia Suppl. Vol. 45, Potrykus, I., Harms, C. T., Hinnen, A., Hütter, R., King, P. J., and Shillito, R. D., Eds. Birkhäuser Verlag, Basel, 1983, 162.
66. Magnien, E., Dalschaert, X., and Devreux, M., Different radiosensitivities of *Nicotiana plumbaginifolia* leaves and regenerating protoplasts, *Plant Sci. Lett.* 19, 231, 1980.

67. Sidorov, V., Menczel, L., and Maliga, P., Isoleucine-requiring *Nicotiana* plant deficient in threonine deaminase, *Nature (London)*, 294, 87, 1981.
68. Sidorov, V. A. and Maliga, P., Fusion-complementation analysis of auxotrophic and chlorophyll-deficient lines isolated in haploid *Nicotiana plumbaginifolia* protoplast cultures, *Mol. Gen. Genet.*, 186, 328, 1982.
69. Bourgin, J.-P., Valine-resistant plants from *in vitro* selected tobacco cells, *Mol. Gen. Genet.*, 161, 225, 1978.
70. Bourgin, J.-P., Goujaud, J., Missonier, C., and Pethe, C., Valine-resistance, a potential marker in plant cell genetics. I. Distribution of two types of valine-resistant tobacco mutants isolated from protoplast-derived cells, *Genetics*, in press, 1984.
71. Muller, J. F., Missonier, C., and Caboche, M., Low density growth of cells derived from *Nicotiana* and *Petunia* protoplasts: influence of the source of protoplasts and comparison of the growth promoting activity of various auxins, *Physiol. Plant.*, 57, 35, 1983.
72. Beadle, G. W. and Tatum, E. L., *Neurospora*. II. Methods of producing and detecting mutations concerned with nutritional requirements, *Am. J. Bot.*, 32, 678, 1945.
73. Lederberg, I. and Lederberg, E. M., Replica plating and indirect selection of bacterial mutants, *J. Bacteriol.*, 63, 399, 1952.
74. Schulte, U. and Zenk, M. H., A replica plating method for plant cells, *Physiol. Plant.*, 39, 139, 1977.
75. Oetiker, J. and King, P. J., unpublished data, 1984.
76. King, J., Horsch, R. B., and Savage, A. D., Partial characterization of two stable auxotrophic cell strains of *Datura innoxia* Mill., *Planta*, 149, 480, 1980.
77. Gebhardt, C., Shimamoto, K., Lázár, G., Schnebli, V., and King, P. J., Isolation of biochemical mutants using haploid mesophyll protoplasts of *Hyoscyamus muticus*. III. General characterization of histidine and tryptophan auxotrophs, *Planta*, 159, 18, 1983.
78. Carlson, P. S., Induction and isolation of auxotrophic mutants in somatic cultures of *Nicotiana tabacum*, *Science*, 164, 487, 1970.
79. Malmberg, R. L., Temperature-sensitive variants of *Nicotiana tabacum* isolated from somatic cell culture, *Genetics*, 92, 215, 1979.
80. Horsch, R. B. and King, J., Isolation of an isoleucine-valine-requiring auxotroph from *Datura innoxia* cell cultures by arsenate counterselection, *Planta*, 159, 12, 1983.
81. Horsch, R. B. and King, J., The isolation of auxotophs from *Datura innoxia* Mill. cell cultures following recovery of arsenate-treated cells on feeder plates, *Planta*, 160, 168, 1984.
82. Shimamoto, K. and King, P. J., Isolation of a histidine auxotroph of *Hyoscyamus muticus* during attempts to apply BUdR enrichment, *Mol. Gen. Genet.*, 189, 69, 1983.
83. Polacco, J. C., Arsenate as a potential negative selection agent for deficiency variants in cultured plant cells, *Planta*, 146, 155, 1979.

Chapter 12

HYBRIDIZATION BY SOMATIC CELL FUSION

Christian T. Harms

TABLE OF CONTENTS

I. INTRODUCTION

Somatic cell fusion leading to the formation of viable cell hybrids has been a tremendously powerful technique in vertebrate cell biology[1] and, particularly, human genetics[2] and immunology.[3] Plant cells with their rigid cellulose wall have long resisted any attempts to fuse them. It was not until the development by Cocking[4] of a technique for the large-scale isolation of plant protoplasts that this major problem, the cell wall, was literally digested away by pectinolytic and cellulolytic enzymes. Although protoplasts were seen to fuse spontaneously under certain conditions, the real breakthrough for somatic plant cell fusion came only with the development of efficient fusion techniques,[5,6] notably through the introduction, in 1974, of polyethylene glycol as a potent fusogen (see Chapter 4).[7,8] With these new tools at hand plant cell biologists have begun to explore what had been termed "alternatives to sex"[9] some 25 years ago. Protoplast fusion has been developed primarily as a method for the genetic manipulation of plant cells. Not only does somatic cell fusion provide us with an opportunity to construct hybrids between taxonomically distant plant species beyond the limits of sexual crossability, but protoplast fusion also creates cells with new genetic, nuclear as well as cytoplasmic, constitutions that cannot otherwise be obtained. The experimental establishment of new combinations of nuclei, chloroplasts, and mitochondria provides a novel and potent tool to study the genetic and physiological interactions between these organelles. Questions as to the complementation, segregation, and recombination of functions encoded by these organelles can now be approached analytically. Somatic incompatibility,[10] resulting from the enforced union of alien nuclear, chloroplastic, and mitochondrial genetic elements is now open to experimental investigation. Cell hybridization might further be exploited for the mapping of genes and linkage analysis in somatic plant cells as it has been so successfully using human-rodent cell hybrids.[2] Further areas of research include basic studies on the regulation of gene expression with respect particularly to the control of differentiation and the realization of morphogenetic programs. The ability to regenerate plants from intraspecific, interspecific, and even intergeneric hybrid cells allows not only the construction of novel plants but it also provides an opportunity to rescue somatic cell markers into heterozygous plants which can then be analyzed by conventional sexual means.

Considering the roughly 10 years of its history, plant protoplast fusion has been applied very successfully to the study of these issues, and progress is well documented.[11-17] Methodological aspects of protoplast fusion, the fusion process itself, and the early stages in the development of fusion products are being discussed elsewhere (see Chapter 4).[5] In this review I shall focus on some aspects regarding the selection and characterization of the products of protoplast fusion as well as on some of the genetic consequences and uses of somatic hybridization.

II. THE NATURE OF PROTOPLAST FUSION PRODUCTS

The induced fusion of protoplasts from diverse sources generates a novel cellular entity in which the protoplasms, organelles, and genetic elements from both partners are put together. This early stage is called a heterokaryon or a heterokaryocyte. It is characterized by the mixing and progressive intermingling of the cytoplasmic components but the two parental nuclei have not fused (Figure 1). Plasmogamy may be followed by karyogamy, either at interphase or in the first mitotic division of the nuclei, to produce a real hybrid cell with one hybrid nucleus. Whether or not the two nuclei will fuse is dependent upon a variety of factors. The frequency of nuclear fusion may be as low as 0.1% or less in remote intergeneric fusion combinations, or close to 100%

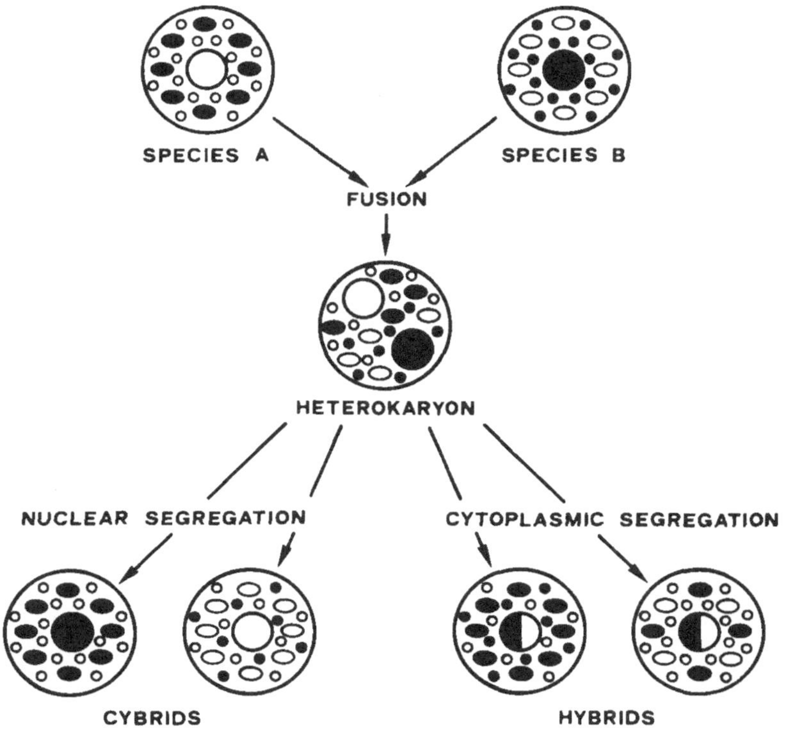

FIGURE 1. Interspecific fusion of protoplasts carrying different genomes, plastomes, and chondriomes, and some of the new genetic combinations that are obtained as a result of segregational events.

in intraspecific fusions. It is evident that the qualitative as well as the quantitative outcome of a fusion experiment is severely determined by the probability at which nuclear fusion occurs.

The genetic amalgam of a newly formed fusion product is usually not maintained for long periods. Rather there is a tendency for segregational events to alter the original genetic constitution of fusion products and hybrid cells. Segregation may affect the cytoplasmic organelles, plastids, and mitochondria, as well as the two nuclear genomes. Segregation of plastids and/or mitochondria gives rise to homoplastidic hybrid cells or cells combining chloroplasts from one parent with the mitochondria from the other (cytoplasmic recombinant types). Segregation of nuclear genomes can be a consequence of the sequential loss of chromosomes from (usually) one parent. Segregation of nuclear genomes can also result from the first cell division of a heterokaryocyte if nuclear fusion did not occur or if one of the parental nuclei has degenerated. Such early segregation, followed by the proliferation of both types of daughter cells, will give rise to chimeric tissue composed of a mixture of genetically diverse cells. Degeneration of one parental nucleus will create a cell which combines both parents' cytoplasms with only one parent's nuclear genome. Such cells are cytoplasmic hybrids (cybrids, heteroplasmons). Further segregation of the two cytoplasms may finally lead to alloplasmic cells containing one parent's nuclear genome combined with the other parent's cytoplasmic constituents. In Figure 1, only a few of the many possible types of cells are depicted to illustrate the kinds and directions of segregational events which may affect protoplast fusion products and hybrid cells. The significance of these segregational events and the conditions under which these diverse types of cells are derived from one initial heterokaryocyte will be discussed later in more detail.

A. Initial Inequality of Fusion Products — a Possible Cause of Segregation

The genetic fate of a fusion product may be determined as early as by its first division. Inequality and genetic imbalances within and among fusion products from their very initiation can be determinative for their further development and for the rate and direction of segregational events that they face during development. Various types of *ab initio* inequality may be distinguished:[15] imbalances due to cell cycle phase, (an)euploidy, ploidy level, structural chromosome rearrangement, differentiative, and physiological state. Experimental evidence as to the effects of these factors is as yet very poor.

1. Cell Cycle Phase

Protoplasts from different sources can differ with respect to the cell cycle phase of their nuclei (see Chapter 5). Leaf cells usually are all in G_1 phase whereas protoplasts isolated from suspension cultured cells are in different phases of their cell cycle. Fusion of heterophasic protoplasts can lead to premature chromosome condensation in the parental nucleus which was in interphase. Even more drastic effects such as chromatin pulverization have been observed in heterophasic fusion products.[18] Segregation and loss of chromatin in subsequent divisions are the likely consequences. Fragmentation of chromatin can, however, be a desirable prerequisite in the transfer of small pieces of genetic material by somatic cell fusion.[19] Ashmore and Gould[20] have shown that protoplasts from cells in different cell cycle phases participate in fusion events at rates different than expected. Changes, according to the cycle phase, of the surface charge of protoplasts have been suggested as a possible cause. To study the effects of different cell cycle phases of parental protoplasts on the genetic behavior of their fusion products in more detail, protoplasts could be isolated from synchronized cultures,[21] fractionated fluorimetrically in a cell sorter[22,23] according to cell cycle phase, then fused in specific combinations and cultured individually after microisolation.[24-33] The fate of each individual product could then be monitored to reveal the beneficial or detrimental effects of certain cell cycle combinations. Fusion of protoplasts which are in controlled phases of their cell cycle has been suggested as a means to obtain more stable genetic constitutions in interspecific and intergeneric hybrid cells.[26]

Frequently, protoplasts from mesophyll cells have been fused with those isolated from suspension cultured cells (cf. Tables 1 to 3). One good reason for doing so is that fusion products of this type are easily recognized during the first hours of culture by the presence of visual markers (chloroplasts and abundant cytoplasm) which are contributed by the two parents. This easy visual distinction of unfused from heterospecifically fused protoplasts has greatly aided in the selection of fusion products by microisolation[24-33] (cf. Tables 1 to 3). Disequilibrium does, however, exist in that the two fusion partners most likely differ with respect to their cell cycle phases, their chromosomal and ploidy constitutions, and the physiological and differentiative state they are in. It is well established that plant cells in culture can become progressively polyploid and aneuploid and are subject to drastic chromosome rearrangements[34] (see Chapter 10). These alterations eventually interfere with differentiation and morphogenesis from these cultures but they may also be maintained and expressed at the regenerated plant level. When incorporated into a fusion product, such abnormalities may well affect its further development in various ways which are as yet not fully understood.[15]

Table 1
INTRASPECIFIC SOMATIC HYBRID PLANTS[a]

Species fused[b]		Selection method	Cytoplasmic constitution[c]	Ref.
Solanaceae				
Datura innoxia (M,r), line A1/5a, chlorophyll-deficient	✱ *D. innoxia* (M,r), line A7/ls, chlorophyll-deficient	Picking of green colonies	n.r.	40
Hyoscyamus muticus (S,r), various combinations of auxotrophic (NR^-, nic^-, his^-, trp^-) and temperature-sensitive (ts) lines	✱ H. muticus (S,r),	Complementation on minimal medium	n.r.	41—43
Nicotiana debneyi (M,R), line TS 233, wt	✱ *N. debneyi* (M,R), line TS 287, wt	No selection	4: TS 287 2: TS 233(cpDNA)	44
Nicotiana plumbaginifolia (M,r), various auxotrophic (ile^-, leu^-, ura^-) and chlorophyll-deficient lines	✱ *N. plumbaginifolia* (M,r)	Complementation on minimal medium; picking of green colonies	n.r.	45
Nicotiana plumbaginifolia (S,r), nia type NR^- line	✱ *N. plumbaginifolia* (S,r) cnx type NR^- line	Complementation on minimal medium	Parental lines are cytoplasmically identical	46
Nicotiana plumbaginifolia (C,r), lines NA9 and NX9, NR^-	✱ *N. plumbaginifolia* (M,R), line A28, n albino	Unilateral medium selection; green shoot regeneration	Parental lines are cytoplasmically identical	47
Nicotiana tabacum (M,r), var. virescent, chlorophyll-deficient	✱ *N. tabacum* (M,r), var. sublethal, chlorophyll-deficient	Picking of green colonies	Parental lines are cytoplasmically identical	39
Nicotiana tabacum (M,R), cv. Hicks-2, wt	✱ *N. tabacum* (C,R), cv. Hicks-2, albino	No selection	Parental lines are cytoplasmically identical	48
Nicotiana tabacum (M,R), cv. Samsun; cv. Xanthi, wt	✱ *N. tabacum* (M,R), cv. Techné, cms-deb	No selection	n.r.	49
Nicotiana tabacum (M,R), cv. Turkish Samsun, cp albino	✱ *N. tabacum* (S,r), cv. Gatersleben cnx-68, cnx type NR^- line	Picking of green autotrophic colonies	22: green	50
Nicotiana tabacum (M,R), cv. Burley 21, cms-sua	✱ *N. tabacum* (S,r) cv. Gatersleben cnx-68 cnx type NR^- line	Unilateral medium selection; green shoot regeneration	16: sua (LSU-RuBCase) 5: tab 2: sua + tab	51

Table 1 (continued)
INTRASPECIFIC SOMATIC HYBRID PLANTS[a]

Species fused[b]		Selection method	Cytoplasmic constitution[c]	Ref.
Nicotiana tabacum (M,R), cv. Burley 21, cms-sua, cms-und, cms-gla (line BP210)	✱ *N. tabacum* (S,r) cv. Gatersleben cnx-68 cnx type NR^- line	Unilateral medium selection; green shoot regeneration	5: tab (LSU-RuBCase + 15: sua tentoxin sens.)	52
			22: und (LSU-RuBCase + tentoxin sens.)	52
			11: tab (LSU-RuBCase + 10: gla tentoxin sens.)	52
Nicotiana tabacum (M,R), SR1, cp streptomycin-resistant	✱ *N. tabacum* (C,r) B6S3 tumour line	Unilateral medium selection; green shoot regeneration	4: SR1 (resistant) 12: ? (sensitive)	53
Nicotiana tabacum (M,R), cv. Xanthi nc, line *PmR1 Gut*, picloram-resistant, glycerol utilizing	✱ *N. tabacum* (S,r) cv. Xanthi nc, line *HuR9* hydroxyurea-resistant	Selective media	Parental lines are cytoplasmically identical	54
Nicotiana tabacum (M,r), cv. Xanthi nc, line *PmR1 Gut*, picloram-resistant, glycerol utilizing	✱ *N. tabacum* (S,r), cv. Xanthi nc, lines clr0, clr15, clr19, clr30, chlorate-resistant	Selective media	Parental lines are cytoplasmically identical	55
Petunia hybrida (M,R), line 64, wt	✱ *P. hybrida* (M,R), line Sfla, cms	No selection	n.r.	56
Umbelliferae *Daucus carota* (S,M,r), line WCH105, cyclohexmide-resistant	✱ *D. carota* (C,R), cv. Nantaise Slendero, line A1, albino	Unilateral inhibition by iodoacetate; green plant regeneration	n.r.	57

[a] As of December 1983.

[b] Abbreviations: In Tables 1 to 4 cultivar and line designations have been used as in the original reports. Further abbreviations: protoplast source — callus (C), suspension culture (S), leaf mesophyll (M); parental protoplast cultures are morphogenic (R), nonmorphogenic (r) under the applied conditions; cms, cytoplasmic male sterile analog containing cytoplasm from *Nicotiana debneyi* (-deb), *N. glauca* (-gla), *N. megalosiphon* (-meg), *N. suaveolens* (-sua), *N. undulata* (-und); cnx, NR^- mutant defective in the molybdenum cofactor of nitrate reductase; cp, chloroplastic; n, nuclear; nia, NR^- mutant defective in the apoenzyme of nitrate reductase; NR, nitrate reductase; wt, wild type.

[c] For legend, see Table 2.

Table 2
INTERSPECIFIC SOMATIC HYBRID PLANTS[a]

Parental species A[b]	× Parental species B	Selection method	Cytoplasmic constitution[c]	Ref.
Solanaceae: Datura				
Datura innoxia (M,R), line A1/5a, chlorophyll-deficient	× *D. candida* (M,r), wt	Picking of largest green colonies	?: inn (cpDNA probes)	58, 59
Datura innoxia (M,R), line A1/5a, chlorophyll-deficient	× *D. discolor* (M,r), wt	Picking of largest green colonies	n.r.	60
Datura innoxia (M,R), wt	× *D. quercifolia* (M,r), wt	Picking of largest colonies	n.r.	37
Datura innoxia (M,R), line A1/5a, chlorophyll-deficient	× *D. sanguinea* (M,r), wt	Picking of largest green colonies	?: san (cpDNA probes)	58, 59
Datura innoxia (M,R), line A1/5a, chlorophyll-deficient	× *D. stramonium* (M,r), var. tatula, wt	Picking of largest green colonies	n.r.	60
Solanaceae: Nicotiana				
Nicotiana glauca (M,r), wt	× *N. langsdorffii* (M,r), wt	Hormone-independent growth	n.r.	61
Nicotiana glauca (M,r), wt	× *N. langsdorffii* (M,r), wt	Hormone-independent growth	8: lan (LSU-RuBCase) 6: gla 1: heteroplastidic	62, 63
Nicotiana glauca (M,r), wt	× *N. langsdorffii* (M,r), wt	Hormone-independent growth	n.r.	64
Nicotiana glauca (S,r), wt	× *N. langsdorffii* (S,r), wt	Hormone-independent growth	n.r.	65
Nicotiana sylvestris (S,r), line KR103, kanamycin-resistant, albino	× *N. knightiana* (M,r), wt	Unilateral medium selection; picking green resistant colonies	n.r.	66
Nicotiana sylvestris (S,r), line KR103, kanamycin-resistant, albino	× *N. knightiana* (M,r), wt	Microisolation of fusion products	2: kni (SR resistant) 2: heteroplastidic	28

Table 2 (continued)
INTERSPECIFIC SOMATIC HYBRID PLANTS[a]

Parental species A[b]	✕ Parental species B	Selection method	Cytoplasmic constitution[c]	Ref.
Nicotiana sylvestris (S,r), wt	✕ *N. rustica* (S,R), cv. chlorotica, chlorophyll-deficient	Differential growth; picking of green colonies	8: rus (LSU-RuBCase)	67
Nicotiana tabacum (M,R), cv. Burley 21, wt	✕ *N. alata* (M,r), wt	No selection	n.r.	68
Nicotiana tabacum (S,R), var. J.W. Broadleaf, Su/Su sulfur mutant	✕ *N. glauca* (M,R), wt	Picking of light green colonies	3: tab (LSU-RuBCase) 1: tab (tentoxin resist.)	69, 75
Nicotiana tabacum (M,R), cv. Bright Yellow, aurea mutant	✕ *N. glutinosa* (M,r), wt	Picking of greenish-white compact calli	n.r.	68
Nicotiana tabacum (S,R), cms-deb	✕ *N. glutinosa* (S,r), wt	No selection	1: glu (LSU-RuBCase)	70, 71
Nicotiana tabacum (S,r), cv. Xanthi nc, 5-methyl-tryptophan-resistant	✕ *N. glutinosa* (M,R), wt	Unilateral medium selection; green plant regeneration	4: glu 1: tab } (LSU-RuBCase)	72
Nicotiana tabacum (S,R), albino	✕ *N. knightiana* (M,r), wt	Green shoot regeneration	n.r.	73
Nicotiana tabacum (C,R), line SR1, cp-streptomycin-resistant	✕ *N. knightiana* (M,r), wt	Microisolation of fusion products	13: tab 15: kni } (cpDNA)	32
Nicotiana tabacum (S,R), var. J. W. Broadleaf, Su/Su sulfur mutant	✕ *N. nesophila* (M,r), wt	Picking of light green shoots	5: nes (tentoxin sensitivity)	74, 75
Nicotiana tabacum (S,R), var. J. W. Broadleaf, Su/Su sulfur mutant	✕ *N. otophora* (M,r), wt	Picking of light green shoots	2: oto (tentoxin sensitivity)	75, 76
Nicotiana tabacum (M,r), cv. Samsun, albino	✕ *N. plumbaginifolia* (C,r), wt	Unilateral inhibition by iodoacetate and irradiation	20: tab (albino) 2: plu (green) 1: heteroplastidic	77
Nicotiana tabacum (C,r), line SR1, cp-streptomycin-resistant	✕ *N. plumbaginifolia* (M,r), wt	Enucleation; selective medium	56: tab (SR resistant)	78

Nicotiana tabacum (M,r), line 70 St-R701, cms-meg, streptomycin-resistant	✱ *N. plumbaginifolia* (M,R), wt	Uniparental lethal irradiation; selective medium	6: tab (streptomycin-resistant	79
Nicotiana tabacum (M,r), cv. Bright Yellow, aurea mutant	✱ *N. repanda* (M,r), wt	Picking greenish-white compact calli	n.r.	80
Nicotiana tabacum (M,r), cv. Bright Yellow, aurea mutant	✱ *N. rustica* (M,r), cv. rustica, wt	Picking greenish-white compact calli	2: tab, 1: rus } (LSU—RuBCase)	81, 82
Nicotiana tabacum (M,r), cv. Burley 21, wt	✱ *N. rustica* (M,r), cv. rustica, wt	No selection	n.r.	81
Nicotiana tabacum (S,R), albino	✱ *N. rustica* (S,R), cv. chlorotica, wt	Picking of green colonies	3: tab, 11: rus } (LSU-RuBCase)	83, 84
Nicotiana tabacum (S,R), var. J. W. Broadleaf, Su/Su sulfur mutant	✱ *N. stocktonii* (M,r), wt	Picking of light green shoots	n.r.	74
Nicotiana tabacum (M,R), var. sublethal, cholophyll-deficient	✱ *N. sylvestris* (M,r), chlorophyll-deficient	Picking of green colonies	n.r.	85
Nicotiana tabacum (M,r), line 92 cms-sua[d]	✱ *N. sylvestris* (M,r), wt	Unilateral lethal irradiation; unilateral selective medium	?: sua (LSU-RuBCase, tentoxin sensitivity, cpDNA	86, 87
Nicotiana tabacum (M,r), cv. Xanthi, wt	✱ *N. sylvestris* (M,r), line f48s, cms-sua[d]	Unilateral lethal irradiation; unilateral selective medium	1: tab (tentoxin resistant) 2: sua (tentoxin sensitive)	88
Nicotiana tabacum (M,r), line SR1, cp-streptomycin-resistant	✱ *N. sylvestris* (M,R), wt	Unilateral inhibition by iodoacetate; selective medium	105: tab (streptomycin-resistant)	89
Nicotiana tabacum (S,R), var. J. W. Broadleaf, Su/Su sulfur mutant	✱ *N. sylvestris* (M,R), wt	Picking of light green shoots	n.r.	76
Solanaceae: Petunia				
Petunia parodii (M,r), wt	✱ *P. hybrida* (M,R), cv. Comanche, wt	Differential growth; shoot regeneration	1: par (cpDNA)	90—92
Petunia parodii (M,r), wt	✱ *P. hybrida* (S,r), wt	Green shoot regeneration	n.r.	93

Table 2 (continued)
INTERSPECIFIC SOMATIC HYBRID PLANTS[a]

Parental species A[b]	✕ Parental species B	Selection method	Cytoplasmic constitution[c]	Ref.
Petunia parodii (M,r), wt	✕ *P. hybrida* (S,r), cv. Comanche, albino	Differential growth; green shoot regeneration	n.r.	94
Petunia parodii (M,r), line 3688, cms	✕ *P. hybrida* (M,r), line 3704, wt	Differential growth in selective medium	3923: fertile 92: sterile 32: heteroplasmic partially sectored, segregating fertile and sterile sexual progeny	95
Petunia parodii (M,r), wt	✕ *P. inflata* (S,r), cp-albino	Differential growth; picking of green colonies	1: par (cpDNA)	92, 96
Petunia parodii (M,r), wt	✕ *P. parviflora* (C,R), n-albino	Differential growth; picking of green colonies	1: par (cpDNA, LSU-RuBCase)	92, 97, 98
Solanaceae: Solanum				
Solanum tuberosum (C,r), wt	✕ *S. chacoense* (M,r), wt	Differential growth; shoot regeneration	n.r.	99
Solanum tuberosum lines HH258 (M,R) HH345 (M,R) HH439 (M,R) Mo9 (M,r)	✕ *S. nigrum* (M,R), cp-atrazine-resistant	No selection	7: tub (atr-sensitive 2: nig (atr-resistant)	100
Umbelliferae: Daucus				
Daucus carota (S,R), cv. Nantaise Slendero, line A1 albino	✕ *D. capillifolius* (S,R), wt	Picking of green embryos	n.r.	101
Daucus carota (S,r), 5-methyltryptophan-resistant	✕ *D. capillifolius* (S,R), wt	Unilateral medium selection; green embryo formation	n.r.	102
Cruciferae: Brassica				
Brassica oleracea (M,r), var. capitata, cv. Early Spring, wt	✕ *B. campestris* (M,r), var. narinosa, wt	Differential growth	n.r.	103

Brassica oleracea (M,r), var. capitata, cv. Savoy King, wt	✱ *B. campestris* (M,r), var. perviridis, cv. Tendergreen, wt	Differential growth	n.r.	103
Brassica oleracea (M,r), var. capitata sabauda, cv. Praeco, wt	✱ *B. campestris* (M,r), var. pekinensis, cv. Early Hybrid, wt	Differential growth	n.r.	103
Fabaceae: Medicago				
Medicago sativa (M,R), line MSR12, cms	✱ *M. falcata* (M,R), line 318, wt	No selection	n.r.	104

[a] As of December 1983.

[b] For abbreviations, see Table 1.

[c] Number of hybrids analyzed; type(s) of plastids found; method of distinction (LSU-RuBCase, large subunit of Ribulose bisphosphate carboxylase oxygenase; cpDNA, restriction fragment analysis of chloroplastic DNA; cpDNA probes, DNA-DNA hybridization using labeled cpDNA probes) mtDNA, restriction fragment analysis of mitochodrial DNA; n.r., not reported, ?, number or constitution not clear. Abbreviation of species names is by 3-letter code. Figures give numbers of plants tested (No. calli, from which plants were regenerated)

[d] Based on recent restriction fragment analysis of cpDNA line 92 has been reclassified as being *Nicotiana tabacum* (cms-undulata).[154]

Table 3
INTERGENERIC SOMATIC HYBRIDS[a]

Parental species A[b]	X Parental speciesB[b]	Selection method	Morphogenetic response	Ref.
Solanaceae				
Atropa belladonna (S,R), cv. Lutea Doll, wt	X *Petunia hybrida* (M,R), cv. Satellite, wt	No selection	Abnormal shoots	105
Atropa belladonna (M,R), wt	X *Datura innoxia* (M,R), lines A1/5a; A7/1—N_L chlorophyll deficient	Picking of hairy green colonies	Abnormal shoots	106, 107
Atropa belladonna (M,R), wt	X *Nicotiana chinensis* (C,r), wt	Microisolation of fusion products	Abnormal shoots	30
Hyoscyamus muticus (S,r), line IVH2, nic^- auxotrophic line	X *Nicotiana tabacum* (S,r), cv. Gatersleben cnx-68, cnx type NR^- line	Selective media	Rooted shoots, flowering (grafted)	41
Solanum tuberosum (C,R), line HH258, wt	X *Lycopersicon esculentum* (M,r), var. cerasiforme, yellow-green yg-6 mutant	Picking of green shoots	Flowering shoots (grafted)	108—110
Solanum tuberosum (M,R), line H_3 703/17/114, chlorophyll deficient	X *Lycopersicon esculentum* (M,r), gilva mutant	Picking of green shoots	Abnormal shoots, sterile flowers	111
Solanum tuberosum (M,R), line H_3 703/14/77, chlorophyll deficient	X *Lycopersicon esculentum* (M,r), gilva mutant	Picking of green shoots	Normal and abnormal shoots, male sterile flowers	112
Solanum tuberosum (M,r), cv. Russet Burbank, clone 774, chlorophyll deficient	X *Lycopersicon esculentum* (M,r), cv. Rugers, cv. Nova, wt	Differential growth, green shoot regeneration	Flowering plants, sterile	113
Nicotiana glauca (M,R)	X *Petunia hybrida* (M,R),	No selection	Flowering plants, sterile	114, 115
Nicotiana tabacum (M,r), cv. Bright Yellow, aurea mutant	X *Salpiglossis sinuata* (S,r), wt	Picking greenish white compact calli	Plants, nonflowering	80
Nicotiana tabacum (C,r), B6S3 tumour line	X Petunia hybrida (M,R), wt	Picking green colonies from hormone-free medium	Shoots	116

Umbelliferae				
Daucus carota (S,R), cv. Nantaise Slendero, line A1, albino	✱ *Aegopodium podagraria* (M,r), wt	Picking green morphogenic colonies	Rooted shoots	117
D. carota (x) Ae. podagraria, somatic hybrid line (M,R)	✱ *Daucus carota* (S,R), cv. Nantaise Slendero, line A1, albino	Picking green morphogenic colonies	Flowering plants	118
Daucus carota (S,R) c.v. Nantaise Slendero, line A1, albino	✱ *Petroselinum hortense* (M,r)	Uniparental lethal irradiation, picking green morphogenic colonies	Shoots	19
Cruciferae				
Arabidopsis thaliana (C,r), wt	✱ Brassica campestris (M,r), cv. Steinacher Winterrubsen, wt	Microisolation of fusion products	Roots, flowering shoots, sterile	119—121

[a] As of December 1983.
[b] For abbreviations see Table 1.

2. Ploidy Level

Gross genomic imbalances will result if the fused cells differ in their ploidy levels or if the fusion event involved more than two protoplasts. The latter case can be rather frequent since multiple fusions usually increase with attempts to increase the fusion efficiency. Yet, the number of somatic hybrids displaying AAB or similar types of genome structure, or chromosome numbers that are suggestive of a multiple fusion origin, seems to be much smaller than one would expect from the frequency of multiple fusion events. This leads us to assume that products of multiple fusion may have to face developmental disadvantages which preclude their manifestation as somatic hybrid plants. On the other hand, a multiple fusion origin has been suggested when a chromosome number close to that expected seemed to support this assumption. In the absence of precise experimental data it is tempting to design an experimental approach which could be used to study the question of how genomic disequilibrium affects the development of protoplast fusion products. Different types of fusion products, varying in the numeric ratio of the two parental genomes, would be produced and then cultured individually after microisolation. This simple and straightforward approach would allow us to monitor the developmental potentialities of each of the various fusion products and to explore the nature and the rate of changes determinative for their development. The lack of experimental data precludes a thorough assessment of the consequences that are induced by unbalanced genomic constitutions of fusion products at this point of time. It is interesting to note, though, that the elimination of *Hordeum bulbosum* chromosomes from interspecific *Hordeum* sexual hybrids was found to be greatly affected by the balance of *H. bulbosum* and *H. vulgare* genomes.[35] Wetter and Kao[26] have observed an increase in chromosome stability when somatic hybrid cells of soybean ⋇ *Nicotiana glauca** were back fused to *N. glauca* protoplasts thus shifting the genomic ratio in these hybrids. Both of these examples seem to suggest a genetic and genomic control of chromosome segregation in hybrid cells.

3. Differentiation and Physiological State

Investigations are also needed to evaluate the effects of various differentiative and physiological states of the parental protoplasts. It is as yet not known how differences in the patterns of gene activity in the parental cells may influence the development of fusion products. It is tempting to assume that the differentiative states and patterns of gene expression of both parental protoplast types are reset in the course of dedifferentiation and rejuvenescence which regularly occur when protoplasts are put into culture. No data are available at present to prove or disprove this notion but experiments could easily be set up to study these questions.

Studies using vertebrate cells seem to suggest that the direction of chromosome loss from hybrid cells can be manipulated simply by chosing the types of cells that are being fused together. Mouse-human cell hybrids were found to eliminate mouse rather than human chromosomes when embryogenic mouse cells had served as one fusion parent.[36] Otherwise, the opposite pattern is usually observed.[1,2] It remains to be investigated with plant cells to what extent the source and the differentiative state of the parental protoplasts have an affect on the segregational events that alter the genetic constitutions of fusion products.

* The terminology of somatic hybrids is controversial in the literature. Most commonly, somatic hybrids have been denoted by "+" (for example *Nicotiana tabacum* + *N. sylvestris*) to contrast somatic from sexual crosses for which the symbol "x" is used. However, the "+" sign has been proposed previously to denote graft combinations. To avoid confusion, Hoffmann and Adachi[120] have suggested "(x)" for somatic hybrids. In my opinion, similarities and dissimilarities of somatic vs. sexual crosses may be best reflected by yet another symbol, "⋇" which is a superposition of the "X" and "+" symbols, and this sign will be used here.

Although the state of experimental analysis is still poor there is evidence to assume that inherent features of the parental protoplasts, as discussed above, can play a triggering role in determining the developmental fate and final genetic constitution of protoplast fusion products. Furthermore, the direction of segregational events is also influenced by various selective pressures that are imposed on a given fusion system intentionally or unintentionally. The consequences are selective advantages which act in favor of maintaining a certain genetic determinant while discriminating against another. Certain variant or mutant types (chloroplastic or nuclear) provide such selective advantage. For instance, fusion between protoplasts carrying functional and albino plastids, respectively, can establish a selective advantage for the wild-type chloroplasts and may thus lead to the recovery preferentially of chlorophyllous hybrids. Such behavior can often be predictable and then be used intentionally for the selection of somatic hybrids (cf. Tables 1 to 3). It is evident, hence, that the final result of a protoplast fusion is nothing but a reflection of the various selective pressures that have been operating. Several examples can be found in Tables 1 to 3 that certain types of hybrids, i.e., those with a specific cytoplasmic constitution, have been obtained more frequently or even exclusively due to the action of selective forces.

III. SELECTION PROCEDURES FOR THE ISOLATION OF SOMATIC HYBRIDS

Despite many efforts to increase the efficiency of protoplast fusion usually not more than 1 to 10% of the protoplasts in a treated population have actually undergone fusion.[5,6] The proportion of viable heterospecific binucleate A+B type fusion products may be even lower. In the absence of systems which would fuse protoplasts specifically to produce heterokaryocytes at a very high rate, there is an obvious need to select the products of fusion from among the unfused and homospecifically fused parental cells. This selection step can be, and often has been, a bottle neck in the somatic hybridization of plant cells although a great variety of different procedures have been described.[6,16,37]

A. Complementation Systems

For several years the apparent lack, in higher plants, of biochemical markers selectable at the cellular level has made it necessary to explore other means of selection (cf. below). As more and more such markers have become available[38] they have been employed successfully for somatic hybrid selection. Various nutritional auxotrophic, temperature sensitive, chlorophyll deficient and dominant drug resistant lines have been used for this purpose. Fusion of two protoplasts each carrying a different recessive selectable biochemical marker will generate a fusion product which happens to be functionally restored since each parent contributes a functional allel which corrects the respective deficiency of the other parent. Parental type protoplasts can thus be selected against when using unsupplemented media or otherwise nonpermissive conditions. Heterologous fusion products, in which both parents' deficiencies are mutually abolished by complementation, possess a unique constitution that allows their survival and proliferation under selective conditions. Recessive auxotrophic and conditional lethal mutants will thus give rise to heterozygous fusion hybrids displaying a wild-type autotrophic phenotype. Chlorophyll deficient lines will complement to produce fully green or intermediate hybrids which can be distinguished visually and then be picked manually from among the parental type colonies or plants. In case of dominant drug resistance, fusion products will carry and express resistance to both drugs. This combination of traits will allow them to proliferate while cells from the parental lines, carrying only one resistance trait, are being knocked out by one or the other of the toxic compounds that are both included in the selection medium.

Complementation of selectable biochemical markers provides powerful selection systems for the enrichment of somatic hybrids from protoplast fusion mixtures. Complementation can be used to study dominance vs. recessiveness of traits in the heterozygous situation of hybrid cells. Complementation, if it occurs after fusion, provides strong evidence for the nonallelic nature of the markers in question. Protoplast fusion and subsequent complementation analysis can thus be employed to genetically characterize markers that are expressed at the cellular level. Such a somatic cell genetical approach can be very helpful when sexual cross analysis is not possible due to the lack of fertile plant regeneration from the cultured cells carrying such markers.

1. Chlorophyll Deficiency Complementation

Intraspecific complementation of chlorophyll deficiencies was first demonstrated by Melchers and Labib[39] in a fusion involving protoplasts from the chlorophyll deficient light-sensitive varieties "sublethal" and "virescent" of tobacco. The same approach led Schieder[40] to isolate green somatic hybrids from among the chlorophyll-deficient parental colonies in *Datura innoxia*. Complementation between several chlorophyll deficient lines derived from irradiated haploid *Nicotiana plumbaginifolia* has shown that they were nonallelic.[45] Sidorov and Maliga[45] also have achieved intraspecific complementation when they fused protoplasts from a chlorophyll-deficient and a nutritional auxotrophic (uracil or isoleucine requiring) line of *Nicotiana plumbaginifolia*.

2. Auxotroph Complementation

Several authors[46,47,55,122-125] have reported on the selective recovery of hybrid cells able to utilize nitrate as their sole source of nitrogen when they fused protoplasts from nitrate reductase (NR) deficient cells. The lack of NR activity causes an absolute requirement for reduced nitrogen and is caused by a deficiency either in the NR apoenzyme (nia-type mutant) or in the molybdenum cofactor (cnx-type mutant). Cnx- and nia-type NR^- mutants regularly complement each other upon fusion whereas complementation within each group of mutants only occurs if they are nonallelic. Fusion complementation can thus be a valuable technique to classify NR^- mutants whose nature is not known and to study allelism of different mutant lines. Fusion complementation within a group of nine NR deficient lines of *Nicotiana plumbaginifolia*[46] has revealed four rather than two complementation groups indicating allelism in some and nonallelism in some other lines. Newly isolated cnx-type mutants of *Nicotiana tabacum* have complemented some previously isolated cnx lines of tobacco as well as nia-type mutants of both tobacco and *Nicotiana plumbaginifolia* but failed to complement each other.[123] Results obtained from interspecific fusion suggested that the cnx-type NR^- mutants of *Nicotiana tabacum* and *N. plumbaginifolia* belong to three different complementation groups.

Fusion of NR deficient tobacco protoplasts with those from a nitrate nonutilizing line of *Hyoscyamus muticus* has demonstrated the successful intergeneric complementation of auxotrophic lines.[124] These experiments have also indicated that the *H. muticus* line used was a cnx-type NR^- mutant. Preliminary evidence has been obtained that fusion of protoplasts from a NR^- line of *Petunia hybrida* and a cnx-type mutant of tobacco has yielded hybrid colonies which were able to grow in selective medium containing nitrate as the sole nitrogen source.[125]

Other mutant traits recently employed for the selection of somatic hybrids are nutritional auxotrophic lines of *Nicotiana plumbaginifolia*[45] and *Hyoscyamus muticus*[41-43] (cf. Table 1). In the latter species a whole series of different lines have been subjected to fusion complementataion tests, including various lines requiring histidine, tryptophan, or nicotinamide. Complementation has been observed in all but one combination which involved two lines independently selected for their nicotinamide requirement. Failure to complement can be taken as evidence for the allelic nature of these nic⁻ mutant lines. Shoot regeneration which was defective in some of the auxotrophs was also restored upon fusion at least when one fusion parent was morphogenic.[41]

Hybrid plants have been selected on the basis of fusion complementation between a cnx-type mutant of tobacco and a nicotinamide-requiring line of *Hyoscyamus muticus*[41] thus demonstrating the utility of complementation selection even in the case of intergeneric fusion combinations. The morphogenetic potential of the *H. muticus* line was dominantly expressed resulting in prolific shoot formation. The regenerated flowering plants are now being analyzed morphologically, physiologically, and if possible, genetically through sexual crosses and anther culture.

3. Complementation of Resistance Markers

Dominant characters such as traits confering resistance to antibiotics, herbicides, or other toxic compounds have recently been recognized as potent selectable markers which can be very useful for the selection of somatic hybrids. In principle, selection is similar to the above described complementation selection using recessive traits in that the resistance to a certain drug in the one parent corresponds to sensitivity to this drug in the other parent. When protoplasts from two resistant lines are being fused together, the sensitivity traits of each parent will be dominated by the respective resistance trait of the other parent. Fusion hybrids are distinguished, hence, by their double resistance from the parental cells each carrying a single resistance. Resistance to the amino acid analogs, 5-methyltryptophan (5MT) and S(2-aminoethyl)cysteine (AEC), has been combined by fusion of protoplasts from cell lines of *Nicotiana sylvestris* exhibiting resistance to 5MT and AEC, respectively.[126] Harms et al.[127] have selected double resistant somatic hybrids in *Daucus carota* following fusion of protoplasts from AEC- and 5MT-resistant parental lines. When exposed to increasing drug concentrations the hybrid lines showed dose-response curves similar to the respective resistant parental line thus indicating the dominant expression of the resistance traits.[128] Resistance to the proline analog, azetidine-2-carboxylic acid (A2C), a trait also carried by the 5MT-resistant parental line and not selected for in these experiments, was also expressed dominantly in some, and semidominantly in some other hybrid lines. The biochemical characteristics correlated with the resistance traits — abolished uptake of AEC and a feedback insensitive form of anthranilate synthase — were found to be retained in the hybrid lines.[129] Despite chromosomal instability and a reduction of chromosome numbers these lines displayed an extraordinary stability of the resistance phenotypes in the absence of selection pressure. The lack of segregating lines thus has precluded studies on the linkage and chromosome assignment of the resistance markers in these hybrids.

Semidominant expression of A2C resistance has been found in intraspecific hybrid cells selected following fusion of protoplasts from A2C-resistant and sensitive carrot cell lines.[130] Overproduction of proline, the causative characteristics of the resistant parental line, was at an intermediate level in the hybrids as expected of semidominant expression.

Complementation of amino acid analog resistance has also been accomplished recently in remote intergeneric fusions between 5MT-resistant carrot and AEC-resistant tobacco protoplasts[131] as well as between AEC-resistant carrot and 5MT-resistant protoplasts of tobacco.[132] In the latter case, hybrid lines selected on medium containing both drugs were found to progressively and preferentially lose carrot chromosomes while the AEC-resistant phenotype was retained in most of the lines examined.

4. *Resistant Auxotrophic Double Mutants*

Many potential fusion partners are of wild-type constitution and do not possess selectable markers. A useful solution to this problem would be to construct one parent in such a way that it carries a negatively together with a positively selectable marker, i.e., an auxotrophic and a resistant trait. As a result, the whole selection system can be placed into just one parent regardless of the genetic constitution of the other parent. When wild-type protoplasts are fused with this double mutant line, they can be eliminated by their sensitivity, whereas exclusion of the necessary auxilliary compounds from the culture medium serves to eliminate the auxotrophic though resistant parent. Only heterologous fusion products with their complemented resistant-autotrophic constitution will be able to survive the selective conditions. Following this line of thought, Hamill et al.[133] have established, through sexual crosses, a double mutant line of tobacco which is nitrate reductase deficient (hence auxotrophic for reduced nitrogen) and which also contains streptomycin-resistant chloroplasts. These authors conclude that this mutant will be valuable in performing somatic crosses between tobacco and other species within and beyond the genus *Nicotiana.*

B. Use of Metabolic Inhibitors

The above examples clearly illustrate the utility of various types of selectable biochemical mutants for somatic hybrid selection using simple complementation systems. Similar systems can be created when specific irreversible inhibitors are employed which block essential metabolic functions in the parental lines. Hence, inactivated parental cells are precluded from proliferation whereas heterologous fusion products possess a reconstituted complement of active metabolic pathways necessary for survival. This concept of mutual biochemical inactivation has first been practiced with mammalian cells.[134] When applied to plant cells in a preliminary way, this system failed to produce viable hybrids.[135] However, inactivation by iodoacetate, a rather unspecific reactant with sulfhydryl groups, has been widely employed to inactivate one of the parental protoplast populations[56,77,89,130] (cf. Tables 1, 2). Although it seems to require special care in establishing the appropriate working conditions (effective inactivation, yet low cross-toxicity, low level of escapes), this selection method may prove useful for fusion combinations particularly when there are no selectable markers present in the parental lines. This method can even be superior to approaches attempting to introduce selectable markers into a given set of parental lines that one might wish to fuse. This general applicability to a wide range of wild-type protoplasts makes this method highly attractive.

C. Use of Physical Characteristics

Provided parental protoplast populations are (or can be made) different from each other with respect to physical properties such as protoplast size, buoyant density, electrophoretic mobility, color, fluorescence, etc. such differences can provide potential selective criteria for the selection of somatic cell hybrids. Systems of this type are in wide use. Fusion of mesophyll and cell culture-derived protoplasts, for instance, yields heterokaryocytes that are easily distinguished from parental protoplasts by the presence of chloroplasts and abundant cytoplasm. Fusion products thus recognized can be handpicked and cultured individually in small drops of culture medium[24-31] or in a suitable nurse culture system[32,33] (cf. Tables 1 to 3). Parental protoplasts that are isolated from the same type of cells and not sufficiently different for visual distinction may be loaded with fluorescent dyes prior to fusion.[23,136-138] The double fluorescent label of fusion products can be recognized in a fluorescence microscope[138] and fusion products then cultured after microisolation. The double fluorescence provides a possibility to separate fusion products from parental protoplasts when fusion mixtures are run through a fluorescence activated cell sorter (FACS).[23,139]

Differences in the buoyant density of different types of protoplasts, and hence their different sedimentation behavior in isoosmotic density gradients,[140] have been used following protoplast fusion to obtain fractions highly enriched for heterokaryocytes.[141] The enrichment factor is dependent upon the existence of sufficient density differences. These can be established by the choice of protoplast types,[140] preselection of a very dense and a very light subpopulation from the parental protoplast populations, or by a density shift experimentally induced in one parental protoplast population.[142] Other possible methods, i.e., use of differences in electrophoretic mobility, viscosity gradients, or magnetic sorting procedures have as yet not seriously been considered for somatic hybrid selection.

D. Compound Selection Systems

Most of the somatic plant hybrids that are known today have been selected using compound selection strategies rather than straightforward marker complementation. Somatic hybrid plants, as well as the major features employed in their selection, are listed in Tables 1 to 3 for intraspecific, interspecific, and intergeneric fusion combinations. Compound selection systems have used differential growth characteristics, regenerability, and morphological features in vitro as well as in vivo to distinguish somatic hybrids from parental cells/plants. Some systems have been based on sequential procedures to eliminate one parental species after the other. Whatever system is employed, the utility of a selective system lies in the fact that it provides enrichment, to whatever degree is possible, for fusion hybrids possessing a certain genetic constitution by reducing or eliminating the nondesired parental-type cells. This enrichment thus greatly reduces the number of colonies or plants among which hybrids must be sought. In the absence of directed fusion systems which would make hybrid selection obsolete, somatic plant cell hybridization is a numbers game. Even when a given selection system is struck with constraints and allows for parental escapes, much work can be saved, work which otherwise would be required to tediously identify somatic hybrids from among large numbers of parental-type cells or plants. Only in exceptional cases, i.e., to avoid action of selective forces as in studies on the randomness of cytoplasmic segregation,[44,143] will it be feasible to renounce a selection system intentionally.

IV. VERIFICATION AND CHARACTERIZATION OF SOMATIC HYBRIDS

Successful passage through a selection system provides the first circumstantial evidence for the somatic hybrid nature of selected materials. Further evidence must be added based on traits not involved in the selection, particularly when the selection system used is prone to some residual leakiness. Proof of hybridity requires a clear demonstration of genetic contribution from both fusion parents, which is either evidenced by the physical presence or the expression of genetic material from both parents.[15]

A. Chromosomal Constitution

Counting chromosomes in presumed hybrid cells can be an easy and reliable method to verify hybrid cells and it also provides information on the ploidy state of the cells. Due to numerical, structural, and size differences of their chromosomes, hybrids between remote species are often easier to identify than are hybrids of more closely related species. The presence of marker chromosomes can greatly facilitate the analysis of genetic events in hybrid cells. More extensive use could be made of chromosome banding techniques in order to identify specific chromosomes and to investigate chromosome rearrangements in somatic hybrid cells.

B. Molecular Biological Techniques

Recent progress in the development of molecular biological techniques has greatly improved our capabilities to analyze the genetic constitution of somatic plant hybrids. While major genetic contributions (full chromosome sets, individual chromosomes, chromosome fragments) are evidenced by cytological methods, more refined techniques are required to detect minor genetic contributions to hybrid cells at the subchromosomal level. Specific restriction patterns of chloroplastic and mitochondrial DNA have been used to great advantage to characterize the nature of the plastoms and chondrioms of somatic hybrids and cybrids.[14] Species-specific restriction fragments of nuclear DNA coding for ribosomal RNA have shown potential to verify somatic hybrids such as *Nicotiana glauca* ✻ *N. langsdorffii*.[65] Another method recently developed by Saul and Potrykus[144] has used species-specific fragments of highly repetitive nuclear DNA as probes to identify parental DNA in intergeneric fusion hybrids between *Hyoscyamus muticus* and *Nicotiana tabacum*. This technique also has revealed recently the presence of reasonable amounts of tobacco DNA in these *Hyoscyamus* ✻ *Nicotiana* hybrids although the tobacco protoplasts had been X-irradiated extensively prior to fusion in order to disintegrate and eliminate their nuclei (see References 12, 88, 154 to 157).[145] DNA specific for tobacco appeared to be scattered over the *Hyoscyamus* nuclear genome as evidenced by Southern blotting and *in situ* DNA hybridization on hybrid cells.

C. Polypeptides

Expression of genetic material from both parents usually gives good indication for hybridity although not all genes of the two parents will probably become expressed in a somatic hybrid. Electrophoretic banding patterns of isoenzymes lend themselves favorably to hybrid verification,[15] but nonenzyme structural polypeptides may be analyzed as well.[128] Somatic hybrids may display isoenzyme bands of certain enzymes specific to one or the other parent, or to both parents simultaneously. In some cases intermediate bands have been detected[31,66,69,74,146] which are indicative of heteropolymers of multimeric enzymes which are composed of subunits coded for by the two fusion parents.

D. Secondary Metabolites

Parental differences in the production of secondary metabolites such as volatile compounds,[147] alkaloids,[77,148] and carotenoids[20,117] have been examined in somatic hybrids. It is obvious that these compounds with their complex multistep biosynthesis are rather far down the line of direct gene action and are usually susceptible to variations as a result of environmental, physiological, and developmental conditions. Their value for hybrid identification must be considered low. Analysis of secondary metabolites may, therefore, be used to provide supportive but never exclusive evidence to the verification of somatic hybrids.

E. Morphology

When plant regeneration has been accomplished from protoplast fusion products (cf. Tables 1 to 3) a wide range of morphological features can be drawn upon for hybrid verification. Characters such as the shape, size, or color of leaves, roots, stems, and floral organs have been employed for this purpose. Care must be taken in the use of such markers since they are often under polygenic control and subject to modifications. Several independent traits must be considered to allow meaningful conclusions. Morphological variation observed with somatic hybrids must not be confused with abnormalities and variation that results from aneuploidy, somatic incompatibility, or as an effect of tissue culture conditions.[10,15] Only euploid hybrids should be considered for morphological comparison, and observations on aneuploid hybrids must be interpreted cautiously. Although more or less perfect intermediate hybrids have been described, the phenotype of somatic hybrids (even if euploid) cannot usually be predicted *a priori*, since dominance/recessiveness and more complex interactions of the parental genomes as well as nuclear-cytoplasmic interactions on the level of gene expression are not at all clear at present. In sexually compatible species combinations a high degree of similarity has been observed when sexual and somatic hybrids were compared. In fusions of taxonomically more distant and sexually noncompatible species there will be morphological distortions and deviations from an intermediate phenotype. For instance, *Nicotiana tabacum* ✱ *Hyoscyamus muticus* somatic hybrids displayed a phenotype which was predominantly *H. muticus*[41] and fusion hybrids *Nicotiana tabacum* ✱ *Salpiglossis sinuata* more closely resembled the *Salpiglossis* parent.[80]

F. Genetic Characterization

Somatic hybrid plants that produce fertile flowers can be analyzed by conventional genetic methods. Fertility can be expected in euploid (diploid, tetraploid, and hexaploid) somatic hybrids. Problems regarding fertility will arise in aneuploid and highly asymmetric hybrids, hybrids with odd genome ratios, and in fusion combinations between more distantly related species which display some sort of incompatibility.[10,15] Genetic analyses by sexual crossing have therefore been limited to intraspecific and interspecific somatic hybrids within the genera *Nicotiana*[44,47,48,54,62,71,74,76,85,88,149] and *Petunia*.[95,150] Results obtained from the segregation of selfed F_2 generations generally were in good agreement with the expected values and have provided a sound confirmation of the somatic hybrid nature of the plants analyzed.

High pollen viability is considered a valuable indication of the compatibility of genomes in sexual as well as somatic hybrids. Indeed, reduced pollen viability has been noted in sexually incompatible, taxonomically more distant fusion combinations as compared to sexually compatible somatic hybrids within the genus *Nicotiana*.[62,68,74,76,80,81,151] Care must be taken, however, in interpreting these findings, since pollen viability is also reduced as a result of aneuploidy, and most of the hybrids examined were in fact not euploid.

Most of the intergeneric somatic hybrid plants did produce flowers[41,108,111-113,117-120] but with the exception of the *Hyoscyamus muticus* ✱ *Nicotiana tabacum* hybrids[41] they were all sterile. Since none of these hybrids was truly euploid, it is difficult to assess whether infertility was due to aneuploidy or caused by genetic incompatibility.

Anther culture has been employed on *Datura*[152] and *Nicotiana*[76,87,88,153] somatic hybrids as another and alternative method for the genetic characterization of somatic hybrids. Segregation of nuclear-coded parental markers among anther-derived progeny provides clear evidence for the hybrid nature of the anther donor plant although the segregation frequencies can be distorted. At the same time, anther culture may prove useful to reduce the ploidy level of somatic hybrids and to recover and stabilize new gene combinations resulting from meiotic recombination.

Quite unexpectedly, two out of nine plants obtained from cultured anthers of a somatic hybrid *Nicotiana rustica* ✱ *N. tabacum* have shown the chloroplast-encoded large subunit of ribulose bisphosphate carboxylase of *Nicotiana rustica* although the hybrid appeared to contain chloroplasts of *N. tabacum* type exclusively.[153] It has been suggested that the hybrid was a cytoplasmic chimera not uncovered by the analytical methods applied.

V. GENETIC CONSEQUENCES OF PROTOPLAST FUSION

As indicated in Figure 1, the genetic amalgam of newly established fusion products is being rearranged and assorted during further development. Segregational events will alter the nuclear as well as the cytoplasmic genetic constitution of fusion products. These processes will be discussed briefly in this section.

A. Segregation of Nuclear Genomes

Karyogamy of the two parental nuclei within the heterokaryocyte establishes what is then a true hybrid cell. Failure of the nuclei to fuse leads to their early segregation but these cells still contain a mixture of cytoplasmic genophores and are therefore termed cybrids. Early segregation of nuclei can be stimulated and directed when a lethal dosage of X- or γ-irradiation is applied to one parental protoplast population.[77,79,86-88,154-157] This treatment renders the protoplasts inactivated and nondividing but they serve as an efficient donor of cytoplasmic genophores when fused with recipient protoplasts. Other inactivating treatments, e.g., by iodoacetate, do not similarly affect the segregation of nuclei of the treated protoplasts.[155]

Maliga et al.[78] have recently reported on another technique which uses enucleated protoplasts (cytoplasts) as donors of cytoplasmic organelles in fusions with recipient protoplasts.

It is generally believed that irradiation at sufficient dosage will completely eliminate the nuclear genome of the treated parent. Indeed cybrid plants have been obtained which have the euploid chromosome number of the recipient species (cf. Table 4) and display normal morphology and fertility (if not, transfer of cytoplasmic male sterility was intended). Minor genetic contributions, at the subchromosomal level, of the irradiated parent cannot, however, be fully ruled out. One plant recovered after fusion of carrot with X-irradiated protoplasts from *Petroselinum hortense*[19] displayed some features characteristic of parsley although the chromosome number was close to that of diploid carrot cells. Imamura et al.[145] using species-specific probes of repetitive nuclear DNA have demonstrated that substantial amounts of tobacco DNA were present in callus material obtained after fusion of heavily irradiated tobacco and nontreated protoplasts of *Hyoscyamus muticus.*

Furthermore, irradiation does not prevent all treated protoplasts from dividing. This leakiness is evidenced by the number of colonies/plants of the type of the irradiated parent which are obtained as "escapes" in each experiment. It is also evidenced by those cases in which cybrids as well as somatic hybrids have been recovered from the same experiment.[49-51,77-79,82,96-89] These cases are indicated by a double entry both in Table 4 (cybrids) and in Tables 1 or 2 (somatic hybrids).

Regeneration of both hybrid and parental type plants from a single callus is indicative of nuclear segregation within the callus.[77-79,89,156] Since no variegated plants have been obtained among the plants regenerated, chloroplast segregation also must have been complete at the time of regeneration.[77]

Cybridization by fusion of protoplasts followed by nuclear segregation thus provides cells containing one parent's nuclear genome with the cytoplasmic, mitochondrial as well as plastidic, genophores from either one or both parents, and cytoplasmic recombinant types are also possible. Cybridization thus opens an exciting perspective to achieve alloplasmic constitutions in one step without the need to perform a series of usually 8 to 12 time-consuming backcrosses. Applications of high potential value would be the directed transfer of cytoplasmic male sterility or herbicide resistance from a donor to a recipient crop plant species. Futhermore, cybrids offer unique prospects for the study of nuclear-cytoplasmic interactions as well as interactions between different cytoplasmic organelles.[158]

Table 4
CYTOPLASMIC HYBRID PLANTS (CYBRIDS)[a]

Cytoplasm donor[b]	Recipient	Constitution[c] Nuclear	Anther type (mt?)	Chloroplastic	Ref.
Solanaceae: Nicotiana					
Nicotiana tabacum, var. Dyubek-44, cms-deb	✱ *N. tabacum*, var. Samsun, cp albino	n.r.	2: sterile	6: variegated (2: tab+deb (varietaged, LSU-RuBCase) 1: deb (green; LSU-RuBCase)	143 143
Nicotiana tabacum, var. Samsun, cp albino	✱ *N. tabacum*, Su/Su sulfur mutant	n.r.	—	19: variegated	143
Nicotiana tabacum, cv. Samsun; cv. Xanthi, wt	✱ *N. tabacum*, cv. Techne, cms-deb	6: tec 3: xan	3: fertile 3: sterile	1: tab, 5: deb (cpDNA) 1: tab, 2: deb (cpDNA)	49 49
Nicotiana tabacum cv. Turkish Samsun, cp albino	✱ *N. Tabacum*, cv. Gatersleben cnx-68, cnx type NR⁻ line	tab	—	1: variegated 4: green	50
Nicotiana tabacum, cv. Burley 21, cms-sua	✱ *N. tabacum*, cv. Gatersleben cnx-68, cnx type NR⁻ line	tab	1: sterile	3:sua (LSU-RuBCase, 1: (tentoxin sensitivity) 2: tab+sua	51
Nicotiana tabacum, line SR1, cp streptomycin resistant	✱ *N. tabacum*, line 92, cms-sua[d]	tab	13: sterile 1: fertile	12: tab (tentoxin resist., SR—resist.) 2: sua (tentox. and SR sens., cpDNA)	154
Nicotiana tabacum, line DPI—VBW, cp albino	✱ *N. tabacum*, cv. Xanthi, line VCMSG, cms-sua	VBW	25: sterile 1: fertile	24: sua (green, tentoxin sensitive 2: sua+tab (variegated, cpDNA; segregating sexual progeny	155
Nicotiana tabacum, var. Samsun, cp albino	✱ *N. debneyi*, wt	n.r.	All sterile	8: variegated (2: tab+deb (varieg.; LSU-RuBCase)	143
Nicotiana tabacum, cv. Samsun, cp mutant pale green	✱ *N. plumbaginifolia*, wt	plu	—	1: deb (green; LSU-RuBCase) 1: plu (green)	77

Table 4 (continued)
CYTOPLASMIC HYBRID PLANTS (CYBRIDS)[a]

Cytoplasm donor[b]	Recipient	Constitution[c]			Ref.
		Nuclear	Anther type (mt?)	Chloroplastic	
				11: tab (pale green; 2: LSU-RuBCase) 3: tab+plu	
Nicotiana tabacum, line SR1, cp streptomycin resistant	✱ *N. plumbaginifolia*, wt	plu	—	8: tab (SR resist., 2:cpDNA) 9: plu (SR sensit., 4:cpDNA)	78
Nicotiana tabacum, line SR1, cp streptomycin resistant	✱ *N. plumbaginifolia*, wt	plu	—	43: tab (SR resist., cpDNA)	154
Nicotiana tabacum, line 70 St-R701, cms-meg, cp streptomycin resistant	✱ *N. plumbaginifolia*, wt	plu	62(30) sterile 3 with fertile and sterile flowers	27: tab (SR resist., 10: cpDNA)	79
Nicotiana tabacum, cv. Bright Yellow, aurea mutant	✱ *N. rustica*, var. rustica, wt	rus	—	1: tab (LSU-RuBCase)	82
Nicotiana tabacum, line 92, cms-sua[d]	✱ *N. sylvestris*, wt	syl	21: sterile	sua (LSU-RuBCase, cpDNA, tentoxin sensitive)	86, 87
Nicotiana tabacum, cv. Xanthi, wt	✱ *N. sylvestris*, line f29s, cms-sua[d]	syl	3: fertile	3: sua (tentoxin sensitive)	88
Nicotiana tabacum, cv. Xanthi, wt	✱ *N. sylvestris*, line f48s, cms-sua[d]	syl	7(3): sterile 28(15): fertile	9(5): sua (tentoxin sensit.) 26(13): tab (tentoxin resistent)	88
Nicotiana tabacum, line SR1, cp streptomycin resistant	✱ *N. sylvestris*, wt	syl	—	23: tab (SR resistant)	89
Nicotiana tabacum, line SR1, cp streptomycin resistant	✱ *N. sylvestris*, wt	syl	—	8(3) (SR resistant) 16(9) (SR sensitive) 18(12) tab+syl (segregation of SR resistant and SR sensitive progeny)	154

Nicotiana tabacum, line DPI—VBW, cp albino	✱ *N. sylvestris*, line f29s, cms-sua[d]	syl	21: fertile 21: sterile	14: albino 35: green 7: variegated	155
Nicotiana rustica, wt	✱ *N. sylvestris*, wt	syl	syl (mtDNA)	8(3) rus (cpDNA) 41(25) syl (cpDNA)	157
Solanaceae: Petunia					
Petunia hybrida, line 64, wt	✱ *P. hybrida* line Sfla, cms	Sfla	3: fertile	—	56
Petunia hybrida, line 2426, cms	✱ *P. axillaris*, line 2785, wt	axi	6 of 290 plants segregate fertile and sterile progeny in F_2 and F_3	—	164, 165
Petunia hybrida, line 2426, cms	✱ *P. axillaris*, line 2785, wt	hyb	1 heteroplasmic fertile segregating homoplasmic sterile and fertile progeny in F_3	—	164, 165
Petunia parodii, line 3688, cms	✱ *P. hybrida*, line 3704, wt	hyb par	2: sterile 2: fertile	— —	95 95
Cruciferae: Brassica					
Brassica napus, type 'C', cms-Raphanus sativus, atrazine sensitive	✱ *B. napus*, var. Brutor, atrazine sensitive, wt	5: 'C' 2: Brutor	5: sterile 2: half-fertile	5: nap (cpDNA, green) 2: nap (cpDNA, green)	166 166
Brassica napus, type 'C', cms-Raphanus sativus, atrazine sensitive	✱ *B. napus*, cv. Tower, cp (B. campestris), cp atrazine resistant	nap	1: sterile	1: cam (green, cpDNA)	166

[a] As of December 1983.
[b,c] For abbreviations see Table 1.
[d] Based on recent restriction fragment analysis of cpDNA line 92 has been reclassified as being *Nicotiana tabacum* (cms-undulata).[154]

B. Chromosome Elimination

While the previously discussed segregation appeared to deal with the assortment of nuclear genomes as a whole, there is plenty of experimental evidence for the sequential loss of individual chromosomes from somatic hybrid cells.[24-26,29-31,107,132,159] Chromosome elimination has been noted preferentially in cells derived from intergeneric fusion combinations, and differences in size and structure of chromosomes have facilitated these studies. Although intergeneric hybrid cells of *Arabidipsis thaliana* ✱ *Brassica campestris*[25] or *Atropa belladonna* ✱ *Nicotiana chinensis*[30,31] were reported to contain remarkably stable chromosome complements at least in some clones, the assumption can be made that chromosome instabilities and elimination more frequently occur in hybrids between remote as compared to more closely related species. Chromosome elimination can be regarded as an indication of incompatibility of the parental genomes[10,15] and it therefore appears to be a general phenomenon not limited to plants[1,2,160] and not limited to remote somatic hybrids either. Intraplant variation of chromosome number has been found in sexual[161] as well as somatic hybrids[73] of *Nicotiana* species. Unidirectional elimination of *Hordeum bulbosum* chromosomes from sexual embryos obtained from interspecies crosses with *H. vulgare* is being utilized to great advantage in the production of monoploid barley for breeding purposes.[35,162]

As far as one can conclude from the data available, there seems to be a tendency for chromosomes of the more slowly growing and more differentiated of the parents to get lost. And like in vertebrate cell hybrids, chromosome elimination from somatic hybrid plant cells appears to be random though directed in the sense that usually only one parent's genome is subject to elimination. Partly due to insufficient distinction of individual chromosomes and plant cell karyotypes there is as yet no thorough study investigating the direction, randomness, rate, and dynamics of chromosome loss from somatic hybrid plant cells, nor is there evidence on the genetic control and the cytological processes that are involved in chromosome instability and elimination. It is true to say that most of what we know of this phenomenon is based on elaborate yet by-chance observations.

Causes as well as mechanisms of chromosome elimination from hybrid cells remain to be elucidated. Previous studies on vertebrate cell hybrids and on the *H. bulbosum* system may serve as assisting guidelines in this task. Chromosome instability and elimination may result from initial disequilibrium within fusion products, as discussed in a previous section, or from an inability of the hybrid cell to coordinate the cell cycles of its contributing parental nuclei, hence, a failure to proceed through mitosis synchronously. Since most intergeneric fusions in which chromosome elimination has been observed were between mesophyll and suspension culture-derived protoplasts whose cycling times are very much different, serious problems can arise in synchronizing the cell cycles. Less chromosome loss would be expected from hybrids composed of parental cells which were growing at similar rates prior to fusion. Much of the initial rapid loss may be attributed to cell cycle asynchrony. The continuing elimination of chromosomes may then be a consequence of the genetic imbalance which was caused by the early loss of chromosomes. In the absence of selective pressure on the cell in favor of maintaining specific chromosomes of the parent under elimination, the remaining chromosomes of this set may constitute redundant genetic information which the cell can afford to lose without losing essential functions. Alternatively, a "modification-restriction" system, a kind of genetic immune and excision system, has been proposed by Davies[163] in order to explain chromosome elimination from the *Hordeum* sexual hybrids though no experimental evidence was provided to support this model.

Expression of species-specific isoenzyme bands has been monitored in several hybrid cell lines which were losing chromosomes,[25,26,29,39,159] but in no case has it been possible to establish a clear correlation between the absence, or presence, of specific isoenzymes and the chromosomal constitutions of the cells examined. Chromosome assignment of cellular biochemical markers and linkage analyses, so successfully performed using mammalian cell hybrids,[1,2] will remain challenges for plant cell biologists to accept.

In addition to chromosome elimination, there can be extensive rearrangement of chromosomes in hybrid cells. Fragmentation, ring and bridge formation, multiconstrictional chromosomes, as well as deletion and translocation of chromosome fragments have been observed.[24-26,29-31,120] Although these examples refer to intergeneric hybrid cell lines between species that are considered sexually incompatible, one can assume similar, though less extensive, rearrangement of chromosomes in hybrids of more closely related species.

C. Cytoplasmic Segregation

Unlike the zygotes of most angiosperm plant species, which in addition to the two parental nuclear genomes contain only maternal type cytoplasm, somatic fusion products harbor plastids, mitochondria, and nuclei from both parental species. This unique amalgam of genetic elements makes protoplast fusion products ideal objects in the analysis of nuclear-cytoplasmic interactions and interorganelle competition. These interactions can be studied when monitoring the genetic behavior of chloroplasts and mitochondria in somatic hybrids (cybrids are even superior since only one nuclear genome is present).

1. Assortment of Chloroplasts

The analysis of the fate of plastids in somatic fusion products has taken great advantage of the fact that various types of markers are available to distinguish plastids of different origin.[14] Some of these markers have mainly diagnostic value, e.g., endonuclease restriction fragment patterns of cpDNA, cpDNA hybridization using radioactive probes, chloroplastic albino mutants, and the large subunit of ribulose bisphosphate carboxylase. Other chloroplast markers have additional potential as selective traits, e.g., resistance to streptomycin, lincomycin, tentoxin, or atrazine.

A great majority of the somatic hybrids and cybrids (cf. Tables 1 to 4) display the cytoplasmic characteristics of one or the other parent, rarely those of both. The initially heteroplasmic constitution of the early fusion products has apparently not been maintained as such. Rather there has been more or less rapid assortment which seemed complete at the time the hybrids were analyzed. The question has emerged whether this sorting out is random. Tables 1, 2, and 4 show examples both for random and nonrandom distribution of chloroplast types among regenerated hybrid plants. Provided chloroplasts from both parents were present in equal numbers initially, and evenly distributed, random assortment should be expected if there is no replicative advantage to one plastid type. Glimelius et al.[51] have suggested that the bias in favor of chloroplasts from *Nicotiana suaveolens*, in their experiments, was caused by the fact that the protoplasts of the other parent were from suspension-cultured cells and most likely contained either defective plastids or proplastids.

In many cases, distorted chloroplast distribution has obviously come about through the action of a selective system applied to recover the hybrids.[50,78,79,89,154,155] The effect of selection for streptomycin resistance has been evidenced by Fluhr et al.,[154] who have obtained a high proportion of heteroplasmic plants together with homoplastic (SR resistant and SR sensitive) ones where no selection was applied, whereas a 12:2 bias for the resistant type was noted as a result of selection using streptomycin. But even in an intraspecific fusion with *Nicotiana debneyi*, where a range of possibly selective influences have been carefully avoided, chloroplast assortment has been somewhat skewed.[44] Only homoplastidic plants, though differing in the type of chloroplasts they contained, have been regenerated from one individual piece of callus which was obviously composed of a mixture of cells in which plastid assortment was already complete.[28,51,77,156]

Distribution of chloroplast types was different among intraspecific somatic hybrids of *Nicotiana tabacum* when chloroplasts from different *Nicotiana* species have been transferred via fusion into the tobacco nuclear background,[51,52] hence suggesting an effect of the genomic environment on chloroplast segregation. These experiments also demonstrate that a (hybrid) genome can be host to different plastomes. The usually low number of hybrids/cybrids analyzed as yet does not permit extrapolation on the preferred direction of segregation nor on the action of genome-plastom incompatibility. Moreover, the interpretation of many findings is very much confounded by obvious experimental constraints such as the use of selection systems or defective mutants, different physiological states of the protoplasts fused, and the presence of ploidy chromosome abnormalities.

The low frequency of heteroplastidic hybrids has been attributed in part to the rapid and complete sorting out of plastids, and in part to insufficient analytical means of detecting a minor "contamination" by a second type of plastids. Recent experiments in which selective pressures have been kept to a minimum have indeed yielded a considerable proportion of heteroplasmic hybrids[28,51,77] and cybrids.[50,51,143,154,155] Best proof for the existence of heteroplasmic as opposed to chimeric hybrids comes from genetic crosses and segregation of both plastid types in the F_2 generation.[95,143,154,164,165]

When different plastid markers have been monitored segregation has never been observed, hence indicating strong linkage of plastid-coded traits. Even with the best of analytical techniques, cpDNA restriction fragment analysis and DNA:DNA hybridization, there was no evidence for recombination of cpDNAs in heteroplastidic cells.[12,14,51,154,155]

2. Assortment of Mitochondria

Much less is known on the genetic behavior of mitochondria in somatic hybrids as compared to plastids. One reason is the paucity, at present, of both diagnostic and selectable markers in mitochondria; another is the small amount of mtDNA, rendering its isolation more difficult and requiring substantial amounts of plant material.

In contrast to cpDNA, mtDNA was found to be highly variable and new types of mtDNA — similar but not identical to either parental pattern — have been detected in somatic hybrids.[167-171] Nagy et al.[170] using endonuclease restriction fragment analysis, have shown mitochondrial DNA rearrangement to occur only when fusion of protoplasts from different species had established a heteroplasmic state, whereas homoplasmic fusions did not induce such rearrangement.

The occurrence of "cytoplasmic recombinant" hybrids and cybrids having cytoplasmic traits from both parents combined in a way not found in the parents[49,51,88,143,154,155,157,166] is clearly indicative of independent assortment of plastid-coded and mitochondria-related traits (cytoplasmic male sterility). No correlation has been found in tobacco hybrids between cytoplasmic male sterility and the type of chloroplasts present, as shown by marker expression or cpDNA restriction fragment analysis.[49,51,143,154,168] Taken together these arguments indicate that cytoplasmic male sterility factors of tobacco are not resident in or related to chloroplasts. Evidence has been presented[167,168] for a correlation between cytoplasmic male sterility and the mitochondria. In contrast to the usually independent assortment of mitochondrial and chloroplastic traits, Menczel et al.[79] have been able to transfer cytoplasmic male sterility from a streptomycin-resistant line of *Nicotiana tabacum* to recipient *N. plumbaginifolia* protoplasts by selection for streptomycin resistance.

It has been a long held view that no heteroplasmic plants exist with respect to cytoplasmic male sterility, and assortment to either sterile or fertile phenotypes always appeared to be complete. Recently, several cybrid *N. plumbaginifolia* plants have been obtained which had both fertile and male-sterile flowers on different branches.[79] In *Petunia* somatic hybrids, somatic segregation of the heteroplasmic state has been observed yielding sterile as well as fertile flowers on the same plant.[95] The heteroplasmic constitution of *Petunia* cybrids has previously been shown to be maintained over several meiotic cycles,[164,165] and homoplasmic fertile and sterile progeny was obtained in F_2 and F_3 generations. Recent analyses of mitochondrial genomes from fertile and sterile somatic hybrids of *Petunia*, as compared to their sterile and fertile parents, have not been able to clearly identify CMS-specific patterns of mtDNA.[171] Unlike in *Nicotiana*, where a strong correlation of cytoplasmic male sterility and the chondriom has been evidenced with the aid of somatic cell fusion, the nature of male sterility in *Petunia* remains to be elucidated.

VI. CONCLUDING REMARKS

In this brief overview it has been possible to spotlight only some of the many unique facets and merits of somatic plant cell hybridization. The last decade has seen the emergence and vigorous flourishing of plant protoplast fusion as a tool of great versatility for cell biological and genetical studies. Basic as well as applied research will continue to benefit from the fascinating perspectives which result from a combination of somatic cell fusion and modern molecular biological techniques.

REFERENCES

1. Morrow, J., *Eukaryotic Cell Genetics*, Academic Press, New York, 1983.
2. Kao, F. T., Somatic cell genetics and gene mapping, *Int. Rev. Cytol.*, 85, 109, 1983.
3. Kennett, R. H., McKearn, T. J., and Bechtol, K. B., *Monoclonal Antibodies Hybridomas: A new Dimension in Biological Analysis*, Plenum Press, New York, 1980.
4. Cocking, E. C., A method for the isolation of plant protoplasts and vacuoles, *Nature (London)*, 187, 962, 1960.
5. Constabel, F. and Cutler, A. J., Protoplast fusion, *Plant Protoplasts*, Fowke, L. C. and Constabel, F., Eds., CRC Press, Boca Raton, in press.
6. Harms, C. T., Fusion of plant cell protoplasts, in *Developments in Cell Biology*, Vol. 2, Dean, R. T. and Stahl, P. D., Eds., Butterworths, London, in press.
7. Kao, K. N. and Michayluk, M. R., A method for high-frequency intergeneric fusion of plant protoplasts, *Planta (Berlin)*, 115, 355, 1974.
8. Wallin, A., Glimelius, K., and Eriksson, T., The induction of aggregation and fusion of *Daucus carota* protoplasts by polyethylene glycol, *Z. Pflanzenphysiol.*, 74, 64, 1974.
9. Haldane, J. B. S., Some alternatives to sex, *New Biol.*, 19, 7, 1955.
10. Harms, C. T., Somatic incompatibility in the development of higher plant somatic hybrids, *Q. Rev. Biol.*, 58, 325, 1983.
11. Evans, D. A., Protoplast fusion, in *Handbook of Plant Cell Culture*, Evans, D. A., Sharp, W. R., Ammirato, P. V., and Yamada, Y., Eds., Vol. 1, MacMillan Publishing, New York, 1983, 291.
12. Galun, E., Somatic cell fusion for inducing cytoplasmic exchange: a new biological system for cytoplasmic genetics in higher plants, in *Plant Improvement and Somatic Cell Genetics*, Vasil, I. K., Frey, K. J., and Scowcroft, W. R., Eds., Academic Press, New York, 1982, chap. 10.
13. Evans, D. A., Agricultural applications of plant protoplast fusion, *BioTechnology*, 1, 253, 1983.
14. Fluhr, R., The segregation of organelles and cytoplasmic traits in higher plant somatic fusion hybrids, in *Protoplasts 1983 Lecture Proc.*, Potrykus, I., Harms, C. T., Hinnen, A., Hütter, R., King, P. J., and Shillito, R. D., Eds., Experientia Supplementum, Vol. 46, Birkhauser, Basel, 1983, 85.

15. Harms, C. T., Somatic hybridization by plant protoplast fusion, in *Protoplasts 1983 Lecture Proc.*, Potrykus, I., Harms C. T., Hinnen, A., Hütter, R., King, P. J., and Shillito, R. D., Eds., Experientia Supplementum, Vol. 46, Birkhäuser, Basel, 1983, 69.
16. Lazar, G. B., Recent developments in plant protoplast fusion and selection technology, in *Protoplasts 1983 Lecture Proc.*, Potrykus, I., Harms, C. T., Hinnen, A., Hütter, R., King, P. J., and Shillito, R. D., Eds., Experientia Supplementum, Vol. 46, Birkhäuser, Basel, 1983, 61.
17. Keller, W. A., Setterfield, G., Douglas, G., Gleddie, S., and Nakamura, C., Production, characterization, and utilization of somatic hybrids of higher plants, in *Application of Plant Cell and Tissue Culture to Agriculture and Industry*, Tomes, D. T., Ellis, B. E., Harney, P. M., Kasha, K. J., and Peterson, R. L., Eds., University of Guelph Press, Guelph, 1982, 81.
18. Szabados, L. and Dudits, D., Fusion between interphase and mitotic plant protoplasts. Induction of premature chromosome condensation, *Exp. Cell Res.*, 127, 442, 1980.
19. Dudits, D., Fejer, O., Hadlaczky, G., Koncz, C., Lazar, G. B., and Horvath, G., Intergeneric gene transfer mediated by plant protoplast fusion, *Mol. Gen. Genet.*, 179, 283, 1980.
20. Ashmore, S. E. and Gould, A. R., Protoplast fusion and the cell cycle, *Plant Cell Rep.*, 1, 225, 1982.
21. Hadlaczky, G., Bisztray, G., Praznovsky, T., and Dudits, D., Mass isolation of plant chromosomes and nuclei, *Planta (Berlin)*, 157, 278, 1983.
22. Galbraith, D. W., Flow cytometric analysis of the cell cycle in higher plants, in *Cell Culture and Somatic Cell Genetics of Plants*, Vasil, I. K., Ed., Academic Press, Orlando, 1984, 765.
23. Redenbaugh, K., Ruzin, S., Bartholomew, J., and Bassham, J. A., Characterization and separation of plant protoplasts via flow cytometry and cell sorting, *Z. Pflanzenphysiol.*, 107, 65, 1982.
24. Kao, K. N., Chromosomal behaviour in somatic hybrids of soybean — *Nicotiana glauca*, *Mol. Gen. Genet.*, 150, 225, 1977.
25. Gleba, Y. Y. and Hoffmann, F., Hybrid cell lines *Arabidopsis thaliana* + *Brassica campestris*: no evidence for specific chromosome elimination, *Mol. Gen. Genet.*, 165, 257, 1978.
26. Wetter, L. R. and Kao, K. N., Chromosome and isoenzyme studies on cells derived from protoplast fusion of *Nicotiana glauca* with *Glycine max* — *Nicotiana glauca* cell hybrids, *Theor. Appl. Genet.*, 57, 273, 1980.
27. Harms, C. T., Kochba, J., and Potrykus, I., Fusion of *Citrus* and tobacco protoplasts — a new system for somatic hybridization studies with remote species, in *Advances in Protoplast Research*, Ferenczy, L. and Farkas, G. L., Eds., Akademiai Kiado, Budapest, 1980, 321.
28. Menczel, L., Lazar, G., and Maliga, P., Isolation of somatic hybrids by cloning *Nicotiana* heterokaryons in nurse cultures, *Planta (Berlin)*, 143, 29, 1978.
29. Chien, Y. C., Kao, K. N., and Wetter, L. R., Chromosomal and isozyme studies of *Nicotiana tabacum* — *Glycine max* hybrid cell lines, *Theor. Appl. Genet.*, 62, 301, 1982.
30. Gleba, Y. Y., Momot, V. P., Cherep, N. N., and Skarzynskaya, M. V., Intertribal hybrid cell lines of *Atropa belladonna* (x) *Nicotiana chinensis* obtained by cloning individual protoplast fusion products, *Theor. Appl. Genet.*, 62, 75, 1982.
31. Gleba, Y. Y., Momot, V. P., Okolot, A. N., Cherep, N. N., Skarzynskaya, M. V., and Kotov, V., Genetic processes in intergeneric cell hybrids *Atropa* + *Nicotiana*. I. Genetic constitution of cells of different clonal origin grown in vitro, *Theor. Appl. Genet.*, 65, 269, 1983.
32. Menczel, L., Nagy, F., Kiss, Z. R., and Maliga, P., Streptomycin resistant and sensitive somatic hybrids of *Nicotiana tabacum* + *Nicotiana knightiana*: correlation of resistance to *N. tabacum* plastids, *Theor. Appl. Genet.*, 59, 191, 1981.
33. Hein, T., Przewozny, T., and Schieder, O., Culture and selection of somatic hybrids using an auxotrophic cell line, *Theor. Appl. Genet.*, 64, 119, 1983.
34. Bayliss, M. W., Chromosomal variation in plant tissues in culture, *Int. Rev. Cytol.*, Suppl. 11A, 113, 1980.
35. Subrahmanyam, N. C., Haploidy from *Hordeum* interspecific crosses. I. Polyhaploids of *H. parodii* and *H. procerum*, *Theor. Appl. Genet.*, 49, 209, 1977.
36. Minna, J. D. and Coon, H. G., Human x mouse hybrid cells segregating mouse chromosomes and isozymes, *Nature (London)*, 252, 401, 1974.
37. Schieder, O., Somatic hybridization: a new method for plant improvement, in *Plant Improvement and Somatic Cell Genetics*, Vasil, I. K., Frey, K. J., and Scowcroft, W. R., Eds., Academic Press, New York, 1982, chap. 12.
38. Bourgin, J. P., Protoplasts and the isolation of plant mutants, in *Protoplasts 1983 Lecture Proc.*, Potrykus, I., Harms, C. T., Hinnen, A., Hütter, R., King, P. J., and Shillito, R. D., Eds., Experientia Supplementum, Vol. 46, Birkhäuser, Basel, 1983, 43.
39. Melchers, G. and Labib, G., Somatic hybridization of plants by fusion of protoplasts. I. Selection of light resistant hybrids of "haploid" light sensitive varieties of tobacco, *Mol. Gen. Genet.*, 135, 277, 1974.
40. Schieder, O., Hybridization experiments with protoplasts from chlorophyll-deficient mutants of some *Solanaceous* species, *Planta (Berlin)*, 137, 253, 1977.

41. Potrykus, I., Shillito, R. D., Jia, J., and Lazar, G. B., Auxotroph complementation via protoplast fusion in *Hyoscyamus muticus* and *Nicotiana tabacum*, in *Genetic Engineering in Eukaryotes*, Lurquin, P. F. and Kleinhofs, A., Eds., Plenum Press, New York, 1983, 253.
42. Fankhauser, H., Gebhardt, C., Jia, J., King, P. J., Laser, M., Lazar, G. B., Potrykus, I., Shillito, R. D., and Shimamoto, K., Fusion complementation tests on a group of independently isolated auxotrophic and temperature-sensitive clones of *Hyoscyamus muticus* and *Nicotiana tabacum*, in *Protoplasts 1983 Poster Proceedings*, Potrykus, I., Harms, C. T., Hinnen, A., Hütter, R., King, P. J., and Shillito, R. D., Experientia Supplementum, Vol. 45, Birkhäuser, Basel, 1983, 112.
43. Gebhardt, C., Shimamoto, K., Lazar, G. B., Schnebli, V., and King, P. J., Isolation of biochemical mutants using haploid mesophyll protoplasts of *Hyoscyamus muticus*. III. General characterization of histidine and tryptophan auxotrophs, *Planta (Berlin)*, 159, 18, 1983.
44. Scowcroft, W. R. and Larkin, P. J., Chloroplast DNA assorts randomly in intraspecific somatic hybrids of *Nicotiana debneyi*, *Theor. Appl. Genet.*, 60, 179, 1981.
45. Sidorov, V. A. and Maliga, P., Fusion-complementation analysis of auxotrophic and chlorophyll-deficient lines isolated in haploid *Nicotiana plumbaginifolia* protoplast cultures, *Mol. Gen. Genet.*, 186, 328, 1982.
46. Marton, L., Sidorov, V. A., Biasini, G., and Maliga, P., Complementation in somatic hybrids indicates four types of nitrate reductase deficient lines in *Nicotiana plumbaginifolia*, *Mol. Gen. Genet.*, 187, 1, 1982.
47. Marton, L., Biasini, G., Sidorov, V., and Maliga, P., Nitrate reductase deficients in the progeny after selfing complemented somatic hybrids of *Nicotiana plumbaginifolia* mutants, in *Protoplasts 1983 Poster Proc.*, Potrykus, I., Harms, C. T., Hinnen, A., Hütter, R., King, P. J., and Shillito, R. D., Experientia Supplementum, Vol. 45, Birkhäuser, Basel, 1983, 10.
48. Kameya, T., Induction of hybrids through somatic cell fusion with dextran sulfate and gelatin, *Jpn. J. Genet.*, 50, 235, 1975.
49. Belliard, G., Pelletier, G., Vedel, F., and Quetier, F., Morphological characteristics and chloroplast DNA distribution in different cytoplasmic parasexual hybrids of *Nicotiana tabacum*, *Mol. Gen. Genet.*, 165, 231, 1978.
50. Glimelius, K. and Bonnett, H. T., Somatic hybridization in *Nicotiana*: restoration of photoautotrophy to an albino mutant with defective plastids, *Planta (Berlin)*, 153, 497, 1981.
51. Glimelius, K., Chen, K., and Bonnett, H. T., Somatic hybridization in *Nicotiana*: segregation of organellar traits among hybrid and cybrid plants, *Planta (Berlin)*, 153, 504, 1981.
52. Bonnett, H. T. and Glimelius, K., Somatic hybridization in *Nicotiana*: behavior of organelles after fusion of protoplasts from male-fertile and male-sterile cultivars, *Theor. Appl. Genet.*, 65, 213, 1983.
53. Wullems, G. J., Molendijk, L., and Schilperoort, R. A., The expression of tumour markers in intraspecific somatic hybrids of normal and crown gall cells from *Nicotiana tabacum*, *Theor. Appl. Genet.*, 56, 203, 1980.
54. Evola, S. V., Earle, E. D., and Chaleff, R. S., The use of genetic markers selected in vitro for the isolation and genetic verification of intraspecific somatic hybrids of *Nicotiana tabacum* L., *Mol. Gen. Genet.*, 189, 441, 1983.
55. Evola, S. V., Chlorate resistant variants of *Nicotiana tabacum* L. II. Parasexual genetic characterization, *Mol. Gen. Genet.*, 189, 455, 1983.
56. Bergounioux-Bunisset, C. and Perennes, C., Transfert de facteurs cytoplasmiques de la fertilité male entre deux lignées de *Petunia hybrida* par fusion de protoplastes, *Plant Sci. Lett.*, 19, 143, 1980.
57. Lazar, G. B., Dudits, D., and Sung, Z. R., Expression of cycloheximide resistance in carrot somatic hybrids and their segregants, *Genetics*, 98, 347, 1981.
58. Schieder, O., Somatic hybrids between a herbaceous and two tree *Datura* species, *Z. Pflanzenphysiol.*, 98, 119, 1980.
59. Muller-Gensert, E., Landsmann, J., Eckes, P., and Schieder, O., Confirmation of chloroplast segregation in somatic hybrids of *Datura* by DNA-DNA hybridization, in *Protoplasts 1983 Poster Proc.*, Potrykus, I., Harms, C. T., Hinnen, A., Hütter, R., King, P. J., and Shillito, R. D., Eds., Experientia Supplementum, Vol. 45, Birkhäuser Verlag, Basel, 1983, 106.
60. Schieder, O., Somatic hybrids of *Datura innoxia* Mill. + *Datura discolor* Bernh. and of *Datura innoxia* Mill. + *Datura stramonium* L. var *tatula* L. I. Selection and characterisation, *Mol. Gen. Genet.*, 162, 113, 1978.
61. Carlson, P. S., Smith, H. H., and Dearing, R. D., Parasexual interspecific plant hybridization, *Proc. Natl. Acad. Sci. U.S.A.*, 69, 2292, 1972.
62. Smith, H. H., Kao, K. N. and Combatti, N. C., Interspecific hybridization by protoplast fusion in *Nicotiana*. Confirmation and extension, *J. Hered.*, 67, 123, 1976.
63. Chen, K., Wildman, S. G., and Smith, H. H., Chloroplast DNA distribution in parasexual hybrids as shown by polypeptide composition of fraction 1 protein, *Proc. Natl. Acad. Sci. U.S.A.*, 74, 5109, 1977.

64. Chupeau, Y., Missonier, C., Hommel, M. C., and Goujaud, J., Somatic hybrids of plants by fusion of protoplasts. Observations on the model system "*Nicotiana glauca — Nicotiana langsdorffii*", *Mol. Gen. Genet.*, 165, 239, 1978.
65. Uchimiya, H., Ohgawara, T., Kato, H., Akiyama, T., and Harada, H., Detection of two different nuclear genomes in parasexual hybrids by ribosomal RNA gene analysis, *Theor. Appl. Genet.*, 64, 117, 1983.
66. Maliga, P., Lazar, G., Joo. F., Nagy, A. H., and Menczel, L., Restoration of morphogenetic potential in *Nicotiana* by somatic hybridization, *Mol. Gen. Genet.*, 157, 291, 1977.
67. Gleddie, S., Keller, W. A., Setterfield, G., and Wetter, L. R., Somatic hybridization between *Nicotiana rustica* and *N. sylvestris*, *Plant Cell Tissue Organ Cult.*, 2, 269, 1983.
68. Nagao, T., Somatic hybridization by fusion of protoplasts. II. The combinations of *Nicotiana tabacum* and *N. glutinosa* and of *N. tabacum* and *N. alata*, *Jpn. J. Crop Sci.*, 48, 385, 1979.
69. Evans, D. A., Wetter, L. R., and Gamborg, O. L., Somatic hybrid plants of *Nicotiana glauca* and *Nicotiana tabacum* obtained by protoplast fusion, *Physiol. Plant.*, 48, 225, 1980.
70. Uchimiya, H., Somatic hybridization between male sterile *Nicotiana tabacum* and *N. glutinosa* through protoplast fusion, *Theor. Appl.* Genet., 61, 69, 1982.
71. Uchimiya, H., Akiyama, T., Ohgawara, T., and Harada, H., Expression of nuclear and cytoplasmic genes in the progeny of a somatic hybrid between male sterile *Nicotiana tabacum* and *N. glutinosa*, *Jpn. J. Genet.*, 57, 343, 1982.
72. Horn, M. E., Kameya, T., Brotherton, J. E., and Widholm, J. M., The use of amino acid analog resistance and plant regeneration ability to select somatic hybrids between *Nicotiana tabacum* and *N. glutinosa*, *Mol. Gen. Genet.*, 192, 235, 1983.
73. Maliga, P., Kiss, Z. R., Nagy, A. H., and Lazar, G., Genetic instability in somatic hybrids of *Nicotiana tabacum* and *Nicotiana knightiana*, *Mol Gen. Genet.*, 163, 145, 1978.
74. Evans, D. A., Flick, C. E., and Jensen, R. A., Disease resistance: incorporation into sexually incompatible somatic hybrids of the genus *Nicotiana*, *Science*, 213, 907, 1981.
75. Flick, C. E. and Evans, D. A., Evaluation of cytoplasmic segregation in somatic hybrids of *Nicotiana*: tentoxin sensitivity, *J. Hered.*, 73, 264, 1982.
76. Evans, D. A., Bravo, J. E., Kut, S. A., and Flick, C. E., Genetic behavior of somatic hybrids in the genus *Nicotiana: N. otophora* + *N. tabacum* and *N. sylvestris* + *N. tabacum*, *Theor. Appl. Genet.*, 65, 93, 1983.
77. Sidorov, V. A., Menczel, L., Nagy, F., and Maliga, P., Chloroplast transfer in *Nicotiana* based on metabolic complementation between irradiated and iodoacetate treated protoplasts, *Planta (Berlin)*, 152, 341, 1981.
78. Maliga, P., Lorz, H., Lazar, G., and Nagy, F., Cytoplast-protoplast fusion for interspecific chloroplast transfer in *Nicotiana*, *Mol. Gen. Genet.*, 185, 211, 1982.
79. Menczel, L., Nagy, F., Lazar, G., and Maliga, P., Transfer of cytoplasmic male sterility by selection for streptomycin resistance after protoplast fusion in *Nicotiana*, *Mol. Gen. Genet.*, 189, 365, 1983.
80. Nagao, T., Somatic hybridization by fusion of protoplasts. III. Somatic hybrids of sexually incompatible combinations *Nicotiana tabacum* + *Nicotiana repanda* and *Nicotiana tabacum* + *Salpiglossis sinuata*, *Jpn. J. Crop Sci.*, 51, 35, 1982.
81. Nagao, T., Somatic hybridization by fusion of protoplasts. I. The combination of *Nicotiana tabacum* and *Nicotiana rustica*, *Jpn. J. Crop Sci.*, 47, 491, 1978.
82. Iwai, S., Nagao, T., Nakata, K., Kawashima, N., and Matsuyama, S., Expression of nuclear and chloroplastic genes coding for fraction 1 protein in somatic hybrids of *Nicotiana tabacum* + *rustica*, *Planta (Berlin)*, 147, 414, 1980.
83. Douglas, G. C., Keller, W. A., and Setterfield, G., Somatic hybridization between *Nicotiana rustica* and *N. tabacum*. II. Protoplast fusion and selection and regeneration of hybrid plants, *Can. J. Bot.*, 59, 220, 1981.
84. Douglas, G. C., Wetter, L. R., Keller, W. A., and Setterfield, G., Somatic hybridization between *Nicotiana rustica* and *N. tabacum*. IV. Analysis of nuclear and chloroplast genome expression in somatic hybrids, *Can. J. Bot.*, 59, 1509, 1981.
85. Melchers, G., Kombination somatischer und konventioneller Genetik für die Pflanzenzüchtung, *Naturwissenschaften*, 64, 184, 1977.
86. Zelcer, A., Aviv, D., and Galun, E., Interspecific transfer of cytoplasmic male sterility by fusion between protoplasts of normal *Nicotiana sylvestris* and X-irradiated protoplasts of male-sterile *N. tabacum*, *Z. Pflanzenphysiol.*, 90, 397, 1978.
87. Aviv, D., Fluhr., Edelman, M., and Galun, E., Progeny analysis of the interspecific somatic hybrids: *Nicotiana tabacum* (cms) + *Nicotiana sylvestris* with respect to nuclear and chloroplast markers, *Theor. Appl. Genet.*, 56, 145, 1980.
88. Aviv, D. and Galun, E., Restoration of fertility in cytoplasmic male sterile (cms) *Nicotiana sylvestris* by fusion with X-irradiated *N. tabacum* protoplasts, *Theor. Appl. Genet.*, 58, 121, 1980.

89. Medgyesy, P., Menczel, L., and Maliga, P., The use of cytoplasmic streptomycin resistance: chloroplast transfer from *Nicotiana tabacum* into *Nicotiana sylvestris*, and isolation of their somatic hybrids, *Mol. Gen. Genet.*, 179, 693, 1980.
90. Power, J. B., Frearson, E. M., Hayward, C., George, D., Evans, P. K., Berry, S. F., and Cocking, E. C., Somatic hybridization of *Petunia hybrida* and *P. parodii*, *Nature (London)*, 263, 500, 1976.
91. Power, J. B., Berry, S. F., Frearson, E. M., and Cocking, E. C., Selection procedures for the production of inter-species somatic hybrids of *Petunia hybrida* and *Petunia parodii*. I. Nutrient media and drug sensitivity complementation selection, *Plant Sci. Lett.*, 10, 1, 1977.
92. Kumar, A., Cocking, E. C., Bovenberg, W. A., and Kool, A. J., Restriction endonuclease analysis of chloroplast DNA in interspecies somatic hybrids of *Petunia*, *Theor. Appl. Genet.*, 62, 377, 1982.
93. Itoh, K. and Futsuhara, Y., Restoration of the ability to regenerate shoots by somatic hybridization between *Petunia hybrida* and *P. parodii*, *Jpn. J. Breed.*, 33, 130, 1983.
94. Cocking, E. C., George, D., Price-Jones, M. J., and Power, J. B., Selection procedures for the production of inter-species somatic hybrids of *Petunia hybrida* and *Petunia parodii*. II. Albino complementation selection, *Plant Sci. Lett.*, 10, 7, 1977.
95. Izhar, S., Schlicter, M., and Swartzberg, D., Sorting out of cytoplasmic elements in somatic hybrids of *Petunia* and the prevalence of the heteroplasmon through several meiotic cycles, *Mol. Gen. Genet.*, 190, 468, 1983.
96. Power, J. B., Berry, S. F., Chapman, J. V., Cocking, E. C., and Sink, K. C., Somatic hybrids between unilateral cross-incompatible *Petunia* species, *Theor. Appl. Genet.*, 55, 97, 1979.
97. Power, J. B., Berry, S. F., Chapman, J. V., and Cocking, E. C., Somatic hybridization of sexually incompatible petunias: *Petunia parodii*, *Petunia parviflora*, *Theor. Appl. Genet.*, 57, 1, 1980.
98. Kumar, A., Wilson, D., and Cocking, E. C., Polypeptide composition of fraction 1 protein of the somatic hybrid between *Petunia parodii* and *Petunia parviflora*, *Biochem. Genet.*, 19, 255, 1981.
99. Butenko, R. G. and Kuchko, A. A., Physiological aspects of procurement, cultivation, and hybridization of isolated potato protoplasts, *Fiziol. Rast.*, 26, 1110, 1979.
100. Binding, H., Jain, S. M., Finger, J., Mordhorst, G., Nehls, R., and Gressel, J., Somatic hybridization of an atrazine resistant biotype of *Solanum nigrum* with *Solanum tuberosum*, *Theor. Appl. Genet.*, 63, 273, 1982.
101. Dudits, D., Hadlaczky, G., Levi, E., Fejer, O., Haydu, Z., and Lazar, G. B., Somatic hybridization of *Daucus carota* and *D. capillifolius* by protoplast fusion, *Theor. Appl. Genet.*, 51, 127, 1977.
102. Kameya, T., Horn, M. E., and Widholm, J. M., Hybrid shoot formation from fused *Daucus carota* and *D. capillifolius* protoplasts, *Z. Pflanzenphysiol.*, 104, 459, 1981.
103. Schenck, H. R. and Röbbelen, G., Somatic hybrids by fusion of protoplasts from *Brassica oleracea* and *B. campestris*, *Z. Pflanzenzüchtg.*, 89, 278, 1982.
104. Téoulé, E., Hybridation somatique entre *Medicago sativa* L. et *Medicago falcata* L., *C. R. Acad. Sci. Paris*, 297, 13, 1983.
105. Gosch, G. and Reinert, J., Cytological identification of colony formation of intergeneric somatic hybrid cells, *Protoplasma (Berlin)*, 96, 23, 1978.
106. Krumbiegel, G. and Schieder, O., Selection of somatic hybrids after fusion of protoplasts from *Datura innoxia* Mill. and *Atropa belladonna* L., *Planta (Berlin)*, 145, 371, 1979.
107. Krumbiegel, G. and Schieder, O., Comparison of somatic and sexual incompatibility between *Datura innoxia* and *Atropa belladonna*, *Planta (Berlin)*, 153, 466, 1981.
108. Melchers, G., Sacristan, M. D., and Holder, A. A., Somatic hybrid plants of potato and tomato regenerated from fused protoplasts, *Carlsberg Res. Commun.*, 43, 203, 1978.
109. Poulsen, C., Porath, D., Sacristan, M. D., and Melchers, G., Peptide mapping of the ribulose bisphosphate carboxylase small subunit from the somatic hybrid of tomato and potato, *Carlsberg Res. Commun.*, 45, 249, 1980.
110. Schiller, B., Herrmann, R. G., and Melchers, G., Restriction endonuclease analysis of plastid DNA from tomato, potato and some of their somatic hybrids, *Mol. Gen. Genet.*, 186, 453, 1982.
111. Melchers, G., The first decennium of somatic hybridization, in *Plant Tissue Culture 1982*, Fujiwara, A., Ed., Japan Association Plant Tissue Culture, Tokyo, 1982, 13.
112. Melchers, G., Topatoes, and pomatoes, somatic hybrids between tomatoes and potatoes, in *Proc. Congr. Eur. Tissue Cult. Soc.*, Budapest, in press.
113. Shepard, J. F., Bidney, D., Barsby, T., and Kemble, R., Genetic transfer in plants through interspecific protoplast fusion, *Science*, 219, 683, 1983.
114. Sun Yongru, Huang Meijuan, Li Wenbin, and Li Xianghui, Regeneration of somatic hybrid plants between *Nicotiana glauca* and *Petunia hybrida*, *Acta Genet. Sinica*, 9, 284, 1982.
115. Li Wenbin, Sun Yongru, Huang Meijuan, and Li Xianghui, Further observations of somatic hybrid plant between *Nicotiana glauca* and *Petunia hybrida*, *Acta Genet. Sinica*, 10, 194, 1983.
116. Li Xianghui, Li Wenbin, and Huang Meijuan, Somatic hybrid plants from intergeneric fusion between tobacco tumor B6S3 and *Petunia hybrida* W43 and expression of LpDH, *Sci. Sinica*, 25, 611, 1982.

117. Dudits, D., Hadlaczky, G., Bajszar, G. Y., Koncz, C., Lazar, G., and Horvath, G., Plant regeneration from intergeneric cell hybrids, *Plant Sci. Lett.*, 15, 101, 1979.
118. Dudits, D., Backfusion with somatic protoplasts as a method in genetic manipulation of plants, *Acta. Biol. Acad. Sci. Hung.*, 32, 215, 1981.
119. Gleba, Y. Y. and Hoffman, F., "Arabidobrassica": a novel plant obtained by protoplast fusion, *Planta (Berlin)*, 149, 112, 1980.
120. Hoffmann, F. and Adachi, T., "Arabidobrassica": chromosomal recombination and morphogenesis in asymmetric intergeneric hybrid cells, *Planta (Berlin)*, 153, 586, 1981.
121. Komarnitsky, I. K. and Gleba, Y. Y., Fraction 1 protein analysis of parasexual hybrid plants *Arabidopsis thaliana* + *Brassica campestris*, *Plant Cell Rep.*, 1, 67, 1981.
122. Glimelius, K., Eriksson, T., Grafe, R., and Muller, A. J., Somatic hybridization of nitrate reductase deficient mutants of *Nicotiana tabacum* by protoplast fusion, *Physiol. Plant.*, 44, 273, 1978.
123. Le Thi Xuan, Grafe, R., and Muller, A. J., Complementation of nitrate reductase deficient mutants in somatic hybrids between *Nicotiana* species, in *Protoplasts 1983 Poster Proc.*, Potrykus, I., Harms, C. T., Hinnen, A., Hütter, R., King, P. J., and Shillito, R. D., Eds., Experientia Supplementum, Vol. 45, Birkhäuser, Basel, 1983, 76.
124. Lazar, G. B., Fankhauser, H., and Potrykus, I., Complementation analysis of a nitrate reductase deficient *Hyoscyamus muticus* cell line by somatic hybridization, *Mol. Gen. Genet.*, 189, 359, 1983.
125. Steffen, A. and Schieder, O., Selection and characterization of nitrate reductase deficient mutants of *Petunia*, in *Protoplasts 1983 Poster Proc.*, Potrykus, I., Harms, C. T., Hinnen, A., Hütter, R., King, P. J., and Shillito, R. D., Eds., Experientia Supplementum, Vol. 45, Birkhäuser, Basel, 1983, 162.
126. White, D. W. R. and Vasil, I. K., Use of amino acid analogue-resistant cell lines for selection of *Nicotiana sylvestris* somatic cell hybrids, *Theor. Appl. Genet.*, 55, 107, 1979.
127. Harms, C. T., Potrykus, I., and Widholm, J. M., Complementation and dominant expression of amino acid analogue resistance markers in somatic hybrid clones from *Daucus carota* after protoplast fusion, *Z. Pflanzenphysiol.*, 101, 377, 1981.
128. Harms, C. T., Oertli, J. J., and Widholm, J. M., Characterization of amino acid analogue resistant somatic hybrid cell lines of *Daucus carota* L., *Z. Pflanzenphysiol.*, 106, 239, 1982.
129. Harms, C. T. and Widholm, J. M., unpublished results.
130. Cella, R., Carbonera, D., and Iadarola, P., Characterization of intraspecific somatic hybrids of carrot obtained by fusion of iodoacetate-inactivated A2CA-resistant and sensitive protoplasts, *Z. Pflanzenphysiol.*, 112, 449, 1983.
131. Harms, C. T. and Oertli, J. J., Complementation and expression of amino acid analogue resistance sudied by intraspecific and interfamily protoplast fusion, in *Plant Tissue Culture 1982*, Fujiwara, A., Ed., Japan Association Plant Tissue Culture, Tokyo, 1982, 467.
132. Hauptmann, R., Kumar, P., and Widholm, J. M., Carrot x tobacco somatic cell hybrids selected by amino acid analog resistance complementation, in *Protoplasts 1983 Poster Proc.*, Potrykus, I., Harms, C. T., Hinnen, A., Hütter, R., King, P. J., and Shillito, R. D., Eds., Experientia Supplementum, Vol. 45, Birkhäuser, Basel, 1983, 92.
133. Hamill, J. D., Pental, D., Cocking, E. C., and Müller, A. J., Production of a nitrate reductase deficient streptomycin resistant mutant of *Nicotiana tabacum* for somatic hybridization studies, *Heredity*, 50, 197, 1983.
134. Wright, W. E., The isolation of heterokaryons and hybrids by a selective system using irreversible biochemical inhibitors, *Exp. Cell Res.*, 112, 395, 1978.
135. Nehls, R., The use of metabolic inhibitors for the selection of fusion products of higher plant protoplasts, *Mol. Gen. Genet.*, 166, 117, 1978.
136. Galbraith, D. W. and Galbraith, J. E. C., A method for identification of fusion of plant protoplasts derived from tissue cultures, *Z. Pflanzenphysiol.*, 93, 149, 1979.
137. Galbraith, D. W. and Mauch, T. J., Identification of fusion of plant protoplasts. II. Conditions for the reproducible fluorescence labelling of protoplasts derived from mesophyll tissue, *Z. Pflanzenphysiol.*, 98, 129, 1980.
138. Patnaik, G., Cocking, E. C., Hamill, J., and Pental, D., A simple procedure for the manual isolation and identification of plant heterokaryons, *Plant Sci. Lett.*, 24, 105, 1982.
139. Galbraith, D. W., Selection of hybrid cells by fluorescence activated cell sorting, in *Cell Culture and Somatic Cell Genetics of Plants*, Vasil, I. K., Ed., Academic Press, Orlando, 1984, 433.
140. Harms, C. T. and Potrykus, I., Fractionation of plant protoplast types by iso-osmotic density gradient centrifugation, *Theor. Appl. Genet.*, 53, 57, 1978.
141. Harms, C. T. and Potrykus, I., Enrichment for heterokaryocytes by the use of iso-osmotic density gradients after plant protoplast fusion, *Theor. Appl. Genet.*, 53, 49, 1978.
142. Harms, C. T., in preparation.
143. Gleba, Y. Y., Extranuclear inheritance investigated by somatic hybridization, in *Frontiers of Plant Tissue Culture 1978*, Thorpe, T. A., Ed., University of Calgary Press, Calgary, 1978, 95.
144. Saul, M. W. and Potrykus, I., Species-specific DNA used to identify interspecific somatic hydrids, *Plant Cell Rep.*, 3, 65, 1984.

145. Imamura, J., Saul, M. W., and Potrykus, I., personal communication.
146. Douglas, G. C., Wetter, L. R., Nakamura, C., Keller, W. A., and Setterfield, G., Somatic hybridization between *Nicotiana rustica* and *N. tabacum*. III. Biochemical, morphological, and cytological analysis of somatic hybrids, *Can. J. Bot.*, 59, 228, 1981.
147. Ninnemann, H. and Jüttner, F., Volatile substances from tissue cultures of potato, tomato, and their somatic fusion products — comparison of gas chromatographic patterns for identification of hybrids, *Z. Pflanzenphysiol.*, 103, 95, 1981.
148. Larkin, P. J. and Harms, C. T., unpublished results.
149. Wullems, G. J., Molendijk, L., Ooms, G., and Schilperoort, R. A., Retention of tumor markers in F_1 progeny plants from in vitro induced octopine and nopaline tumor tissues, *Cell*, 24, 719, 1981.
150. Power, J. B., Sink, K. C., Berry, S. F., Burns, S. F., and Cocking, E. C., Somatic and sexual hybrids of *Petunia hybrida* and *Petunia parodii*. A comparison of flower color segregation, *J. Hered.*, 69, 373, 1978.
151. Evans, D. A., Flick, C. E., Kut, S. A., Reed, S. M., Comparison of *Nicotiana tabacum* and *Nicotiana nesophila* hybrids produced by ovule culture and protoplast fusion, *Theor. Appl. Genet.*, 62, 193, 1982.
152. Schieder, O., Genetic evidence for the hybrid nature of somatic hybrids from *Datura innoxia* Mill., *Planta (Berlin)*, 141, 333, 1978.
153. Iwai, S., Nakata, K., Nagao, T., Kawashima, N., Matsuyama, S., Detection of the *Nicotiana rustica* chloroplast genome coding for the large subunit of fraction 1 protein in a somatic hybrid in which only the *N. tabacum* chloroplast genome appeared to have been expressed, *Planta (Berlin)*, 152, 478, 1981.
154. Fluhr, R., Aviv, D., Edelman, M., and Galun, E., Cybrids containing mixed and sorted-out chloroplasts following interspecific somatic fusions in *Nicotiana*, *Theor. Appl. Genet.*, 65, 289, 1983.
155. Fluhr, R., Aviv, D., Galun, E., and Edelman, M., Generation of heteroplastidic cybrids by protoplast fusion: analysis for plastid recombinant types, *Theor. Appl. Genet.*, 67, 491, 1984.
156. Menczel, L., Galiba, G., Nagy, F., and Maliga, P., Effect of radiation dosage on efficiency of chloroplast transfer by protoplast fusion in *Nicotiana*, *Genetics*, 100, 487, 1982.
157. Aviv, D., Bleichman, S., Arzee-Gonen, P., and Galun, E., Intersectional cytoplasmic hybrids in *Nicotiana*. Identification of plastomes and chondriomes in *N. sylvestris* + *N. rustica* cybrids having *N. sylvestris* nuclear genomes, *Theor. Appl. Genet.*, 67, 499, 1984.
158. Eberhard, W. G., Evolutionary consequences of intracellular organelle competition, *Q. Rev. Biol.*, 55, 231, 1980.
159. Power, J. B., Frearson, E. M., Hayward, C., and Cocking, E. C., Some consequences of the fusion and selective culture of *Petunia* and *Parthenocissus* protoplasts, *Plant Sci. Lett.*, 5, 197, 1975.
160. Shaw, D. D., Wilkinson, P., and Coates, D. J., Increased chromosomal mutation rate after hybridization between two subspecies of grasshoppers, *Science*, 220, 1165, 1983.
161. Yang, S. J., Numerical chromosome instability in *Nicotiana* hybrids. II. Intraplant variation, *Can. J. Genet. Cytol.*, 7, 112, 1965.
162. Jensen, C. J., Monoploid production by chromosome elimination, in *Plant Cell, Tissue, and Organ Culture*, Reinert, J. and Bajaj, Y. P. S., Eds., Springer, Berlin, 1977, 299.
163. Davies, D. R., Chromosome elimination in interspecific hybrids, *Heredity*, 32, 267, 1974.
164. Izhar, S. and Power, J. B., Somatic hybridization in *Petunia*: a male sterile cytoplasmic hybrid, *Plant Sci. Lett.*, 14, 49, 1979.
165. Izhar, S. and Tabib, Y., Somatic hybridization in *Petunia*. II. Heteroplasmic state in somatic hybrids followed by cytoplasmic segregation into male sterile and male fertile lines, *Theor. Appl. Genet.*, 57, 241, 1980.
166. Pelletier, G., Primard, C., Vedel, F., Chetrit, P., Remy, R., Rousselle, P., and Renard, M., Intergeneric cytoplasmic hybridization in Cruciferae by protoplast fusion, *Mol. Gen. Genet.*, 191, 244, 1983.
167. Belliard, G., Vedel, F., and Pelletier, G., Mitochondrial recombination in cytoplasmic hybrids of *Nicotiana tabacum* by protoplast fusion, *Nature (London)*, 281, 401, 1979.
168. Galun, E., Arzee-Gonen, P., Fluhr, R., Edelman, M., and Aviv, D., Cytoplasmic hybridization in *Nicotiana*: mitochondrial DNA analysis in progenies resulting from fusion between protoplasts having different organelle constitutions, *Mol. Gen. Genet.*, 186, 50, 1982.
169. Nagy, F., Török, I., and Maliga, P., Extensive rearrangements in the mitochondrial DNA in somatic hybrids of *Nicotiana tabacum* and *Nicotiana knightiana*, *Mol. Gen. Genet.*, 183, 437, 1981.
170. Nagy, F., Lazar, G., Menczel, L., and Maliga, P., A heteroplasmic state induced by protoplast fusion is a necessary condition for detecting rearrangements in *Nicotiana* mitochondrial DNA, *Theor. Appl. Genet.*, 66, 203, 1983.
171. Boeshore, M. L., Lifshitz, I., Hanson, M. R., and Izhar, S., Novel composition of mitochondrial genomes in *Petunia* somatic hybrids derived from cytoplasmic male sterile and fertile plants, *Mol. Gen. Genet.*, 190, 459, 1983.

Chapter 13

PLANT PROTOPLAST TRANSFORMATION BY *AGROBACTERIUM* IN RELATION TO PLANT BIOTECHNOLOGY

G. J. Wullems and R. A. Schilperoort

TABLE OF CONTENTS

I. INTRODUCTION

Since the beginning of this century, plant breeders have helped increase the productivity and quality of many important plants. However, the lack of knowledge concerning the nature of genes and their mode of action is still a limitation for optimal use of the potentials of genes of higher plants.

The past decennium of molecular biology research has widened perspectives and provided opportunities to change this situation. New technologies have been developed that may provide potentially useful tools in combination with classical plant breeding techniques. Modern molecular biology has focused attention on the structure and organization of genes and to the regulation of their expression leading to a certain phenotype. The development of recombinant DNA technology has made it possible to isolate individual genes and the development of single-cell transformation methods has made it possible to introduce foreign DNA into various organisms. These procedures have also been established now for plant cells.[1-3]

These new technologies, which are potentially useful tools in plant biotechnology and classical plant breeding are not mutually exclusive. They have to be used in combination with each other to be effective in manipulating genetic information through methods that have been adapted from genetic recombination that occurs in nature. If, for instance, by genetic engineering natural barriers such as incompatibility can be overcome, the new plant must still be selected and evaluated under field conditions to ensure that the genetic change is stable and the attributes of the new variety meet commercial requirements.

II. GENETIC ENGINEERING OF PLANTS

The development of genetic engineering in general involves three different phases:

Phase 1 — The isolation of viable single cells, preferably protoplasts because the cell wall may act as a barrier for the genetic exchange

Phase 2 — The engineering of those protoplasts to alter the genetic composition of the cells. Desired traits are selected at this stage if possible

Phase 3 — The regeneration of the altered single cells, so that they can grow into fertile plants

A. Cell and Tissue Culture

Tissue culture involves growing plant material under defined conditions in vitro. Chapters 1 and 2 deal with some of the methods of plant tissue culture. Protoplasts derived from various parts of the plants, but in general from leaves, by incubating with enzymes that degrade the cellulose walls and intercellular compounds like pectins, can also be cultured.

Protoplasts are single cells that lack the cell wall. This absence of a cell wall is crucial for studies involving the uptake of macromolecules like nucleid acids. During the early stages of culture the protoplasts start to regenerate a new cell wall. This regenerating cell wall is important for cell wall specific interactions with microorganisms such as *Agrobacterium.* Based on these properties of protoplasts we have developed two methods to introduce foreign DNA into protoplasts, namely DNA transformation[3] and cocultivation.[1,2] Both methods will be described further in this chapter.

Protoplasts begin tissue culture as single cells, but after division they develop into solid callus or they can be kept in a liquid medium as a cell suspension. In either case they may begin to differentiate and can grow into the well-organized tissues and organs of a normal plant.

ASEXUAL MIX OF GENOMES

NUCLEAR CHARACTERS

- FUSION OF *VIABLE* PARENTAL PROTOPLASTS
  - *PARENTS NON TREATED* (+)
  - *ONE PARENT IRRADIATED TO STIMULATE CHROMOSOME ELIMINATION* (+)
- FUSION OF PROTOPLASTS WITH :
  - *MINIPROTOPLASTS* (VIABLE SUBPROTOPLASTS CONTAINING NUCLEUS AND LITTLE CYTOPLASM)
  - *NUCLEI*
  - *CHROMOSOMES*
  - *MINICHROMOSOMES* (OBTAINED VIA RECOMBINANT DNA TECHNOLOGY)

↓

SOMATIC HYBRIDS

NEW COMBINATION OF CHROMOSOMES

(+) reproducible positive results

FIGURE 1. Approaches for plant genetic engineering by asexual mixing of nuclear genomes. Only a few have resulted in reproducible positive results.

A serious problem in using plant cell culture that is not yet understood is the fact that plant cell cultures cause an enormous genetic disturbance in the cell due to genetic instability, i.e., polyploidy, aneuploidy, chromosomal rearrangements, and various kinds of small mutations[5-7] (see also Chapter 10). In addition, plant cells in tissue culture tend to lose their regenerative or morphogenetic capacity after being subcultured for a prolonged time. Other tissues derived from many crops, in particular monocots, cannot yet be stimulated to regenerate plants.[8] In any case, when using procedures for genetic manipulations where plants must be regenerated, the tissue culture phase should be kept as short as possible, or preferably avoided. In addition it is essential that more be understood about the molecular processes which control cellular differentiation and regeneration.

1. Genetic Manipulation of Cells in Tissue Culture

The second phase involves the genetic manipulation of cells in tissue culture and the selection of desired traits. These new technologies fall into two categories, those involving genetic modification through mixing of complete genomes of different cells and/or organelles[15,16] and those involving the insertion or modification of genetic information through DNA and DNA vectors. This second category can be called DNA transformation. Examples of the first category are shown in Figures 1 and 2. The first of these gives examples of asexual mixing of nuclear genomes resulting in somatic hybrids.[9-12] This might also involve fusion of irradiated nuclei[13] or fusion with miniprotoplasts containing only one or a few chromosomes[14] to eliminate most of the genomic properties.

As indicated in Figures 2 and 3 the target for the introduction of well-defined genes via the molecular approach is not only the nucleus but also the chloroplasts or mitochondria in order to modify cytoplasmic characters. The interest in these cytoplasmic organelles comes from the fact that chloroplasts are centers of photosynthesis and

ASEXUAL MIX OF GENOMES

CYTOPLASMIC CHARACTERS

- FUSION OF *VIABLE* PARENTAL PROTOPLASTS (+)
- FUSION OF *INACTIVATED* PARENTAL PROTOPLASTS (+)
  - ONE PARENT IRRADIATED, THE OTHER IODOACETATE TREATED; *CHLOROPLAST* TRANSFER FROM IRRADIATED PARENT (*MITOCHONDRIA ?*) WITH INACTIVATED NUCLEUS
- FUSION OF PROTOPLASTS WITH :
  - *CYTOPLASTS* (ENUCLEATED, NON VIABLE SUBPROTOPLASTS)
  - *ORGANELLES* (MITOCHONDRIA, CHLOROPLASTS)

↓

CYBRIDS

NEW COMBINATION OF NUCLEUS AND CYTOPLASM

(+) reproducible positive results

FIGURE 2. Cybrids can be obtained via asexual mixing of genomes using various approaches and can be used to manipulate the cytoplasmic characters of a cell.

INTRODUCTION OF WELL DEFINED GENES (RECOMBINANT DNA)

NUCLEAR CHARACTERS

I TRANSFORMATION OF *PROTOPLASTS*
- NAKED DNA (+)
- ENCAPSULATED DNA
  - *LIPOSOMES* ($\pm$)
  - *PROTEIN COAT*
  - *BACTERIAL SPHEROPLASTS* ($\pm$)
- MICROINJECTION
- COCULTIVATION WITH *A.TUMEFACIENS* (+) (DICOTS ONLY)

II TRANSFORMATION OF *EGG CELLS*
- FERTILIZATION WITH TRANSFORMED GERMINATED POLLEN
- MICROINJECTION OF FERTILIZED *EGG CELLS* VIA POLLEN TUBE ENTRANCE OF OVULE (MICROPYLE)

CYTOPLASMIC CHARACTERS

TRANSFORMATION OF
- CHLOROPLASTS
- MITOCHONDRIA

(+) reproducible positive results
($\pm$) reproducibility not well established yet

FIGURE 3. The recombinant DNA technology can be used to change either nuclear or cytoplasmic characters of a cell. Various approaches are followed to transform cells, but only a few of these have given reproducible results in the stable transformation of plant cells for nuclear characters. No data proving transformation of chloroplasts or mitochondria are yet available.

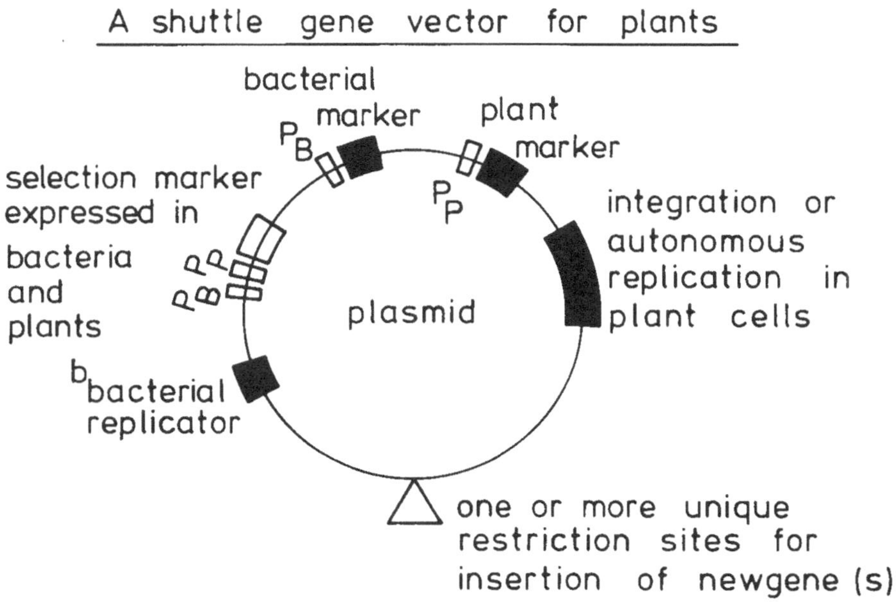

FIGURE 4. Schematic representation of a shuttle gene vector for plants. P_B bacterial promotor; P_P, plant promotor.

mitochondria are the energy power plants in the cell. Both organelles also play a role in certain disease resistance, while mitochondria are known to play a role in cytoplasmic male sterility (CMS).[17] In addition, chloroplasts synthesize compounds which participate in certain biosynthetic pathways involved in the production of secondary metabolites. Since chloroplasts are also centers of production of new proteins, they could be interesting targets for the molecular approach.

The second category of methods to transfer genetic information is shown in Figure 3. This involves the introduction of well-defined genes which can be from both nuclear or cytoplasmic origin. Only a few methods so far have been applied successfully for the stable transformation of nuclear characters.[2,3,18] For dicotyledons, *Agrobacterium tumefaciens* has been shown to be the most efficient vector for the introduction of DNA into plant cells.

B. The Recombinant DNA Technique

Well-defined nuclear and cytoplasmic genes are candidates for transfer into plant cells. In fact, using recombinant DNA every gene from microbial origin also might be transferred into plant cells. However, to achieve this, a number of preconditions would have to be fulfilled. First, one should have a shuttle gene vector (Figure 4). Since many of the recombinant DNA manipulations are performed in *E. coli*, a cloning vehicle for plants should preferably be a "shuttle gene vector". This means that it must contain DNA sequences for replication and selection in bacteria as well as in plant cells. For both stable maintenance of introduced genes during cell division and sexual transmission to progeny, a vector containing DNA sequences which can force the integration of foreign DNA into the genome is preferred over vectors showing autonomous replication. Various possibilities exist to accomplish autonomous replication of a plant vector. With the recombinant DNA technique minichromosomes can be constructed which still carry a chromosomal centromere and sequences for autonomous replication (ARS). Such AR-sequences are present in the DNA of eukaryotic chromosomes at regular distances. Minichromosomes as well as Ar-sequences inserted into a plasmid have been used successfully in yeast.[19,20] Also plant viruses can be applied for the development of vectors showing autonomous replication. Plant viruses containing RNA as genetic material might be used via cDNA for construction of vectors, but not much has been achieved as yet with these viruses. Currently most attention is paid to the DNA-containing plant virus called cauliflower mosaic virus (CaMV).[21] Attempts are being made to insert foreign DNA into the CaMV genome and to introduce this DNA into plant cells through infection of susceptible host plants. CaMV has a limited host range. Progress has been made with the encapsidation of inserted DNA fragments. However, stability, size limit, and expression of foreign DNA inserted into the CaMV genome need further study in order to establish whether or not this system can be used for plant transformation. No plant viruses have been found which are able to integrate their genetic information into the host genome. But it has been established now that the plant pathogens, *Agrobacterium tumefaciens* and *Agrobacterium rhizogenes*, are able to introduce a particular part of their DNA into plant cells. *A. tumefaciens* induces tumors called crown galls, while *A. rhizogenes* causes the hairy root disease on dicotyledons. Large plasmids in these agrobacteria, which are called tumor-inducing plasmid or Ti plasmid and root-inducing plasmid or Ri plasmid, confer on their hosts the pathogenic capacity. Both sorts of diseases result from transfer and functional integration into plant chromosomes of a particular part of the Ti of Ri plasmid. The bacteria do not penetrate into the cells which are being transformed. They firmly attach to the cell wall of cells in an infected area of a wound in order to transfer DNA. Here, only a few relevant aspects about the Ti plasmid and its usefulness as plant vector will be discussed.

If a good vector system is available, the second precondition for gene transfer is the isolation and characterization of genes of interest. Although quite a number of genes have been isolated directly from fragments of viral and bacterial chromosomes, the isolation of specific genes from fragmented eukaryotic chromosomes is still rather difficult. One of the problems is that although various genes exist in multiple copies or multiple gene families, many others do not and are thus such tiny parts of the large amount of DNA in a plant cell. Many complex traits that are of agricultural importance, such as yield and resistance to biotic and abiotic factors, are polygenic traits which are neither well identified nor biochemically characterized.

With a suitable plant vector and the appropriate DNA, methods are available for introduction of this DNA into the genome of plant cells, where it is expressed in RNA which can be translated into stable, active proteins. The expression of such genes can be achieved by making the vector construction in such a way that the gene of interest is controlled by the regulatory sequences (promotor region) of a plasmid gene. This can be the marker gene of the vector.

III. THE Ti PLASMID

Of the transformation systems which exists in nature the Ti plasmid present in the soil-bacterium *Agrobacterium tumefaciens* meets the conditions described here and might be a good candidate as a vector. *A. tumefaciens* is the bacterium that causes crown gall disease on dicotyledonous plants.

Crown gall is characterized by unlimited plant cell proliferation (gall formation). Crown gall tumor tissue differes from nontransformed tissue by its ability to grow autonomously on synthetic media in the absence of hormones, i.e., auxins and cytokinins. Moreover, tumor-specific compounds, called opines, are produced by these cells as a direct result of the genetic transformation by the bacterium. Opines are unusual amino acid derivatives which accumulate in the plant cells. They appear not to

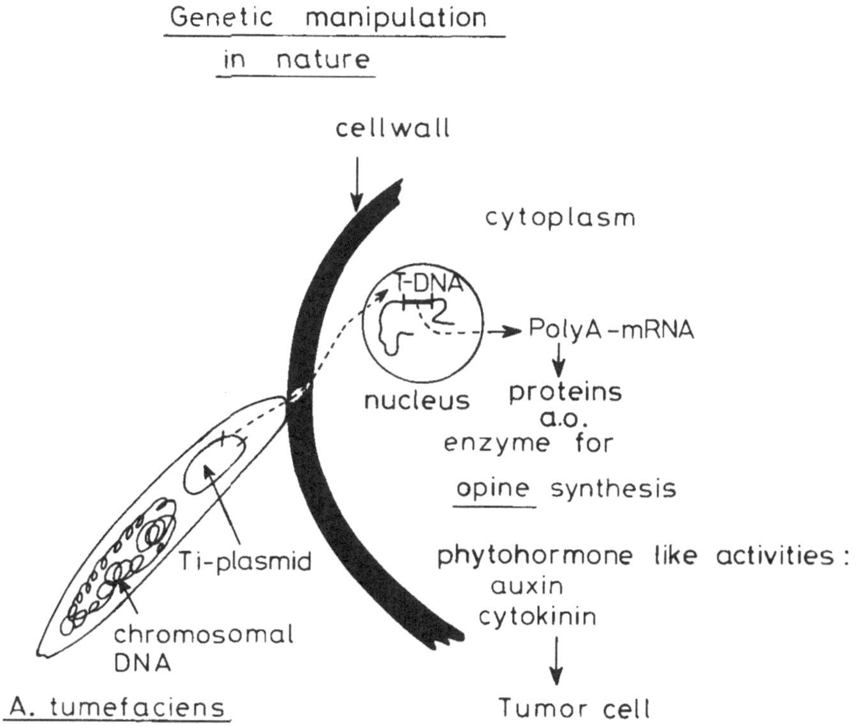

FIGURE 5. Schematic representation of the events that occur upon infection of a plant cell by an octopine type virulent *Agrobacterium tumefaciens*.

have any known useful function in the plant cells and are not responsible for the tumorous state of the cells. However, opines can be useful to agrobacteria, since they can serve as a source of carbon and nitrogen, while some of the opines also induce conjugation between the bacteria. The Ti plasmid of *A. tumefaciens* carries the genetic information for tumor formation, synthesis of opines in crown gall cells, catabolism of these compounds by the bacterium, and for conjugation. The type of opine produced in the tumors defines the Ti plasmids as octopine, nopaline, or leucinopine Ti plasmid.

Crown gall is a clear example of genetic manipulation via recombinant DNA in nature (Figure 5). *Agrobacterium* introduces a part of its Ti plasmid, called the T-region, into plant cells via a still unknown mechanism. The bacterial DNA recombines with chromosomal DNA in the plant nucleus. As part of the plant genome, this piece of foreign DNA is called transferred DNA or T-DNA. The T-DNA can be subdivided into a T_L- and a T_R-DNA. The T_R DNA can be absent in the tumors. The size of the T-DNA varies between 8 and 16 Mdaltons. The T-DNA is transcribed into various RNA molecules carrying a poly A tail, which is characteristic for eukaryotic mRNA. The enzymes for opine synthesis have been shown to be products of T-DNA transcripts. Other T-DNA genes, namely the oncogenes (*onc*-genes), are responsible for the tumorous or oncogenic state of crown gall cells. Through the activity of *onc*-genes the plant cells are stimulated to divide continuously and so provide the bacteria a "niche" with sufficient food in the form of opines. Knowing how agrobacteria are able to manipulate plant cells genetically for their own benefit makes it possible to exploit the system for our own objectives. To this end more must be known about those genes and sequences of the Ti plasmid which are essential for transfer, integration, and expres-

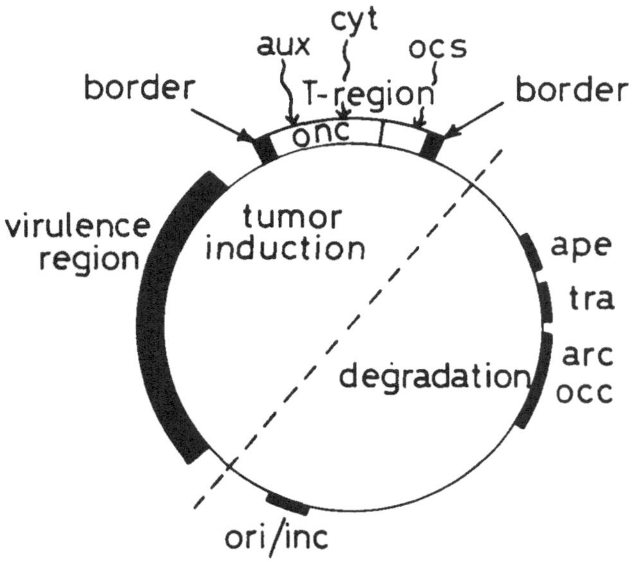

FIGURE 6. Representation of the most important markers on the octopine-type Ti plasmid. The octopine Ti plasmid can be separated into two halves of which only one is required for tumor induction. This half carries the T-region and the *Vir*-region. The other half carries genes for degradation or catabolism of octopine (occ) and arginine (*arc*) as well as genes for conjugation (*tra*), exclusion of a phage (ape), incompatibility (*inc*), and the origin of replication (*ori*). The T-region carries genes for auxin (*aux*) and cytokinin (*cyt*) activity and a gene for octopine synthesis (ocs) in tumor cells. It is flanked by direct repeats of 24 base pairs, which are called border- or junction-sequences. The virulence region carries genes involved in an unknown mechanism by which the T-region becomes transferred and integrated as T-DNA in plant cells.

sion of T-DNA in plant cells. Here, we will deal with the knowledge gained for the octopine Ti plasmid only. A schematic drawing showing a number of phenotypic traits is shown in Figure 6. For detailed information the reader is referred to some recent reviews.[22,23]

Molecular genetic studies have demonstrated that about half the plasmid does not carry any genes involved in tumor induction. It harbors mainly catabolic functions for utilization of octopine (occ) and arginine (*arc*); genes for conjugation (*tra*) and the exclusion of a phage (*ape*) are also located on this catabolic or degradation region of the Ti plasmid. All genes responsible for tumor induction are positioned on the other half in two regions that are called the virulence or *Vir*-region and the T-region (Figure 6).

The T-region carries the *onc*-genes, which are active in the plant cells as part of T-DNA. Mutagenesis of the T-region has revealed that it consists of a number of loci and that it is bordered by particular base sequences or repeats that are called border sequences. Besides a gene for octopine synthase, at least two loci have been identified on T-DNA which control plant cellular growth and differentiation. Inactivation of one locus triggers shoot formation (shooter mutants), while disruption of the other results in root formation (rooter mutants) from relatively small tumors when the mutants are tested on *Kalanchöe* or tobacco plants. Infection with a mixture of a shooter and a rooter mutant gives rise to normal tumor development. The same result is obtained when either a shooter mutant is supplemented with auxin or a rooter mutant with cytokinin. This indicates that the shooter mutant is affected in a locus for cytokinin-like activity (cytokinin-locus).[24] The combined activity of both loci leads to unorganized tumors. They probably are directly responsible for the phytohormone autotrophic growth of the tumor tissue. Study of the function of these T-DNA loci and of the targets of their products in plant cells might shed light upon molecular processes which control regeneration of plant cells.

The *Vir*-region of the Ti plasmid has been mapped on a short distance of the T-region. In contrast to mutations in the T-region, which give rise to tumors with altered morphology, mutations in *vir*-genes abolish or attenuate tumor formation. As the virulence region of the Ti plasmid has never been detected in transformed plant cells obtained after infection with *Agrobacterium*, and as mutations in genes in this region can be complemented *in trans* in *Agrobacterium* strains, it is likely that *vir*-genes are expressed in the bacterium for virulence.[25] Some of them could play a role in the processing of the Ti plasmid DNA, that is found as T-DNA in transformed cells, via recognition of T-region border sequences. Other *vir*-genes are supposed to be important in establishing a mechanism for transfer of Ti plasmid DNA into plant cells.

A. Ti plasmids as Plant Gene Vectors

Since *Agrobacterium* is so efficient in transformation of plant cells much work has been focused on the construction of Ti plasmid-derived vehicles, which can be used in *Agrobacterium* for the introduction of foreign genes into plant cells. Obviously, genes that are to be introduced into the plant genome have to be inserted into the T-region of the Ti plasmid first. Because of the large size of the Ti plasmid and the various functions needed for virulence, it seemed impossible until recently to develop a simple cloning vector derived from the T-region. One wants to introduce only useful genes, avoiding the presence of large remnants of T-DNA, with the use of a small plasmid which is easy to handle for recombinant-DNA work in *E. coli*. Therefore, in order to construct a more advanced plant gene vector derived from the Ti plasmid, we investigated whether it was possible to disconnect the T-region from the *Vir*-region without loss of their functions. To this end, two compatible plasmids have been constructed, one carrying the T-region, the other harboring the *Vir*-region of the octopine Ti plasmid. The results obtained with such plasmids are summarized in Figure 7 showing a binary plant vector strategy.[45] *Agrobacterium* strains carrying only one plasmid with either the T-region or the *Vir*-region were not able to induce tumors. However, when both sorts of plasmid were combined within one cell, an *Agrobacterium* strain was obtained with normal tumor inducing capacity and octopine was detected in the tumors. This result means that the T-region and *Vir*-region can be separated physically without loss of oncogenicity.

Inserts in T-region vectors can now be constructed in *E. coli* without the necessity of later transferring the insert into an intact Ti plasmid via homologous recombination. Vectors carrying a genetically manipulated T-region can be introduced directly into an *Agrobacterium* strain which already has a *Vir*-plasmid. Without any further manipulations the agrobacteria can subsequently be used for infection and transformation of plant cells. However, it still would be desirable not to have segments of the T-region present as part of the foreign DNA that becomes integrated into the plant genome. it has already been said that the *onc*-genes can be eliminated without affecting transfer and integration of T-DNA in plant cells. This indicates that a construction carrying only that part of the T-region which is essential for its transfer, such as the border sequences, could be transferred in the presence of a *Vir*-plasmid. The development of these vectors is in progress, but it is quite likely that apart from the border sequences

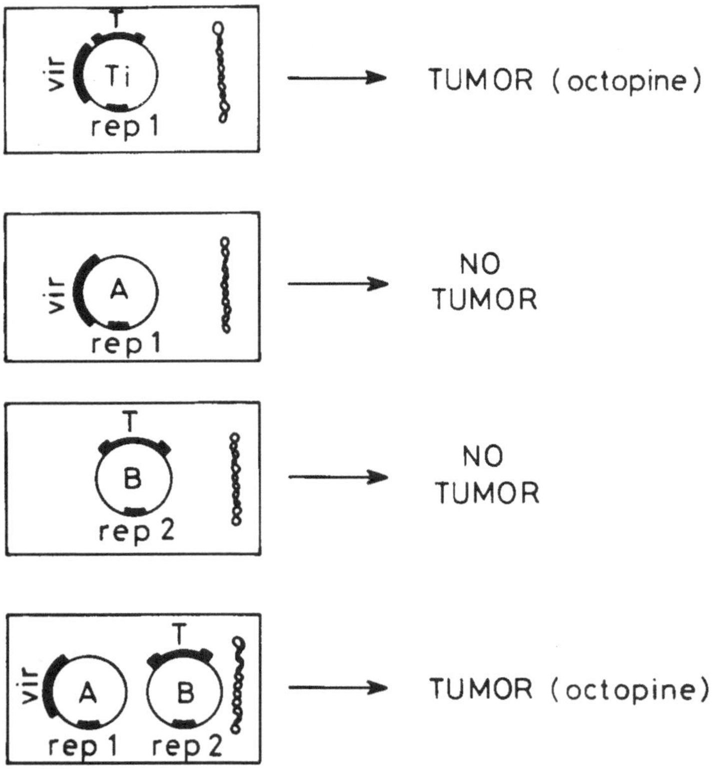

FIGURE 7. The binary plant vector strategy[45] is based on a functional separation of the T-region and *Vir*-region, each on a different plasmid. The T-region plasmid (B) and *Vir*-region plasmid (A) alone do not confer tumorigenity to their host. Agrobacteria acquire a normal tumor-inducing capacity when both compatible plasmids A and B are present within one cell. The replicator of these plasmids is indicated by rep 1 and rep 2.

of about 24 bases, no DNA needs to be present other than those useful genes which have to integrated into the plant. With this knowledge a gene vector has been constructed which, in the presence of the *vir*-region, can transfer the T-DNA region to the plant cells. This gene vector contains only the border sequences and the marker gene for opine synthesis (Figure 8). In addition it has selectable markers for manipulation in the bacteria and restriction sites for the insertion of foreign sequences. The absence of the *onc*-genes does not limit the transformation capacity of this plasmid and is favorable because now transformation can occur without disturbing the phytohormone balance and so without tumor formation. A disadvantage of this development is that selection of transformed cells based on this phytohormone independency can no longer occur. Therefore, an additional selection marker has to be introduced. With this construction we have an important tool to introduce genes into plants at will. When a plant is infected with such a construction some cells in the wound area will be transformed. However, since tumors do not form, the transformed cells cannot be detected. Therefore, it is necessary to develop systems that allow the transformation of individual cells that can be grown in tissue culture and from which plants composed entirely of transformed cells can be regenerated.

Two approaches have been followed to develop such systems. The first is designed particularly to avoid the natural cellular interaction between agrobacteria and dicoty-

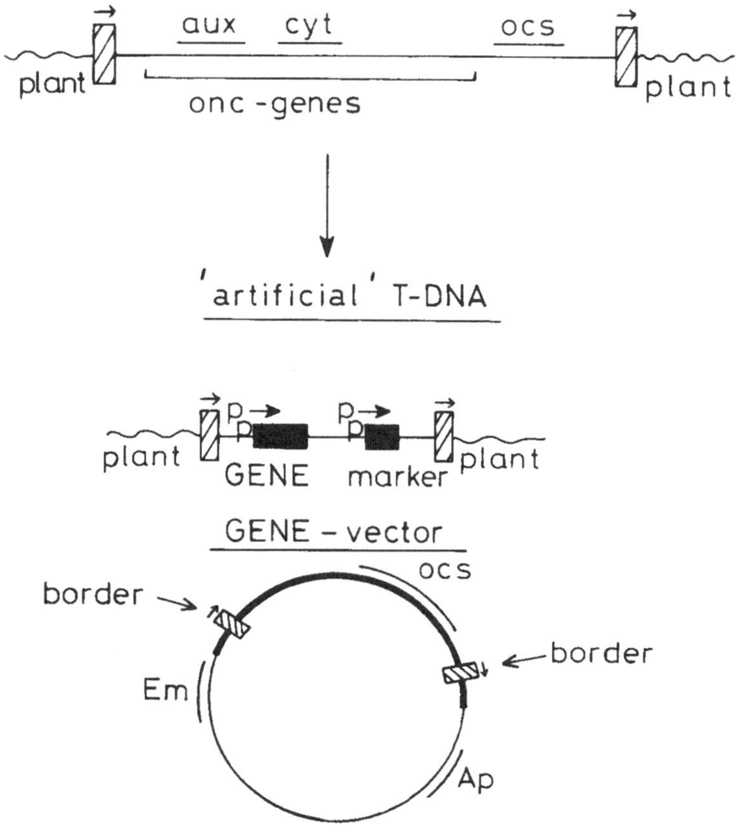

FIGURE 8. "Artificial" T-DNA is constructed by placing the gene to be introduced into plant cells, together with a marker gene for selection or identification of transformed cells, between fragments containing the T-region border sequence (▨). *Onc*-genes present on the normal T-DNA are not required for its transfer and integration into plant chromosomes. *Aux* and *cyt* stand for auxin- and cytokinin-loci. Promotor sequences for expression in plant cells are indicated by P_P. *Ocs* stands for octopine synthase used as a marker gene.

ledonous plants and is called the DNA transformation method.[3] The second approach is based on this naturally occurring transformation capacity of *Agrobacterium* and is called the co-cultivation procedure.[1,2]

B. Transformation of Plant Protoplasts with Ti plasmid DNA

Recipient cells that are used for DNA transformation experiments vary from bacteria to eukaryotic cells derived from animals or plants. Each system requires its own particular conditions for the uptake of DNA, but some basic features are similar in several methods. DNA can be offered to the recipient cells in four different ways: (1) as pure DNA,[26] (2) as calcium phosphate-DNA coprecipitate,[27] (3) as DNA encapsulated in liposomes,[28] and (4) by direct microinjection of the DNA.[29] DNA uptake can be enhanced by a heat shock treatment, by a high pH treatment, and/or by addition of polyethylene glycol (PEG), poly-L-ornithine (PLO), or polyvinyl alcohol (PVA). Recipient cells usually must be made "competent"; for bacteria this can be achieved by a $CaCl_2$ treatment. For transformation of plant and yeast cells protoplasts must be produced by enzymatic removal of the cell wall. The combination of $CaCl_2$ and PEG is typically used for many procedures of DNA transformation applied to different organisms. It has been shown that transformation via liposomes can be as efficient as the "calcium phosphate DNA coprecipitate" method.[30] The latter method was initially developed to transform human cells with human adenovirus 5 DNA. A modification of this procedure was used in further experiments with other cell types and DNAs.[31] The basic principle of this method is that a fine calcium phosphate precipitate is allowed to form in the presence of DNA. Cell monolayers are then incubated with the DNA-calcium phosphate precipitate and postincubated in a medium containing a high concentration of Ca^{2+}. This is currently a well-established procedure for the transformation of mammalian cells.

The final method to be mentioned here is microinjection. In general micropipets are made with the aid of a mechanical puller. Micromanipulators are used to guide the pipet into the cell; the injection solution is then delivered by pressure from a syringe. This method was initially applied to systems consisting of cells of sufficient size, e.g., oocytes, and at present also for mammalian tissue culture cells that are well immobilized in a monolayer. The frequency of transformation must be high to overcome the relatively small number of cells which can be manipulated.

In order to achieve uptake of Ti plasmid DNA by *Petunia* protoplasts a few methods have been published.[32] Protoplasts were incubated with either PLO and DNA or with $CaCl_2$ and DNA for 15 sec at 25°C followed by addition of PEG/Ca^{2+} and incubation for 2 min at 34°C. The reported results are somewhat ambiguous. Transformants were found to be unstable and although by Southern blot hybridization one small fragment was observed with DNA sequence homology to the T-region of the Ti plasmid, the main internal restriction fragments of T-DNA were not detected and therefore genetic information that is essential for the tumorous character of crown gall cells seemed to be absent. Another peculiar point is that the same fragment having homology to T-DNA was found in every transformant. In our laboratory a method was developed which has proven to be reproducible and which results in stable transformants carrying large, but variable stretches of T-DNA.

Many transformation procedures, in various combinations, have been tried at our laboratory in order to achieve stable transformation of tobacco protoplasts with octopine Ti plasmid DNA. We were successful with one procedure that is based on the use of PEG/Ca^{2+} (Table 1). In this procedure 1 $m\ell$ of a suspension of tobacco SR1 leaf protoplasts (5×10^5 protoplasts per milliliter in K3 medium with phytohormones) is mixed with 0.5 $m\ell$ F medium (among other salts containing 125 $m\ell$ $CaCl_2$)[12] supplemented with 40% (w/v) PEG 6000. The final concentrations of Ca^{2+} and PEG are thus approximately 40 m*M* and 13%, respectively.[3] During the preparation of F medium a fine calcium phosphate precipitate is formed, which is not removed. To the mixture is added 10μg Ti plasmid DNA and 50 μg calf thymus DNA as a carrier. After incubation for 30 min at 26°C, the PEG is decreased and the Ca^{2+} concentration is increased stepwise with F medium during a postincubation period, which in total lasts about 20 min. Higher concentrations of PEG result in more cell death and appearance of fusion bodies. No transformants were obtained at lower concentrations of Ti DNA or when carrier DNA was omitted. Lower concentrations of Ca^{2+} during postincubation or a one-step addition of F medium followed by 20 min postincubation also gave negative results.

The medium in which the protoplasts are cultured does contain phytohormones. However, the hormone level is rather reduced when compared to hormone levels used in culture media normally employed for the growth of callus tissue. Although non-transformed and therefore normal SR1 tobacco protoplasts are able to grow in the medium we are using, it is necessary also for recovery of transformed protoplasts to add phytohormones at the start.[33] The chosen level of phytohormones, however, al-

Table 1
A STEPWISE SCHEME OF THE IN VITRO DNA TRANSFORMATION PROCEDURE

1 mℓ (5×10^5) protoplasts in K_3 ↘ 0.5 mℓ 40% (w/v) PEG 6000 in F-medium ↙

Mix and add 10μg Ti-plasmid DNA
and 50 μg calf thymus DNA

↓

Mix and incubate for 30 min at 26°C with occasional shaking

↓

5 × [add 2 mℓ F-medium, mix and incubate for 5 min, at 21°C]

↓

Centrifuge at 600 rpm

↓

Remove supernatant and resuspend protoplasts in 10 mℓ K_3 medium containing 0.4 *M* sucrose, 250 μg/mℓ carbenicillin and phytohormones

↓

Plate and leave in the dark at 26°C for 24 hr

↓

Place in 2000 lx for 12 hr/day and culture for ca. 2 weeks

↓

Add fresh K_3 medium containing 0.4 *M* sucrose
and phytohormones and culture ca. 2 weeks

↓

Dilute the suspension 1:10 with K_3 medium containing 0.3 *M*
sucrose, hormones and 0.3% agar and culture ca. 4 weeks

↓

Take colonies and place them on K_3 medium-H containing
0.2 *M* sucrose and 0.5% agar; culture ca. 4 weeks

↓

Take surviving colonies and put them on fresh K_3-H,
0.2 *M* sucrose, 0.5% agar and culture ca. 4 weeks

↓

Putative transformed calli are placed on
LS medium and tested for LpDH activity

lows a more rapid selection against nontransformed cells, when later on they are cultured in the absence of phytohormones. In the in vitro DNA transformation procedure the tobacco cells are kept 1 month longer on medium containing the reduced phytohormone level than is usually done in the co-cultivation procedure.

After the dark period (see Table 1) the petri dishes with protoplasts are brought into the light (2000 lx for 12 hr/day) and further cultured at 26°C; 2 weeks later the cultures are divided into two (about 5 mℓ each) and 5 mℓ of fresh K_3 medium are added to both. Again 2 weeks later the suspensions of actively dividing cells are diluted 1:10 with K_3 medium still containing phytohormones, but with a reduced sucrose concentration, (0.3 *M*). The medium is solidified with 0.3% agar. After 1 month of growth on these plates small colonies are transferred to plastic petri dishes containing a layer of phytohormone-free K_3 medium (K_3-H), 0.2 *M* sucrose which is solidified with 0.5% agar. They are cultured for 1 month and then the surviving calli are placed on fresh K_3-H, 0.2 *M* sucrose, 0.5% agar plates. After a culture period of 1 month the calli which have survived are called putative transformants. They are transferred and maintained on LS medium lacking phytohormones until they are examined biochemically for their transformed nature. Compared to K_3 medium the LS medium lacks in nicottinic acid, pyridoxine-HCl, and xylose, while thiamine-HCl and sucrose concentrations are much lower. In this respect LS medium is a more minimal medium than K_3 medium. The mineral salts are about the same with slight variations.

Criteria for the transformed nature of calli obtained after the selection procedure are the continued growth on LS medium without phytohormones, the presence of lysopine dehydrogenase (LpDH) activity, and the presence of T-DNA. The fact that calli are obtained with a capacity to grow continuously on medium lacking hormones is an indication of the transformed nature of these calli; however, habituation might have occurred spontaneously. Habituation is the phenomenon in which plant cells may acquire a phytohormone-independent growth trait during subculturing in the presence of phytohormones. So far, no habituated tissues have been isolated using SR1 protoplasts and the described transformation procedure. An additional tumor specific marker is the presence of LpDH. If calli have reached a size of approximately 0.5 cm^3, a piece can be used to assay the enzyme activity.[34] Although the presence of LpDH is a clear indication for the transformed state of a callus, its absence does not mean that a tissue is not transformed. Direct and conclusive evidence for transformation is obtained if the exogeneously supplied DNA or fragments of it are detected as part of the plant DNA, i.e., if T-DNA is present. This is done by Southern blot hybridization using restriction endonuclease digested DNA of the plant material and ^{32}P-labeled cloned T-region fragments.[35]

The described method has been applied in our laboratory to transform tobacco protoplasts reproducibly with octopine-type Ti plasmid DNA, nopaline-type Ti plasmid DNA, and with cloned octopine T-region DNA. The transformation frequency appears to be rather low: on average two to three transformants are obtained if one starts with 5×10^5 protoplasts. However, for a more accurate calculation of transformation frequency, cell death, clumping, and the number of colonies subjected to selection have to be taken into consideration. Usually, the indicated number of transformants are found among about 3000 colonies that are placed on plates without hormones from K_3 medium containing 0.3 *M* sucrose, 0.3% agar, and phytohormones.

Based on phenotypic properties the DNA transformants derived from two independent transformation experiments with octopine Ti plasmid DNA were subdivided into three classes as listed below.

Class A — LpDH positive, hormone autotrophic, no regeneration, 8 transformant lines
Class B — LpDH negative, hormone autotrophic, regeneration, 5 transformant lines
Class C — LpDH positive, hormone auxotrophic, no regeneration, 1 transformant line

Class C was taken from K_3-H medium where it reached a sufficient size for the measurement of LpDH activity. Growth ceased on LS medium and could only be stimulated again by maintaining the callus on medium with phytohormones. T-DNA analysis showed the presence of sequences homologous to the T-region of the Ti-plasmid in the nuclear DNA of all transformants. It revealed that the T-DNA varied among the transformants suggesting that the transformants have originated from independent transformation events. In transformed tobacco tissues either obtained by in vivo infection or by the co-cultivation method it has been demonstrated that usually with only a few exceptions a piece of T-DNA of a well-defined size is present.[35,36] This suggests that the T-region might contain special sequences, called border sequences, which are used preferentially for its integration into plant DNA in cases when the bacterium is involved. In DNA transformation these sequences appear to be used less preferentially.

As can be seen in Figure 9, representing some of the most remarkable results, the T-DNA in transformants differs largely from that found in tumor tissue obtained via *A. tumefaciens.* Some DNA transformants carry a continuous stretch of T-DNA extending more to the left as well as to the right of T_L-DNA. Some contain T_R-DNA present

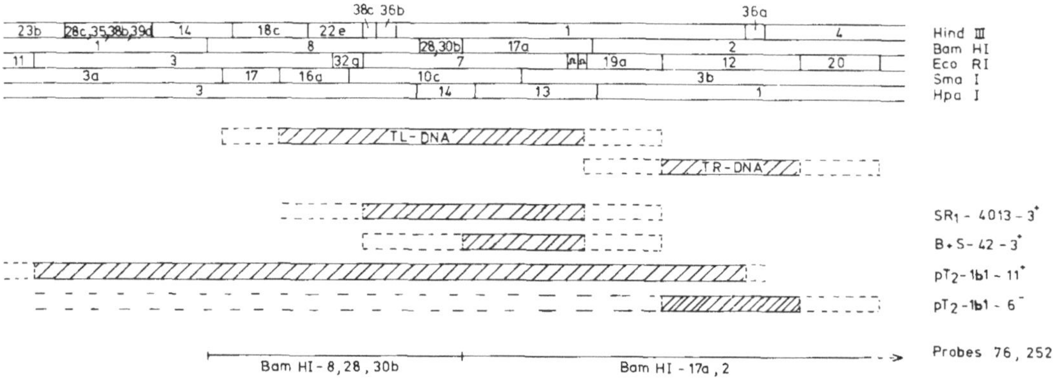

FIGURE 9. Representation of the T-DNA present in transformed tissues obtained via DNA transformation (pT2-lbl-11, pT2-lbl-6). The T-DNA architecture is compared with the consensus T-DNA (TL-DNA and TR-DNA) and with the T-DNA found in a tumor tissue obtained after co-cultivation (SR1-4013-3) and after somatic hybridization (B + S-42-3).

in a considerably higher copy number than T_L-DNA in the same transformant. Also, transformants are isolated showing a very complex banding pattern in which all or most of the known internal T-DNA restriction fragments are lacking. In this case various rearrangements in T-DNA as well as integration of parts of T-DNA at different sites in the plant genome might have occurred.

It has been reported that the octopine and nopaline T-region has at its ends nucleotide sequences that are preferred for T-DNA integration into plant DNA.[37,38] Such border sequences, determining a well-defined fragment of T-DNA, seem to be used less frequently when Ti plasmid DNA instead of *A. tumefaciens* carrying the same Ti plasmid is used in transformation experiments. This might indicate that these border sequences play a specific role in the bacterium by which a more fixed piece of T-DNA is integrated into tumor cells via *A. tumefaciens*. It has been shown that virulence genes exist far outside the T-region on the left half of the Ti plasmid. Genetic complementation studies recently have indicated that these genes act *in trans* and, therefore, most likely have to be expressed in the bacterium for virulence.[25,39] It can be envisaged that this set of genes is involved in determining the fixed size of T-DNA through the recognition of border sequences, as well as in its transfer into plant cells.

C. Transformation of Plant Cells via the Co-cultivation Method

The co-cultivation procedure involves tumor induction due to the natural infection process of *A. tumefaciens* (Figure 5) under in vitro conditions. Virulent agrobacteria are temporarily co-cultivated with protoplasts from the moment the protoplasts have regenerated a new cell wall.[1] Such protoplasts are comparable to conditioned cells that arise from differentiated cells after wounding a plant stem. The protoplasts are treated according to the scheme in Table 2.

The protoplasts are suspended at a density of 10^4 to 10^5 protoplast per milliliter in K_3 culture medium,[46] containing a comparatively reduced level of hormones (0.1 mg NAA and 0.2 mg kinetin/ℓ). Petri dishes containing 10mℓ of protoplast suspension are kept at 26°C in the dark for the first 24 hr, thereafter at 2000 lx for 48 hr. At day 3, i.e., well before cell division takes place and when new cell wall material has formed as indicated by Calcofluor white staining, the cells are mixed with agrobacteria in a 400:1 bacteria/protoplast ratio. After a co-cultivation period of 32 hr, during which heavy aggregation of protoplasts and bacteria occurs, free bacteria are removed by

Table 2
SCHEME FOR TRANSFORMATION OF *NICOTIANA TABACUM* PROTOPLASTS BY *AGROBACTERIUM TUMEFACIENS* AND SELECTION OF TRANSFORMED CELL STRAINS AND SHOOTS

N. tabacum SR1 shoots → Leaf protoplasts(pps,3 days old)

A. tumefaciens → Log phase culture

↓

Mixed incubation in K_3 medium + hormones
10^5 pps/mℓ
10^7 bacteria/mℓ
32 hr at room temp.

↓

Removal of the bacteria by centrifugation and washing

↓

Culture in K_3 medium + hormones and antibiotics

↓

Transfer to solid K_3 medium without hormones

↓

Transformed SR1 callus

↓

Transfer to solid LS medium without hormones

↓

Transformed SR1 shoots

repeated washing. The plant cells are further cultured in K_3 medium supplemented with hormones. Carbenicillin (250 μg/mℓ) is added to kill remaining bacteria; 3 to 4 weeks later small calli are plated at a 1:10 dilution on hormone-free medium solidified with agar. Continued selective pressure allows only calli consisting of transformed cells to survive; habituated tissue consisting of cells that spontaneously have acquired hormone autotrophy is not obtained with the SR1 protoplasts.

When virulent agrobacteria are added to freshly isolated protoplasts or to 1-day-old protoplasts neither bacterial attachment nor transformation is observed. The presence of newly formed cell wall constituents, as present on 3-day-old protoplasts, could be one of these essential conditions for transformation via agrobacteria. This assumption is supported by the results of experiments in which inhibitors of cell wall formation (coumarin, colchicine, 2,6-dichlorobenzonitrile) were used. We also found that EDTA, a chelating agent of divalent cations, but not EGTA, an agent that only binds Ca^{2+}, inhibits both attachment and transformation. Since a cell wall already had formed before application of these chelators, the result suggest that certain divalent cations, such as Mg^{2+} but not Ca^{2+}, may influence the binding of bacteria to cells. As has been suggested for other bacterium-plant interactions. lectins, which are hydroxyproline-rich glycoproteins, might play a role in the binding process. Their activity seems to depend on the presence of divalent cations. We found that substantial amounts of hydroxyproline are present both in cell wall material extracted from 3-day-old protoplasts and in material isolated from medium in which protoplasts are grown. If lectins, which interact with sugar groups, play a role in the bacterial attachment they perhaps could function as a bridge between the polygalacturonic acid component in the cell wall of dicotyledons and the lipopolysaccharide (LPS) component of the agrobacterial wall on the other side. Since agrobacteria often show a wide host range, the function of any compound which controls effective attachment can be neither plant species nor

Agrobacterium species specific. It might, however, have a specific function in establishing an effective attachment only between *Agrobacterium* and dicotyledonous plants. In this respect it is worth nothing that the LPS fraction of *E. coli* does not inhibit tumor induction by *A. tumefaciens* and that *E. coli* does not show attachment to 3-day-old protoplasts. This, moreover, indicates that the attachment observed for agrobacteria in vitro is mainly a specific phenomenon. Also, bacteria related to *Agrobacterium*, such as *Rhizobium trifolii*, attach to protoplasts regenerating cell walls, and earlier it was shown that when they acquire a Ti plasmid they were able to induce tumors.[40] However, *R. meliloti*, carrying a Ti plasmid, does not have the capacity to induce tumors or to attach to protoplasts with a regenerated cell wall. Agrobacteria that have been cured of their Ti plasmid, and therefore are avirulent, still have the capacity to attach to such protoplasts. Their LPS fraction also inhibits tumor induction by virulent agrobacteria. Therefore, it is likely that genes on the *Agrobacterium* chromosome determine the bacterial attachment.

When *A. tumefaciens* is co-cultivated with aggregates of normal tobacco cells, that have developed from dividing protoplasts in 16 days, attachment is observed, but no transformants are found. In this case either the Ti plasmid (or T-DNA) transfer mechanism does not function completely or plasmid DNA is transferred, but tobacco cells in culture are not "competent" or "conditioned" for stable transformation. Similar results obtained from tumor induction on plants led to the conclusion that cells at the wound site are "conditioned" during a period before division takes place, whereafter no transformation occurs. Whether this is generally true for all plant species has still to be established.

Crown gall cells are different from normal cells in that they do not show attachment of *A. tumefaciens*. We observed, however, that 3-day-old protoplasts, isolated from tobacco crown gall cells or from leaves of T-DNA containing transformed plants, do attach *A. tumefaciens*. This capacity appears to be lost again when tumor protoplasts have divided and cell aggregates are tested. This indicates that a component required for attachment, which is probably in the cell wall, is synthesized at the protoplast stage but modified, masked, or even absent at the cellular stage.

Via the co-cultivation procedure we have isolated hormone autotrophic calli using agrobacteria of the nopaline type. Transformants could also be isolated with agrobacteria harboring a cointegrate plasmid consisting of an octopine and nopaline Ti plasmid. The frequency of transformed tissues obtained with this method lies between 0.1 and 1% of the starting population of protoplasts. A characterization of transformants based on expression of the tumor markers, hormone autotrophy, and opine synthesis shows that these markers segregate among the various independent isolated tumors. This is shown in Table 3. Such a situation is not observed for crown galls induced on plants with the same bacterial strains. This difference can be explained by assuming that transformants derived in vitro via co-cultivation of protoplasts are homogeneous with respect to T-DNA structure and organization while this is not the case for crown gall tumors derived in vivo on the plant.[41] This last category of tumors consists of a mixed population of tumor cells and segregation of tumor markers, if it exists, will escape detection due to complementation. The homogenicity of in vitro tumors is also demonstrated by subcloning of transformant lines derived from co-cultivation via protoplast isolation. The subclones that can be obtained without selection all have the same phenotype and harbor an identical T-DNA structure as the primary transformant.

The architecture of T-DNA in octopine type transformants obtained via co-cultivation is not significantly different from that of T-DNA found in the calli derived in tumors on plants.[36] They, in general, contain only the T_L-DNA of a fairly fixed size. The observed phenotypic variation might be due either to deviations from normal T_L-

Table 3
VARIATION IN PHENOTYPE AMONG TRANSFORMANTS OBTAINED AFTER CO-CULTIVATION OF TOBACCO PROTOPLASTS WITH DIFFERENT *A. TUMEFACIENS* STRAINS

	Phenotype[a]			
Type of bacteria	Aut	Ocs	Nos	Frequency of occurrence[b]
Octopine	+	+		69
	+	−		30
	−	+		1
Nopaline	+		+	77
	+		−	23
	−		+	0
Octopine nopaline	+	+	+	13
Cointegrate	+	+	−	33
	+	−	+	33
	+	−	−	17
	−	+	−	4
	−	−	+	0

[a] Aut, phytohormone autotrophy; Ocs, octopine synthesis; Nos, Nopaline synthesis.
[b] With each bacterial type about 200 different transformants have been screened.

DNA or to partial expression of T-DNA genes. For instance, some octopine-negative transformants do contain the T-DNA gene for octopine synthesis although the RNA transcript cannot be detected. In such cases the absence of octopine presumably is a matter of expression. These observations are supported by the finding that application of the cytosine analog 8-azacytidine to the culture medium in some cases results in the restoration of octopine synthesis.[42] It is known that the base cytosine can be methylated, resulting in the absence of the expression of the DNA. 8-Azacytidine is a base-analog that cannot be methylated. If this is incorporated in the DNA, these sequences might be expressed in RNA. We think that this phenomenon might be the reason for the absence of octopine in certain transformed tissues.

Another interesting feature of the co-cultivation method is that it often gives rise to octopine-type transformants with a high shooting capacity. Regeneration of shoots has not been observed for wild-type octopine tumors on plants or for callus tissues derived from them. In the cases that we have studied so far, it is observed that shooting tissue contains T_L-DNA which lacks part of the left side of normal T_L-DNA. The deletions concerned affect at least part of the auxin-locus and this could well explain the spontaneous shooting capacity of this octopine-type transformant. One of the characteristics of the shoots is that they do not develop roots. In analogy with tumor-morphology mutants carrying an affected auxin-locus, this property might be due to the activity of the cytokinin-locus that is still present.

D. T-DNA and T-DNA Expression in Plants and Progeny

Transformed shoots, both from octopine- and nopaline-type transformants obtained after co-cultivation, were separated from the callus tissues and cultured separately. None of them formed roots; they produced opines and were phytohormone independ-

FIGURE 10. Mature flowering tobacco plant obtained from co-cultivation of protoplasts with *Agrobacterium tumefaciens.* Note the aberrant flower morphology shown in the inset. The flowers have an outgrowth of the pistil and are male-sterile which is characteristic for these transformed plants.

ent when brought back in tissue culture. Due to the absence of root formation, shoots were grafted on top of healthy tobacco stems, where they developed into mature flowering plants with retention of the ability to synthesize opines. The flowers of such plants were characterized by an outgrowth of the style (heterostyly) and were male-sterile[18] (Figure 10). Seeds could be obtained from such plants by cross-pollination with pollen from control tobacco plants. When seeds derived from transformants were germinated on culture medium, all seedlings initially developed a root tip.

Two classes of seedlings could be distinguished. One class representing about 50% of the seedlings showed a normal morphological appearance and were opine negative. The other 50% were opine positive and had a transformed phenotype: thick, sprouting shoots and lack of normal root development. The root tips of this class of seedlings had developed into callus. The results demonstrate that the T-DNA must have been transmitted and that it behaved as a single dominant Mendelian trait.

To establish the stability of the T-DNA during regeneration and meiosis, the DNA from different tissue types and seedlings derived from both nopaline and octopine tumor lines were analyzed for the presence of T-DNA.[41] From these experiments we conclude that the T-DNAs that were present in the original callus tissues are retained without detectable changes during regeneration and development of cells into mature plants and are transmitted intact into the progeny plants. Southern blot hybridization revealed the presence of identical internal T-DNA fragments and plant DNA/T-DNA junction fragments in tissue lines representing successive developmental stages, including transformed seedlings.

T-DNA sequences coding for opine synthases are expressed in both parental tissues and in seedlings. Of the F1 plants 50% were opine positive and developed no roots. As the seeds were obtained from cross-pollination of regenerants with pollen from normal tobacco plants, this indicates that the regenerants are heterozygous concerning the T-DNA and that T-DNA sequences are expressed in a dominant way. This is supported by the fact that about 50% of the seedlings obtained from back-crosses showed the transformed phenotype.

No T-DNA sequences were detected in opine-negative rooting seedlings. Furthermore, opine-negative F1 plants are also susceptible to tumor induction by *Agrobacterium tumefaciens* strains. However, heterostyly and male sterility as observed on the parental plants and on all opine$^+$ F1 and BC1 plants were also observed on all normal appearing opine-negative F1 and BC1 plants that were tested for flower development. This indicates that these newly acquired properties are not directly due to the presence of T-DNA. On the other hand, we found that normal plants induced on control callus tissue by culturing it on hormone-free medium did not express these traits. Therefore, we assume that the transformation event could have induced an additional genomic change in the cytoplasm leading to the observed male sterility. Maternal inheritance via the cytoplasm would explain the observation of heterostyly and male sterility on all progeny plants. The nature of the cytoplasmic change is the subject of further investigations.

It has been established that foreign DNA sequences that are inserted into the T-DNA are cotransferred and integrated into the plant genome.[24] This together with our observation that large T-DNA segments are transmitted to progeny plants demonstrates that, in principle, it is possible to integrate stable genes of interest in plants by using the Ti plasmid as a vector. Since the loci on the T-DNA involved in morphogenesis are known, mutations of these loci on the Ti plasmid, followed by the transformation of single cells might lead to a complete restoration of the morphogenic potential of the transformed tissues with retention of the (mutated) T-DNA in regenerants and progeny.

E. Crosses between "Hairy Root" Plants and "Crown Gall" Plants

A. rhizogenes harbors an Ri plasmid which is responsible for the induction of the hairy root disease, which is characterized by an excessive root formation on the site of infection. The infection and transformation process of *A. rhizogenes* is similar to the infection process by *A. tumefaciens.* Part of the Ri plasmid, the Ri T-DNA, is transferred to the plant genome, where it is expressed.[43,44] The roots that grow out of the wound can be brought into tissue culture and from such tissues plants can be regenerated. These plants are characterized by the opine agropine, an overdeveloped root system, bubbled leaves, and have a reduced male fertility. Crosses have been carried out between such hairy root tobacco plants and control plants and between hairy root plants and nopaline- and octopine-type crown gall tobacco plants derived from our cocultivation experiments. Table 4 shows the results from such crosses. The transmission pattern of the hairy root phenotype is similar to the transmission pattern of the crown gall phenotype as described above. About 50% of the seedlings from the cross between

Table 4
PHENOTYPIC PROPERTIES OF PROGENY PLANTS FROM CROSSES WITH NORMAL (n), HAIRY ROOT (hr), AND CROWN GALL (cg) PLANTS

Cross ♂ × ♀ % of seedlings cg × n	Phenotype[a]					Assumed (%)
	Normal	Hr	Ags	Cg	Nos	
56	+	–	–	–	–	50
44	–	–	–	+	+	50
hr × n						
53	+	–	–	–	–	50
47	–	+	+	–	–	50
cg × hr						
32	+	–	–	–	–	25
24	–	+	+	–	–	25
22	–	–	+	+	+	25
22	–	–	–	+	+	25

[a] Hr, excessive roots, bubbled leaves, reduced apical dominance;
Ags, agropine synthesis;
Cg, no roots, teratomata;
Nos, nopaline synthesis.

hairy root plant and control plant have the hairy root phenotype, indicating that also the Ri T-DNA is inherited according to Mendelian laws.

Since wild-type crown gall tumors have a T-DNA with active auxin and cytokinin loci resulting in undifferentiated tumor tissue, whereas the T-DNAs in the transformed tobacco used in these crosses have only an active cytokinin-locus, resulting in shoots, and the hairy root tobacco plant has an active auxin-locus resulting in excessive root formation, it is interesting to see how these properties influence each other when these plants are crossed. According to Mendelian laws, we would expect that 25% of the seedlings would be completely normal, 25% of the seedlings would have the hairy root phenotype, 25% of the seedlings would have the crown gall phenotype, and 25% of the seedlings would have a combination of the hairy root and the crown gall phenotypes. The appearance of the phenotype of this last category is not predictable. If we look at the figures for the nopaline crown gall x hairy root cross we see numbers that deviate from the prediction. This deviation is similar to the deviation of Mendelian segregation found earlier with different crosses between hairy root plants and healthy plants and between crown gall plants and healthy plants. In these crosses, in general, 45% of the seedlings have the transformed phenotype. From the cross between hairy root and crown gall plants we would expect 75% of the seedlings to have a transformed phenotype, either of hairy root type, crown gall type, or a combination of these. We observed 67%. If we look at the number of seedlings without a root system we would expect 50% and we found about 44%. The discrepancy between our observation and the expectation is likely due to the fact that the germination percentage of transformed seedlings is lower than that of nontransformed seedlings.

The major conclusion that we can draw from these experiments is that indeed one category of seedlings harbors both Ti T-DNA and the Ri T-DNA. These seedlings contain both agropine and nopaline. If we look at the phenotype of these seedlings (Figure 11) we conclude that the auxin-locus of the Ri T-DNA is complementing the cytokinin-locus of the Ti T-DNA resulting in seedlings that lack a root system and that

FIGURE 11. Representatives of the four classes of seedlings derived from a cross between hairy root (♂) and crown gall (♀) tobacco plants. (1) normal seedling; (2) seedling with crown gall phenotype; agropine negative, nopaline positive; (3) seedling with crown gall and hairy root phenotype, both agropine positive and nopaline positive; (4) seedling with hairy root phenotype, agropine positive.

degenerate into tertomatous callus, similar to the wild-type nopaline callus. The category of seedlings with only agropine synthesis is similar to the hairy root parent, whereas the seedlings with only nopaline synthesis resemble the nopaline crown gall parent. T-DNA analysis and T-DNA transcript analysis must be carried out to study the segregation pattern of the Ri genes for agropine synthesis and the Ri-onc-gene(s).

IV. CONCLUDING REMARKS

In order to obtain optimal results with plant cell transformation experiments yielding a high frequency of transformants, the DNA of interest must be incubated, with or without agrobacteria as microinjector, with a large population of target cells, in order to select those cells that have incorporated desired genes in a stable fashion. In this respect, the methods described here based on single plant protoplasts as tartet cells are similar to those used in bacterial, yeast, and mammalian cell genetics. Millions of single plant cells can be exposed to the transforming bacteria or DNA by starting with protoplast populations.

DNA transformation and co-cultivation are different systems, both with advantages and disadvantages. The advantage of the co-cultivation method lies in the relatively high frequency of transformation and the incorporation of a nonscrambled, distinct size of transferred DNA. If a gene is introduced in the T-region, it therefore will be incorporated with the T-DNA in the plant genome. A disadvantage of this system is the fact that it is only applicable to dicot plants. DNA transformation, on the other hand, presumably is the most promising method for monocot plants because it does not depend on specific cell-to-cell interactions occurring between agrobacteria and dicot plants. Furthermore, this method might be very useful if DNA other than T-DNA sequences have to be introduced into the cells which can occur via co-transformation. In fact, DNA transformation might turn out to be independent of the crown gall system and, if adequate selection markers are available, might also be used with DNA sequences from other sources.

At the end of this chapter we are aware of the fact that, although the title indicates otherwise, the real application of genetic engineering for biotechnological and agricultural purposes still has not yet begun. However, we strongly believe that the necessary tools to do so are available and they they can be tested on dicotyledonous plant systems. These tools are the gene vectors for plants that have been constructed based on the Ti plasmid, and the single-cell tranformation methods starting with individual single cells and ending with mature, flowering plants. For dicot plants, of course, these tools have to be refined where necessary; for monocot plants a very great deal of experimental work must be carried out before one can begin to think of applying these kinds of techniques.

ACKNOWLEDGMENTS

All members of the Molbas-research group are acknowledged for their contribution to the results presented here. Our research was supported in part by the Netherlands Foundation for Fundamental Biological Research (BION) and Chemical Research (SON) with financial aid from the Netherlands Organization for Advancement of Pure Scientific Research (ZWO) and the Dutch Programme Committee on Biotechnology (PCB) within the framework of the Innovation Oriented Research Programm on Biotechnology (I0P-b). It was also supported by Research Contract no. GB]-4-021-NL of the Biomolecular Engineering Programme of the Commission of the European Communities.

REFERENCES

1. Marton, L., Wullems, G. J., Molendijk, L., and Schilperoort, R. A., In vitro transformation of cultured cells from *Nicotiana tabacum* by *Agrobacterium tumefaciens, Nature (London)*, 277, 129, 1979.
2. Wullems, G. J., Molendijk, L., Ooms, G., and Schilperoort, R. A., Differential expression of crown gall tumor markers in transformants obtained after in vitro *Agrobacterium tumefaciens*-induced transformation of cell wall regenerating protoplasts derived from *Nicotiana tabacum, Proc. Natl. Acad. Sci. U.S.A.*, 78, 4344, 1981.
3. Krens, R. A., Molendijk, L., Wullems, G. J., and Schilperoort, R. A., In vitro transformation of plant protoplasts with Ti-plasmid DNA, *Nature (London)*, 296, 72, 1982.
4. Vasil, I. K., Ahuja, M. R., and Vasil, V., Plant tissue cultures in genetics and plant breeding, *Adv. Genet.*, 20, 127, 1979.
5. Lörz, H. and Scowcroft, W. R., Variability among plants and their progeny regenerated from protoplasts of *Su/su* heterozygotes of *Nicotiana tabacum, Theor. Appl. Genet.*, 66, 67, 1983.
6. Evans, D. A. and Sharp, W. R., Single gene mutations in tomato plants regenerated from tissue culture, *Science*, 221, 949, 1983.
7. Austin, A. and Cassells, A. C., Variation between plants regenerated from individual calli produced from separated potato stem callus cells, *Plant Sci. Lett.*, 31, 107, 1983.

8. Vasil, I. K., Regeneration of plants from single cells of cereals and grasses, in *Genetic Engineering in Eukaryotes*, Nato ASI series, Series A: Life Sciences, Vol. 61, Lurquin, P. F. and Kleinhofs, A., Eds., Plenum Press, New York, 1983, 233.
9. Carlson, P. S., Smith, H. H., and Dearing, R. D., Parasexual interspecific plant hybridization, *Proc. Natl. Acad. Sci. U.S.A.*, 69, 2292, 1972.
10. Maliga, P., Lazar, G., Joo, F., Nagy, A. H., and Menczel, L., Restoration of morphogenetic potential in *Nicotiana* by somatic hybridization, *Mol. Gen. Genet.*, 157, 291, 1977.
11. Kao, K. N. and Michayluk, M. R., A method for high frequency intergeneric fusion of plant protoplasts, *Planta*, 115, 355, 1974.
12. Wullems, G. J., Molendijk, L., and Schilperoort, R. A., The expression of tumour markers in intraspecific somatic hybrids of normal and crown gall cells from *Nicotiana tabacum*, *Theor. Appl. Genet.*, 56, 203, 1980.
13. Dudits, D., Fejer, O., Hadlaczky, G., Koncz, G., Lazar, G. B., and Horvath, G., Intergeneric gene transfer mediated by plant protoplast fusion, *Mol. Gene. Genet.*, 179, 283, 1980.
14. Lörz, H., Paszkowski, I., Dierks-Ventling, C., and Potrykus, I., Isolation and characterization of cytoplasts and miniprotoplasts derived from protoplasts of cultured cells, *Physiol. Plant.*, 53, 385, 1981.
15. Maliga, P., Lörz, H., Lazar, G., and Nagy, F., Cytoplast-protoplast fusion for interspecific chloroplast transfer in *Nicotiana*, *Mol. Gen. Genet.*, 185, 211, 1982.
16. Cocking, E. C., Somatic hybridization by the fusion of isolated protoplasts. An alternative to sex, in *Plant Cell and Tissue Culture*, Sharp, W. R., Larsen, P. O., Paddock, E. F., and Raghaven, V., Eds., Ohio State University Press, Columbus, 1979, 353.
17. Leaver, C. I., Hack, E., Dawson, A. I., Isaac, P. G., and Jones, V. P., Mitochondrial genes and their expression in higher plants, in *Nuclear Mitochondrial Interaction*, Kaudewitz, E., Ed., de Gruyter, in press.
18. Wullems, G. J., Molendijk, L., Ooms, G., and Schilperoort, R. A., Retention of tumour markers in F1 progeny plants from in vitro induced octopine and nopaline tumour tissues, *Cell*, 24, 719, 1981.
19. Clarke, L. and Carbon, I., Isolation of a yeast centromere and construction of functional small circular chromosomes, *Nature (London)*, 287, 504, 1980.
20. Stinchcomb, D. R., Thomas, M., Kelly, I., Selker, E., and Davis, R. W., Eukaryotic DNA segments capable of autonomous replication in yeast, *Proc. Natl. Acad. Sci. U.S.A.*, 77, 4559, 1980.
21. Howell, S. H., Walker, L. L., Dudley, R. K., Cloned cauliflower mosaic virus DNA infects turnip *Brassica rapa*, *Science*, 208, 1265, 1980.
22. Hooykaas, P. J. J. and Schilperoort, R. A., The molecular genetics of crown gall tumorigenesis, in *Advances in Genetics*, Vol. 22, Caspari, E. W. and Scandalios, I. G., Eds., Academic Press, New York, 1983, 209.
23. Chilton, M.-D., A vector for introducing new genes into plants, *Sci. Am.*, 248, 36, 1983.
24. Ooms, G., Hooykaas, P. J. J., Molenaar, G., and Schilperoort, R. A., Crown gall plant tumors of abnormal morphology, induced by *Agrobacterium tumefaciens* carrying mutated octopine Ti-plasmids; analysis of T-DNA functions, *Gene*, 14, 33, 1981.
25. Hille, J., Klasen, I., and Schilperoort, R. A., Construction and application of R prime plasmids, carrying different segments of an octopine Ti plasmid from *Agrobacterium tumefaciens*, for complementation of *vir* genes, *Plasmid*, 7, 107, 1982.
26. Chang, S. and Cohen, S. N., High frequency transformation of *Bacillus subtilis* protoplasts by plasmid DNA, *Mol. Gen. Genet.*, 168, 11, 1979.
27. Graham, F. L. and Van der Eb, A. J., A new technique for the assay of infectivity of human adenovirus 5 DNA, *Virology*, 52, 456, 1973.
28. Fraley, R. and Papahadjopoulos, D., New generation liposomes: the engineering of an efficient vehicle for intracellular delivery of nucleic acids, *Trends Biochem. Sci.*, March 1981.
29. Diacumakos, E. G., Methods for micromanipulation of human somatic cells in culture, in *Methods in Cell Biology*, Prescott, D. M., Ed. Academic Press, New York, 1973.
30. Fraley, R., Strawbinger, R. M., Rule, G., Springer, E. L., and Papahadjopoulos, D., Liposome mediated delivery of DNA to cells: enhanced efficiency of delivery related to lipid composition and incubation conditions, *Biochemistry*, 20, 6978, 1981.
31. Wigler, M., Pellicer, A., Silverstein, S., Axel, R., Urlaub, G., and Chasin, L., DNA mediated transfer of the adenine phosphoribosyl transferase locus into mammalian cells, *Proc. Natl. Acad. Sci. U.S.A.*, 76, 1373, 1979.
32. Draper, J., Davey, M. R., Freeman, I. P., Cocking, E. C., and Cox, B. I., Ti plasmid homologous sequences present in tissues from *Agrobacterium* Ti-plasmid-transformed *Petunia* protoplasts, *Plant Cell Physiol.*, 23, 451, 1982.
33. Wullems, G. J., Márton, L., Molendijk, L., Krens, F., Ooms, G., Würzer-Figurelli, L., and Schilperoort, R. A., Genetic modification of plant cells by transformation and somatic hybridization, in *Advances in Protoplast Research*, Ferenczy, L. and Farkas, G. L., Eds., Pergamon Press, Elmsford, N.Y., 1979, 407.

34. Otten, L. A. B. M. and Schilperoort, R. A., A rapid microscale method for the detection of lysopine dehydrogenase activities, *Biochem. Biophys. Acta*, 527, 497, 1978.
35. Tomashow, M. F., Nutter, R., Montoya, A. L., Gordon, M. P., and Nester, E. W., Integration and organization of Ti plasmid sequences in crown gall, *Cell*, 19, 729, 1980.
36. Ooms, G., Bakker, A., Molendijk, L., Wullems, G. J., Schilperoort, R. A., Gordon, M. P., and Nester, E. W., T-DNA organization in homogeneous and heterogeneous octopine-type crown gall tissues of *Nicotiana tabacum*, *Cell*, 30, 589, 1982.
37. Zambryski, P., Holster, M., Kruger, K., Depicker, A., Schell, J., Van Montagu, M., and Goodman, H. M., Tumor DNA structure in plant cells transformed by *A. tumefaciens*, *Science*, 209, 1385, 1980.
38. Simpson, R. B., O'Hara, P. J., Kwok, W., Montoya, A. L., Lichtenstein, C., Gordon, M. P., and Nester, E. W., DNA from the AGS/2 crown gall tumor contains scrambled Ti plasmid sequences near its junctions with plant DNA, *Cell*, 29, 1005, 1982.
39. Klee, H. J., Gordon, M. P., and Nester, E. W., Complementation analysis of *Agrobacterium tumefaciens* Ti plasmid mutations affecting oncogenicity, *J. Bacteriol.*, 150, 327, 1982.
40. Hooykaas, P. J. J., Klapwijk, P. M., Nuti, M. P., Schilperoort, R. A., and Rörsch, A., Transfer of the *Agrobacterium tumefaciens* Ti plasmid to avirulent Agrobacteria and to *Rhizobium ex planta*, *J. Gen. Microb.*, 98, 477, 1977.
41. Memelink, J., Wullems, G. J., and Schilperoort, R. A., Nopaline T-DNA is maintained during regeneration and generative propagation of transformed tobacco plants, *Mol. Gen. Genet.*, 190, 516, 1983.
42. Van Slogteren, G. M. S., Hoge, J. H. C., Hooykaas, P. J. J., and Schilperoort, R. A., Clonal analysis of heterogeneous crown gall tumor tissues induced by wild type and shooter mutant strains of *Agrobacterium tumefaciens*. Expression of T-DNA genes, *Plant Mol. Biol.*, 2, 321, 1983.
43. Chilton, M.-D., Tepfer, D. A., Petit, A., David, C., Casse-Delbart, F., and Tempé, J., *Agrobacterium rhizogenes* inserts T-DNA into the genomes of the host plant root cells, *Nature (London)*, 295, 432, 1982.
44. Spano, L. and Costantino, P., Regeneration of plants from callus cultures of roots induced by *Agrobacterium rhizogenes* on tobacco, *Z. Pflanzenphysiol.*, 106, 87, 1982.
45. Hoekema, A., Hirsch, P. R., Hooykaas, P. J. J., and Schilperoort, R. A., A binary plant vector strategy based on separation of Vir- and T-region of the *Agrobacterium tumefaciens* Ti plasmid, *Nature (London)*, 303, 179, 1983.
46. Nagy, J. I. and Maliga, P., Callus induction and plant regeneration from mesophyll protoplasts of *Nicotiana sylvestris*, *Z. Pflanzenphysiol.*, 78, 453, 1976.

INDEX

A

B

C

D

E

F

G

H

I

K

L

M

O

P

Q

R

T

U

V

W

X

Y

Z